MW01033842

HOPE DIES LAST

HOPE

VISIONARY PEOPLE
ACROSS THE WORLD, FIGHTING
TO FIND US A FUTURE

DIES

ALAN WEISMAN

LAST

DUTTON

An imprint of Penguin Random House LLC
1745 Broadway, New York, NY 10019
penguinrandomhouse.com

Portions of this book have appeared in different form in the *Los Angeles Times*, *The Boston Globe Magazine*, *Salon*, and *Pacific Standard*; on CNN.com; and as an essay in *Eden Turned on Its Side*, by photographer Meridel Rubenstein, University of New Mexico Art Museum, 2018.

MAPS BY DAVID LINDROTH INC.

BOOK DESIGN BY LORIE PAGNOZZI

LIBRARY OF CONGRESS CATALOGING-IN-PUBLICATION DATA
has been applied for.

ISBN: 9781524746698 (hardcover)
ISBN: 9781524746711 (ebook)

Printed in the United States of America
1st Printing

The authorized representative in the EU for product safety and compliance is Penguin Random House Ireland, Morrison Chambers, 32 Nassau Street, Dublin D02 YH68, Ireland, https://eu-contact.penguin.ie.

To sister Rochelle,
who taught me how much words matter;

and to wife Beckie,
who taught me what matters even more.

CONTENTS

One should, for example, be able to see
that things are hopeless and yet be determined
to make them otherwise.

—F. SCOTT FITZGERALD

Yo he preferido hablar de cosas imposibles
porque de lo posible se sabe demasiado.

I prefer to speak of impossible things
because we know all too well what's possible.

—SILVIO RODRÍGUEZ

CHAPTER ONE

CRADLE TO GRAVE

i. Origin

How, Azzam Alwash wonders, is it possible that he's never been here before? Born in this cradle land to a family from Babylon, taught by his father how to fish for shad and barbel in marshes where people still lived like the ancients in reed houses on floating islands, he knows his country's storied waterways like his own veins and arteries. But now he is seeing something for the first time.

Strange, because except for his years in exile, all his life it's been so near, just downriver from the city of Basra, where he attended university—although needing a police escort to get here might explain why he didn't come earlier. It's a tense spot, Iraq's tiny, 36-mile wedge of coastline between its wealthier, powerful neighbors at the top of what the Kuwaitis call the Arabian Gulf and the Iranians call the Persian Gulf.

"We just call it the Gulf," he murmurs, unable to tear his gaze away.

Azzam, still trim in his mid-60s, clean-shaven with buzz-cut silver hair that narrows to a widow's peak, his blue hoodie unzipped to the breeze, is considered a national hero in Iraq—internationally as well: winner of the 2013 Goldman Environmental Prize, the "green Nobel," for miraculously resuscitating the Mesopotamian marshes of his boyhood, the biggest wetland in the Middle East. In 1993, that lush convergence of the Tigris and Euphrates floodplains, believed by many to be the biblical Eden, had shriveled to barren, dust-blown expanses after Iraqi dictator Saddam Hussein intentionally diverted the fabled rivers, draining thousands of verdant square miles in order to flush Shi'ite rebels out from hiding in the marshes' impenetrable reed and papyrus thickets.

The damage was irreversible, Azzam kept hearing from biologists, as he begged the US State Department for funds to reflood the marshes after US-led forces deposed Saddam in 2003. The roots and desiccated seed banks were over 10 years old, their chemistry irreparably changed, a CIA environmental contractor told him. "It's probably impossible to revive them."

Azzam was an engineer, not an ecologist, but navigating exile and building a career had taught him that *impossible* often masks a lack of imagination. If there was one lesson he'd learned, it was that nature is resilient: she's seen all of this before.

"How will we know unless we try?" he asked.

Today, with the fate of not just a wetland but his entire planet in play, Azzam Alwash's stubborn persistence that proved experts wrong is needed everywhere, he believes. For our civilization to continue—if not our very existence—will demand vision that transcends conventional wisdom and boundaries.

Humanity's astonishing success as a species has resulted in a dangerously lopsided world. With most unfrozen land now devoted to the needs and nourishment of just one species, our own, we've literally pushed most of our Earthly companions off the planet: all but 4 percent of mammalian weight is either us or our livestock. Nature, the fossil record warns, doesn't permit such imbalance for long. Already, our exhaust has packed the sky with heat-trapping gases that wither our crops, trigger whirlwinds, and alternately and increasingly rain flood or hellfire.

Yet we can't seem to stop, so poles keep shrinking and plastic keeps coming. As an existential ultimatum threatens, what, realistically, are our hopes? What or who can get us through this?

Amid pervasive, chilling uncertainty, Azzam Alwash is among a worldwide legion of kindred visionaries, often unknown to one another, toiling in separate arenas to find humanity a path through this treacherous century to a livable future. Neither in denial nor naively optimistic, knowing full well how daunting the odds are, they nevertheless refuse to quit trying.

Sometimes their local challenges turn out to have global implications. Azzam Alwash went from saving his cherished fishing spots to designing a plan to revitalize all that was once Mesopotamia into the world's bridge from an oil-based past to a sun-blessed future. Though repeatedly thwarted, he keeps on, banging into hard realities until something budges.

The last half kilometer to Iraq's southernmost point had taken them over a single-lane sand causeway atop an embankment barely higher than the mudflats and tide pools on either side. A tattered, wind-whipped black flag marked the road's end, beyond which the tide pools continued, dissolving into an unreadable gray horizon.

On his iPhone Azzam checks Google Maps, which shows him surrounded by blue.

"It says we're underwater," he tells Ameer Naji.

Ameer, 31, with short wiry hair and a trimmed brown beard, is an environmental engineer at Iraq's sole port, Umm Qasr, on a nearby inlet by the Kuwaiti border. He's long wanted to meet Azzam and, thanks to an uncle who knows the police captain, was happy to arrange security for this excursion. "The tide is still out," he explains, indicating a faint band where open water shimmers a few hundred meters distant. He then points east, to a line of silhouettes: cargo

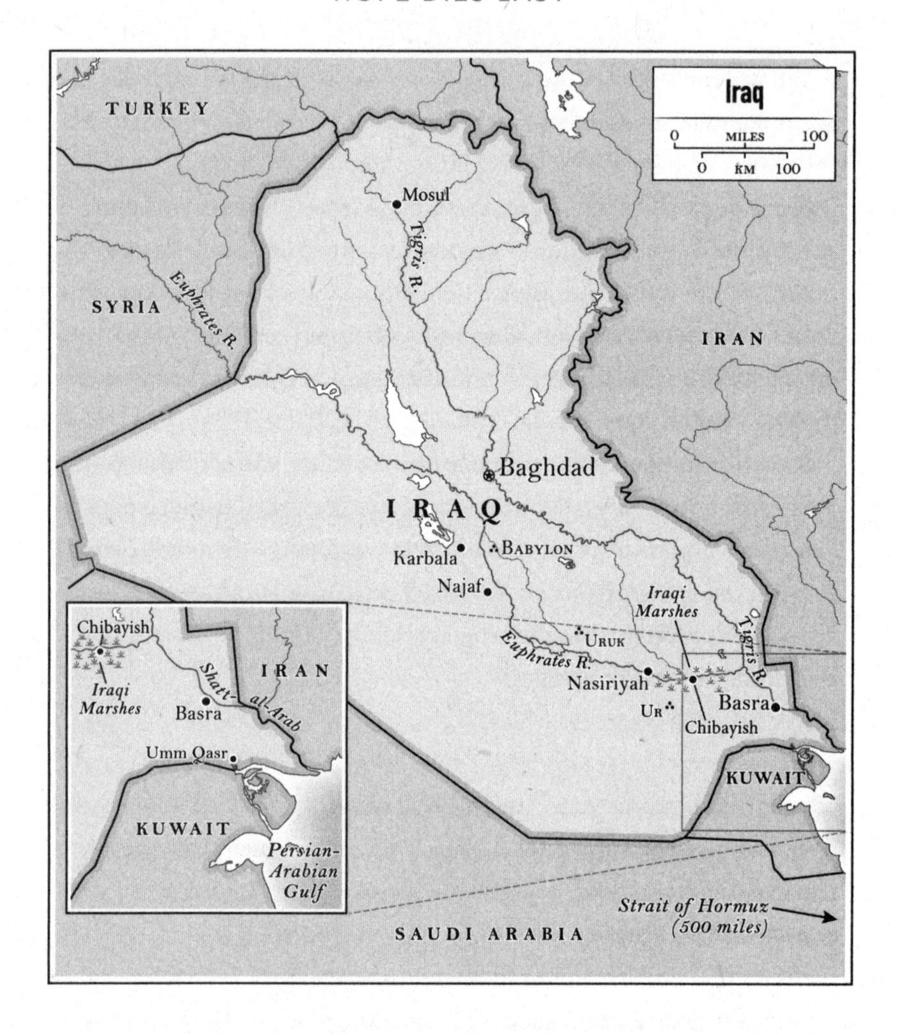

vessels lumbering up the Shatt al-Arab, the river that forms where the Tigris and Euphrates merge, 200 kilometers north. Halfway down its length, the Shatt al-Arab becomes Iraq's fraught border with Iran. The ships hug the Iranian shore.

Heedless of the March wind in a light cotton pullover, Ameer is from Iraq's second-biggest city, Basra, an hour upriver. The world's largest date palm forest once lined the Shatt al-Arab's banks there. Then, in 1980, Saddam Hussein attacked Iran to annex its oil fields. The murderous war that ensued—Saddam used nerve gas; Iran laid waste wherever it counterattacked—ended in an exhausted truce after eight years and a half million deaths on each side. In Basra, on the front lines, millions of date palms also fell victim.

In January 1991—the year Ameer was born—a few months after Saddam had invaded Kuwait for the same reason came the Americans to rescue US oil interests. For five weeks they bombed Baghdad and Basra nonstop before storming Kuwait and pushing out the dictator's Republican Guard troops. As they fled, on his orders they set fire to 600 Kuwaiti oil wells. When US forces left without deposing Saddam, a nationwide rebellion erupted to finish the job, but the Republican Guard that cowered before the Americans had no qualms about mauling its own citizenry into submission.

A decade later, Afghanistan-based jihadis, most of them Saudi, crashed hijacked planes into Manhattan skyscrapers and the Pentagon. Under a crackpot pretext that Saddam was somehow responsible, Americans returned with British allies. The 2003 Battle of Basra was among that invasion's first, and worst. This time, the Americans stayed nearly 10 years. Between bomb attacks by guerrillas resisting their occupation and US military reprisals, Iraq, once deemed the flower of Arab culture for its poetry, became one of the most dangerous places on Earth.

After the Americans finally pulled out in 2011, giving a justification as fanciful as the one that had brought them—that Iraqi forces were now capable of defending their own homeland—ISIS leapt into the breach. The Islamic State's savagery made it possible to miss the hated American occupiers. It took four years and the intervention of Iran to finally drive ISIS out. Except now, Iran has stayed: with Iraq's weak government often locked in parliamentary paralysis, in much of the country Iranian militias are the tacit, corrupt local authority.

Ameer endures their highway checkpoints daily, going from his city, now rubbled by repeated wars, to his various jobs at the port.

Sometimes he's a cargo inspector; sometimes a health, safety, and environmental supervisor for foreign petroleum companies operating here. "Everybody here works for them," he tells Azzam. Southern Iraq has three of the world's biggest oil fields. Dozens of tall flares there burn constantly, because neither Saddam nor any government after his invested in infrastructure to capture the methane.

When summer temperatures here reach 51°C (nearly 124°F), oil pipeline welders collapse. Ameer takes them to the clinic, where medics elevate their legs and give them ice and IV liquids. Once he saw 15 drop, one after another. He and coworkers shuttled them for treatment in a pickup truck bed.

"We must stop," he implored the Korean manager.

"We must continue," the manager told him.

What Ameer calls his real work begins after hours, as the Basra director for Humat Dijlah, a national river preservation network run by young activists. It almost feels like a lost cause, he tells Azzam. With Turkey, where 90 percent of Iraq's fresh water originates, building more upstream dams on both the Tigris and Euphrates, there's even less fresh water to resist the wedge of salinity pushing up the Shatt al-Arab from the Gulf, weakening what remains of the palms. The river, Basra's source of drinking water, already reeks from toilets flushing into urban canals that were lovely for swimming when he was young.

In 2018, as Turkey began filling a huge dammed reservoir upriver on the Tigris, Basra's tap water turned black. More than 120,000 were hospitalized.

On his fingers, Ameer ticks off people he's lost. "My cousin, brain cancer. My best friend, stomach cancer."

After 20 colleagues were killed in protests they'd organized, the Shatt al-Arab got a desalinization plant, but so inadequate that the water is still undrinkable. Trucks jam the streets bearing flats of plastic water bottles to slake the endless thirst of Basra's 3 million people, alongside more trucks stacked with air conditioners. Each summer, temperatures approach 54.5°C (130°F), far beyond where sweating can cool the human body, though with the humidity rising off the nearby Gulf, that point is exceeded here at even lower temperatures. So many windows have AC units that the grid crashes daily.

When the summers get that lethal, Ameer takes his wife and son north to stay with relatives in Turkey, then returns to work.

His wife, terrified he'll be killed or jailed like his protester friends, begs him to leave, or threatens to leave herself. Their son's severe ADHD, they're told, owes to toxicity from the water and the flares ringing the city. The air is greasy, every breath tastes oily. Frequent dust storms render everything invisible except the flares.

Even friends tell him to leave. "But if I go, who will lead? Who will protect Basra for my son's generation?" He carries a picture of his son, Muhammad, holding a sign that reads "Make Basra Green!" He loves the old city's mud-brick houses with ornate second-story cantilevered wooden facades, loves his environmental group that literally raises flowers amid the ruins.

Azzam claps him on the shoulder. "Failure is not an option," he says.

The tattered black flag here at the road's end marks where each year, Shi'ite Muslims gather to begin Arba'een, a two-week, 686-kilometer trek to the holy Iraqi city of Karbala. By the time they arrive, their numbers swell to 22 million: the biggest pilgrimage on Earth, nearly 10 times larger than the annual hajj to Mecca. Arba'een commemorates the martyrdom of Imam Hussain, whose tomb is their destination.

Imam Hussain, who was beheaded while defending the faith, was the son of Imam Ali, who was also assassinated, and whose tomb in Najaf, another holy city, the pilgrims will stop to venerate en route to Karbala. Shi'ites believe that Imam Ali, married to Fatima, daughter of his cousin Muhammad, was the Prophet's chosen heir to his spiritual mantle. Sunnis, however, believe the lineage passed through Muhammad's closest companion, Abu Bakr, who was also one of Muhammad's 13 fathers-in-law. The disagreement has sadly continued the bloodshed that anoints Muslim history.

"A deadly feud between a couple of families," Azzam sighs. But although Jewish history's ceaseless Old Testament battles and Christianity's Crusades, inquisitions, and a messiah tortured to death are no less grisly, there's one thing, he notes, that Shi'ites, Sunnis, Jews, and Christians have in common.

"They all believe it began right here," he says, gesturing at the view before him.

In all the world, he reckons, there may be no better vantage point than the top of this Gulf to peer simultaneously into humanity's past and its future.

What became known as Eden, Azzam knows, wasn't just his beloved Iraqi marshes, although they're relics of the same ecosystem. Descriptions of Eden in the Judeo-Christian Bible, and in similar Sumerian and Akkadian stories predating it, don't correspond with the creation of a universe 13.6 billion years ago. Rather, they resemble the end of the Pleistocene.

The epoch of Ice Ages, the Pleistocene represents one of the most drastic shifts in geologic history. For much of the previous 550 million years, this had been a far warmer planet. Then, around 2.5 million years ago, just as the hairy biped *Australopithecus* was morphing into tool-wielding *Homo*, great sheets of polar ice began advancing, receding, and returning. This happened more than 20 times. The last retreat ended just 8,000 years ago.

The glacial-interglacial fluctuations were due to periodic quirks in Earth's orbit that alter how much solar radiation reaches the surface. But what provoked the Pleistocene's great refrigeration in the first place began in the previous geologic epoch, the Pliocene, when turbulence deep in the planet's magma sent its tectonic plates—twelve vast slabs of rocky crust and several smaller ones that drift like soup crackers atop the semi-molten mantle below—colliding with each other.

As the Indian plate rammed the Eurasian plate to the north, it began to push up the Himalayas. The African plate, crunching into what is now Europe, likewise pushed up the Alps and squeezed the dregs of an ancient ocean into today's Mediterranean. What doomed Earth's balmier climate, however, was when the Pacific plate began to slide under the smaller Caribbean plate. About 2.8 million years ago, this touched off volcanic flows that turned a string of islands into the isthmus of Panama.

We mostly think of that isthmus connecting North and South America, but equally momentous was how it disconnected the Atlantic from the Pacific. With these oceans no longer sharing warm tropical currents at the Earth's midriff, the planet began to cool, setting the stage for the Pleistocene's Ice Ages.

During the previous epoch, several *Homo* species had appeared, but after repeated glaciations, by 40,000 years ago only one remained: our own, *sapiens.* Well before that, we modern humans had begun dispersing from our African origins. Fifty thousand years ago, we'd spread all the way to Australia, and 15,000 years ago, we'd arrived to stay in the Americas. What we call civilization thus had various cradles: in India, China, Mexico, Peru, and Egypt. But the one that emerged at the eastern tip of a fertile crescent our African forebears followed through the Levant into Mesopotamian Asia has special resonance for more than half the world's population, whose Abrahamic cultures hearken to the place called Eden.

Which, it turns out, was more than just a metaphor.

Genesis locates the Garden near the confluence of four rivers: the Hiddekel, the Pirat, the Gihon, and the Pison. Linguists and biblical scholars widely agree that Hiddekel, derived from the Akkadian *Idiqlat*, is Hebrew for the Tigris. Pirat, from *Purattu* in Akkadian, is the Euphrates.

The 1611 English-language King James Bible placed the Gihon far from Mesopotamia, across the Red Sea in Ethiopia. Many modern scholars, however, believe its translators confused Ethiopia's ancient name, Cush, with a similar Hebrew name for the land of the Kashshites, today Iran's Zagros Mountains. A large river, the Karun, flows southward from the Zagros, joining the Shatt al-Arab just above the Gulf at a point almost directly opposite the mouth of a now-intermittent river that flows north through Saudi Arabia and Kuwait, the Wadi al-Batin.

That point today is 60 miles below where the Tigris and Euphrates meet, but back then—as noted by archaeologists who contend that the Karun and Wadi al-Batin are the Bible's Gihon and Pison—local topography was far different.

Until around 14,000 years ago, Pleistocene glaciations locked much of the world's water into ice sheets, and oceans were hundreds of feet lower than they are today. When the first *Homo sapiens* reached today's Arabian-Persian Gulf, its upper third was well above sea level. Had Azzam and Ameer been standing here then, they would have seen a verdant Mesopotamian valley stretching hundreds of miles farther south, its converged rivers finally reaching the sea above today's Strait of Hormuz.

British archaeologist Jeffrey Rose dubbed this the Gulf Oasis: a luxuriant delta as big as Great Britain, with freshwater springs welling up from its limestone basin, continuously fed by nutrient-rich sediments borne by the Tigris, Euphrates, Karun, and Wadi al-Batin. For hunter-gatherers, it would surely have been a paradise of date palms, wild wheat, legumes, fish, mollusks, fallow deer, gazelles, river dolphins, water buffalo, boars, and aurochs: the wild ancestors of today's cattle.

But was this Eden?

The Judeo-Christian Bible begins with two conflicting creation tales. In the first, God creates the world. Within six days, there are stars overhead, sun by day, moonlight by night, seas and sea creatures, land, plants, land animals, and finally people: male and female simultaneously.* It's a world we'd basically recognize, other than that it is entirely vegetarian: fruit, seeds, and green herbs shall be for meat, God tells humans and beasts alike. All of that is described in Genesis, Chapter 1—where, incidentally, there's no mention of any garden: the entire Earth is bountifully blessed.

But after the seventh day, when God rests, suddenly in Chapter 2 the story starts over. This time, there's no workweek plus a Sabbath; instead, God conjures up Heaven and Earth in a single day. And the sequence changes: again, He invents plants early on, but sets them aside in some celestial nursery because, unlike before, He hasn't created water yet. Next, He does so—then, just as the soil begins to moisten, He selects some dust to mold into a human being. A male.

Only after breathing life up man's nostrils does God add the foliage—but this time, to a garden. However gorgeous and fertile, it isn't all bliss and leisure. Here, God intends for man to "dress," "keep," and "till the ground." He sets guidelines about what can or can't be harvested, designates a network of rivers for irrigation, and finally, lest man be lonely, creates animals.

* Genesis 1:27.

Adam gets to name these, and to identify which are domestic and which are wildlife, yet they fail to ease his solitude. So, borrowing one of his ribs, God clones him a more tempting companion. And you know the rest.

As the syntax and even names for God differ (Elohim in Chapter 1; Yahweh in Chapter 2) in these two conflicting, coexisting creation tales, it's widely assumed that they were written at different times by different authors, drawing on different versions of earlier Assyrian and Sumerian legends that were later conflated, often awkwardly, into a single testament by anointed redactors who thrashed out what was canonical and what wasn't.

Small wonder religion breeds so much discord. To literalists who claim that the Bible is divinely inspired and thus entirely true, Chapter 2 simply details Chapter 1's sixth day, when man appears, to underscore God's special relationship with us. (Thus, presumably, it's more fitting that humans should precede the flora and fauna over which they hold dominion.)

Babylonian Talmudists tried reconciling another notable discrepancy, over whether woman was born together with man or later ripped from his rib cage like a bloody afterthought, by declaring that both were true. Appropriating a Sumerian legend of a she-demon, they concluded that Chapter 1's female was Adam's rebellious first wife, Lilith, who disdained his authority and abandoned him. In Chapter 2, God tries to ensure that this doesn't happen again. The woman fashioned from Adam's own bone is merely his "helpmeet"—so subservient she doesn't even get a proper name until after she also proves too disobedient and seductive to control and gets them booted from the Garden.

And us, too, so the tale goes—but that may well be why both creation myths are included in Genesis. Taken together, they reflect an evolutionary truth sensed by whoever compiled the Bible from collective memories inherited through their forebears' stories and perhaps even through their genes. In one version, we had all Earth's bounty to pluck for our nourishment. (Archaeology, littered with spear points, choppers, and blades, belies the vegetarianism, which probably seeped into the account much later, during the time of Isaiah, who condemned animal sacrifice and who prophesized a post-Messianic paradise where wolves would dwell among lambs and leopards lie peacefully with kids.)

But in version two, we are no longer hunter-gatherers. We're stewards of orchards, tillers of soil, and pastoralists of the cattle that Adam named. Thus was born agriculture.

Yet by giving us a garden, was God setting us up for a Fall?

Eventually, the first Mesopotamians noticed that wild wheat and barley would germinate where they spilled them, so they didn't have to roam as much. Likewise, instead of just killing animals, they could capture some to eat later—and, like those crops, breed even more. Once farming and pastoring were established, for the first time in our or any species' history, a significant portion of the population didn't have to spend all day finding food. Gradually people devised new things to do for a living: from artisanship to architecture to trade, from prostitution to priesthood. Soon many of us lived farther from crops and livestock and nearer to each other, in settlements, then villages, then cities. Whatever pining early urban dwellers might have sustained for some fabled, lost Eden gave way to the thrill of something entirely new: civilization.

But post-Pleistocene warming also meant that to the north, vast glaciers were melting. As the thawing accelerated, worldwide coastal flooding commenced. At its peak, 8,000 years ago, Persian-Arabian Gulf waters encroached so fast, suggests American archaeologist and Middle East specialist Juris Zarins, that with each generation they advanced several kilometers farther inland. By the time civilization fully awakened in its cradle, paradise was lost beneath rising seas.

Fall and flood: memories passed from prehistoric ancestors to Mesopotamia's Sumerians, Akkadians, Assyrians, and Babylonians, who recorded them on cuneiform tablets and passed them to the Hebrews, and to us. Today, the last traces we have of the oasis that inspired them are some lush wetlands in Iraq's Tigris-Euphrates floodplain, upstream of that submerged Eden.

And now, the seas are rising again, pushing salinity up the Shatt al-Arab toward the marshes. Rising with them are temperatures that approach the limits that humans can tolerate.

ii. Reeds

Just past dawn, Abu Haider's mashouf, its narrow prow curved skyward, its bow flattened to accommodate a 25-horse outboard motor, glides up the canal in front of Nature Iraq in Chibayish, the central town of the marshes. Cofounders Azzam Alwash and Jassim Al-Asadi are already waiting. Abu Haider's corrugated face beams beneath the black-and-white keffiyeh twisted around his head. *"As-Salaam-Alaikum!"*

"And unto you, peace," they reply. It's still March; temperatures have yet to begin their inexorable rise. Jassim, a ball cap jammed over his leathery features, his corduroy jacket buttoned over a wool shirt, and Azzam, fleece hoodie pulled tight, face each other on the rug lining the mashouf's wooden hull, leaning on pillows propped against the bench seats. Abu Haider, a thick black robe over his gray caftan, yanks the starter. They skirt Saddam's old Euphrates deflection embankment, then enter the river itself, its broad surface molten with sunrise. They pass under a now-useless steel bridge that Russians built for Saddam, where fishermen are already gathered with buyers.

When Jassim was young, Chibayish was called Iraq's Venice: hundreds of palm-covered islands separated by canals where neighbors poled gondola-like mashoufs to each other's reed houses. After Saddam drained the marshes, the canals became dirt streets. Few trees now remain—Saddam had the date palms decapitated—and bricks have supplanted reeds. But as they turn into Hammar, the southwestern marsh, things again look like they did when Jassim was growing up.

As Abu Haider weaves them down channels walled on either side by tall green reeds, bevies of coots, night herons, and egrets rise to join black-and-white kingfishers whizzing by. Before Saddam Hussein dug the Glory Canal, Mother of Battles Canal, Saddam River, and Loyalty to the Leader Channel to divert the Tigris and Euphrates in 1991, these marshes had been home to hundreds of native and migratory birds, as well as otters, hyenas, boars, honey badgers, gazelles, jackals, wolves, frogs, and soft-shelled Euphrates turtles more than two feet long. Not all have returned, but this is again a functioning ecosystem, circulating nutrients, filtering the water, and freshening the air.

Around 60,000 people—the Ma'dan, also known as Marsh Arabs—had lived here, too, raising herds of water buffalo that swam off after morning milking but scrambled back to snug reed barns on their home islands each night. The Ma'dan fished, too: dozens of different species, feeding themselves and netting enough to supply 60 percent of Iraq's fish protein. But after Saddam's ecocide, only 6,000 Ma'dan remained here, most crowded into Chibayish. Many had left for menial jobs in Iran; others scattered as far as Australia, Sweden, or California. Abu Haider, his two wives, and 10 kids had to abandon the island where his people had lived since 1161 for a village near Baghdad.

His vibrato tenor soaring over the motor's drone, he sings a lament about leaving, but soon they're passing tiny islands filled with waving children. Since water returned to the marshes, so have most of the Ma'dan. Their reclaimed ancestral islands are actually mounds of layered reeds, cut and piled by countless previous generations until they compressed into soil. Each has at least one house and a bigger, carpeted mudhif for receiving guests. Both are sided with woven mats fastened across parallel rows of columns of bundled reeds that join into arches. Some mudhifs have columns nearly 4 feet thick, strong as hardwood trunks, forming arches 12 feet high or more. Their design dates to reed structures depicted in 5,000-year-old Sumerian clay tablets—and probably to an earlier drowned Eden.

"I have a surprise," Abu Haider tells Azzam as they arrive at his island: something else has returned that Iraq hasn't seen for decades. Outside the mudhif where they slept the night before, a dozen tourists from seven different countries—Germany, the US, New Zealand, China, Italy, Poland, and the UK—are seated cross-legged around a blanket. They're enjoying a breakfast of buffalo-milk cheese and tahini swirled with date syrup, spread on wheels of hot flatbread that one of his granddaughters bakes by slapping thinly rolled dough on the sides of a hot clay tannur. Azzam is amazed. "How did you find this place?" he asks them.

They'd found each other first, explains a young woman from New Mexico in a purple stocking cap, on a Facebook group for intrepid travelers. Someone had posted something about Ur, Abraham's birthplace. Before Saddam started attacking his neighbors, you could climb the ziggurat he'd restored there and visit Nebuchadnezzar's palace in nearby Babylon. Google coughed up a Baghdad tour guide, who assured them that in the 2020s bombs have stopped exploding, Iraq was safe again, and the famous marshes had miraculously come back to life. "So we came. It's so great here."

"I would not equate Iraq with tourism," says Azzam. "I'm glad that you do."

"People think we're mad to come," agrees a Belfast lad standing behind her in a thick wool sweater, swaying on the spongy footing. "We'll prove them wrong."

"You see, brother?" Jassim says to Azzam. "There's hope."

In 1978, Azzam Alwash was studying civil engineering in Basra when Saddam's Ba'ath Party began demanding eight years of government service from college graduates—in essence, extended military conscription. He left for the US and began all over again, ultimately earning a doctorate at Southern Cal and an offer at an engineering firm. Grateful to be designing pilings for Brentwood's J. Paul Getty Museum instead of being slaughtered like so many of his generation after Saddam attacked Iran, he eventually became a partner in his company and a US citizen.

Then, in the 1990s, he saw NASA images of southern Iraq that he could barely believe: as if setting ablaze all Kuwait's oil fields after the first US invasion weren't enough, by emptying the Mesopotamian marshes, Saddam had perpetrated one of the biggest environmental catastrophes of a century that already had known too many, from passenger pigeon extinction to the 1984 Bhopal, India, pesticide gas leak disaster that killed thousands and injured a half million, to a nuclear meltdown at Chernobyl. When the US invaded Iraq after 9/11, Azzam joined an Iraqi expatriate opposition that lobbied Congress with satellite evidence of what they called Saddam's real weapon of mass destruction: destroying a 7,000-year-old civilization by denying it water. Their attempts to convince the State Department to bomb Saddam's levees proved unsuccessful, but once he was toppled, Azzam decided to see for himself if anything could be done.

Jassim Al-Asadi's mother had been out collecting reeds one day in 1957 when her water broke early, so he was born in her mashouf. Because the Asadi tribe were latecomers, arriving merely 1,300 years ago when an old enemy forced them out of Babylon, they didn't consider themselves Ma'dan. Nevertheless, Jassim grew up in a reed house without electricity, poling his mashouf to one of Iraq's best primary schools because its teachers were freethinking troublemakers exiled from Baghdad to Chibayish—then Iraq's equivalent of Siberia, because it was accessible only by boat.

Jassim earned a scholarship to high school and then to Baghdad University, where he became a freethinking troublemaker himself. For refusing to join the Ba'ath Party, he was jailed for nine months by Saddam's secret police, who tortured him by wadding newspaper between his toes and lighting it. Upon his release, he earned his engineering degree and a post with the Ministry of Water Resources. He was working there when Azzam Alwash appeared in Baghdad in early 2003 with a US Agency for International Development delegation.

During the previous decade, Azzam had personally lobbied governments and conservation groups in Washington, London, Geneva, and Rome and at the UN for a project he called Eden Again, raising exploratory funds but no actual commitments. He'd put so much time into his dream of rewatering the marshes that when he asked his engineering firm for a leave of absence, he was fired instead. Through the expatriate Iraq Foundation, he cobbled together enough donations to set up a Baghdad office, and kept looking for money.

USAID, he learned, offered funding for agricultural rehabilitation in the post-Saddam era. He arranged site visits for agency personnel; as this was still a war zone, they traveled by military helicopter and armored vehicle. A pair of engineers from Iraq's Ministry of Water Resources accompanied them. One was Jassim Al-Asadi.

The only significant vegetation and wildlife left were at a marsh that Saddam had avoided because it spanned the Iranian border. Azzam didn't even recognize the Central and Hammar marshes of his boyhood, two of the biggest, until Jassim told him where they were—all was bare desert and charred stumps. After the water was gone, Jassim explained, Saddam's forces pulled remaining trees up by their roots and burned everything.

He took them to his family's mudhif, where they sat on carpets and drank tea with his uncle Fadhel, a prosperous fish merchant before his world went dry. Azzam remembered Chibayish's charming canals from when his father, irrigation director for Dhi Qar, the regional governorate, brought him here to fish. He promised Fadhel he'd be back.

Months passed. No USAID funding appeared. An Italian grant was delayed. Azzam, nearly broke, was in Baghdad in December 2003 when he received a call from Jassim.

"Can you please come to Chibayish?" People who had been expecting something to happen were desperate, he said.

A day later, Azzam, Jassim, and the irrigation director with Azzam's father's old job stood atop the seven-meter-high, five-meter-wide embankment Saddam had built with fill from hundreds of kilometers away, to divert the Euphrates from its former course through the marshes. Below them, Ma'dan with picks and shovels were trying to hack through it. Suddenly Azzam realized they didn't need some big internationally funded project.

"We just need an excavator."

They turned to the irrigation director. "If the minister hears that I sent a machine to cut this embankment—"

"No one will know whose machine it was," Jassim promised him.

The excavator, a Chinese amphibious backhoe with a jumbo bucket, was a hundred kilometers away in Nasiriyah, Azzam's hometown. While Jassim rented a trailer for 100,000 dinar,* Azzam called his father, who knew Dhi Qar water topography better than anyone. The next day, December 19, they made three cuts, exactly where his father said the embankment should be breached. Hundreds of Ma'dan rejoiced as water gushed into the marsh for the first time in 10 years.

The water, mixed with parched soil, turned a lifeless metallic rust-red. For a few awful weeks Azzam dreaded that a decade of working to restore the marsh had been all for naught. But after a month, sprouting cattails lined the shores. Belying many biologists' predictions, up through the water came reeds: fast-growing phragmites, thick as bamboo.

A few months later, the rehydrated marshes were bright green. The arrival of kingfishers, herons, egrets, and pygmy cormorants proved that fish were back. For a decade, birds migrating between Africa and Siberia had packed into the sole remaining wetland on the Iranian border, competing for limited food. Now, spoonbills, ibis, coots, godwits, marbled teals, and two species of eagles again spread through the reconstituted marshes. The endemic Basra reed warbler was again spotted in the only known place it breeds. Mammals, too: boars, badgers, smooth-coated otters.

Within a year, tens of thousands of Ma'dan and their water buffalo returned.

* $31 US. Including diesel, the cost to rewater the marshes totaled under $100 US.

iii. Future

"We trusted that one day Allah would return our marshes to us," Abu Haider tells the tourists, Jassim translating. "But there's a different shape to them now."

Raising his bearded chin, he gestures toward a case of plastic water bottles. "The marshes are now too salty to drink. We have to buy water in Chibayish."

They rarely catch Mesopotamian whitefish or barbel anymore, he adds. Mostly they get introduced carp, or salt-tolerant African tilapia so small they fry them whole to eat like chips. He doesn't mention that to get enough, they go out illegally at night to electroshock.

After wishing the tourists safe travels, Abu Haider motors Azzam and Jassim north, into the Central Marsh. They pass alongside a landfill spit that conceals a buried Chibayish sewage line; at its end, black effluent gushes from a three-foot pipe. For a decade, they've planned a natural treatment plant here that would also be a public park: a 6.5-acre "living machine" using reeds, flowering shrubs, and fruit trees to filter the sewage, patterned to resemble an Iraqi blanket. From the garden's farthest end, clean water would flow to the marsh. Inevitably dubbed Eden in Iraq, it was designed by American artist Meridel Rubenstein, who since Vietnam has created artworks in US war zones. But like much in Iraq, it's been bogged in permitting requirements, entailing endless presentations to the mayor, the regional governorate, and the Ministry of Water Resources—only to have the process start anew after each election when new officials take over.

Azzam notes that reeds are flourishing around the reeking outfall, suggesting that they're already filtering the discharge. But he supports Rubenstein's project because the war-battered people here would have something beautiful to feel proud of, and it could attract new international attention to this last slice of Eden. That's needed, because 20 years after they rewatered most of the marshes with plans to revive them all, now all is imperiled again.

"Okay, let's go," he tells Abu Haider. Via a maze of trails cut by water buffalo between walls of 12-foot reeds, navigable only by boatmen whose families have been here for centuries, they arrive at the northernmost extreme of the reflooded marshes. Here, vegetation gives way to an open expanse of water, and the wind chops the surface where they're bobbing.

Abu Haider stands at the mashouf's bow swaddled in his keffiyeh, humming one of his dirges and contemplating the bleak shore. When he and Jassim were younger, this, too, was thick with green reeds. For nearly three decades it's been just bare dirt, dark as a cup of chai before adding buffalo milk. A village of 500 Ma'dan with a school and a clinic was here. All are gone, because during Saddam's time, Turkey began damming the Tigris and Euphrates upstream. After Saddam finally fell, there was only enough flow to revive two-thirds of the former wetlands, about 5,500 square kilometers total.

After Turkey completed its 22nd and second-biggest dam in 2020 on the Tigris, there's now even less fresh water: Iraq is getting barely a third of what it used to receive. Syria, the other country these rivers pass through, is in the same predicament. Turkey increasingly treats the Tigris and Euphrates as domestic waters and ignores pleas for a water-sharing agreement.

"It's not just Turkey," says Azzam. "Farmers use way more than they're allotted. There's no enforcement of the regulations." Barley and wheat farmers upstream are still flooding their fields like ancient Sumerians did when they invented irrigation—but 5,000 years ago, no one upstream was impounding 80 percent of Tigris and Euphrates water, and the snowpack in eastern Turkey's Taurus Mountains, where they originate, was twice what it is today.

"It's not just farmers," adds Jassim, "it's drought." From 2013 to 2018, it got so dry they channeled drainage runoff from farmland to keep part of the marshes alive. Then a blessedly wet 2019 diluted the pesticides and fertilizers. For the first time since Saddam, the marshes were almost 100 percent full. Mashoufs were piled with fresh-cut reeds for fodder and island repair. Then the cycle reverted in 2020, and again the marshes were only 20 percent full, which continued into 2021.

With decent winter rains, 2022 started well—then turned into the worst. For three straight months, water levels were sucked down by daytime temperatures averaging 46.1°C (115°F), sometimes reaching 51.1°C (124°F).

When numbers overwhelm, there's a danger of masking the reality of what is happening. Part of that reality is a searing orange light on the horizon behind Jassim's head, rivaling the noon sun. Thirty-seven kilometers away, a Russian company, Lukoil, is flaring methane from Iraq's West Qurna oil field. Nothing in the Bible mentions that beneath the Garden of Eden lay immense quantities of petroleum. For 300 million years, an ancient ocean of which the Mediterranean is a remnant covered today's Middle East. During Jurassic and Cretaceous times, when plate tectonics positioned this region over the equator, vast amounts of algae, phytoplankton, zooplankton, and bacteria settled to the ocean's floor, forming layers miles deep that compressed to black ooze.

Beginning 55 million years ago, as Arabia and Asia collided, tectonic forces pushed up the seabed. The exposed land grew parched. Through its cracking soil, that pressurized ooze seeped to the surface. The Sumerians, building boats from reeds because trees were scarce, used the black goo to waterproof their hulls—as did Noah, attest ark-building instructions in the Epic of Gilgamesh, Genesis 6:14, and a Babylonian flood narrative translated in 2013. Abu Haider's mashouf is still coated with it. The Sumerians used tar and bitumen to mortar the mud bricks of ziggurats from Ur to the infamous one at Babel. Mesopotamia, literally, was the first civilization built on fossil hydrocarbons.

Today, the remaining fragments of that civilization are oil-soaked, crude or its derivatives being nearly Iraq's only exports. That all began in 1953, when the southern part of Hammar Marsh was drained to develop Rumaila, the world's third-largest oil field. Norway's Equinor, ExxonMobil, BP, Shell, Korea Gas Corporation, and China's CNPC are all here. Decades of war plus corrupt government, alternately puppeteered by Americans or Iranians, have kept most Iraqis impoverished despite their country's petro-wealth—but not untouched by it. Each day, petroleum companies inject 4 million barrels of their precious water to push out even more oil, and flare even more gas into their acrid air. While the rest of the world frets over passing 1.5°C, land temperature data show Iraq is already 2.3°C warmer.

When civilization's cradle is deep-frying in its own oil, what does it tell us? Atop everything else, Iraq's population of 40 million, growing at more than twice the global rate, will double by 2050 (barring catastrophe, which seems increasingly inevitable). None of that is any secret to Azzam and Jassim, floating at the brink of their dream of resurrecting something thought lost forever. Yet still they continue on.

Since founding Nature Iraq, the country's first environmental nongovernmental organization, early this century, they've gotten the Mesopotamian marshes designated as Iraq's first national park and a UNESCO World Heritage Site. Besides the Goldman Environmental Prize, Nature Iraq received one of the Arab world's highest honors: the Takreem Award for sustainability and environmental leadership. They've trained 500 field ecologists to collect and catalog the entire country's biodiversity, preserved wildlife corridors, and teamed with an Iranian NGO to protect the rare and elusive

Persian leopard. The majority of Ministry of Environment employees have passed through Nature Iraq.

All this has come despite personal cost. Azzam's California wife loyally supported and commemorated in a memoir his obsession to revive one of Earth's key wetlands—but his obsession never abated, and their marriage couldn't survive his prolonged absences. Jassim's wife refuses to live in the marshes anymore; he now commutes twice a week from a gated community in Babylon, four hours away.

And while they've given so much, it's still not enough. "It's time," says Azzam, "for bigger, unorthodox ideas." A chance came in a 2020 phone call from President Barham Salih after Joe Biden's election. With Trump gone, Iraq's president asked him, what will America now support? Azzam sent him a briefing paper he'd just written for an American think tank, The Century Foundation, imagining a radical climate approach for not just Iraq but the entire Middle East: the world's hottest region, its oil-drenched economy financing its own immolation. It would take unprecedented cooperation among often hostile neighbors—for instance, Iraq storing water in Turkish reservoirs, where there's less evaporation. In exchange, Iraq would ship gas to Istanbul instead of flaring it, and beyond into Europe. Gradually, the gas would be replaced by solar electricity or solar-electrolyzed hydrogen from photovoltaic infrastructure begging to be built in one of Earth's sunniest countries.

His scheme would expand to Kuwait, Saudi Arabia, Qatar, and Iran: connecting first their gas fields and hydroelectric, then their exploding solar output in a region-wide grid, to share excess capacity with each other and reduce each country's cost of generating electricity. Using new storage media like hydrogen and ammonia, they could be a transport corridor of solar electricity to Turkey and Europe. Iraq's strategic geography could become a "dry canal": roads and rail lines linking all the Gulf ports to Europe, coordinating with Egypt to handle freight unsuitable for the Suez Canal. To deal with rising seas pushing salinity upstream, Iran and Iraq together could build something like the retractable flood barrier on the Thames, to close the Shatt al-Arab during high tide and open it when the tide recedes.

It was as if Azzam would perform cultural brain surgery, ablating lobes where nationalism, tribalism, sectarianism, and a few thousand years of unrequited vengeance reside. But his plan made enormous sense: it would save water—including phasing out the oil industry, one of the thirstiest users—and provide the Middle East a post-carbon map for remaining a profitable energy exporter.

Dreaming on, Azzam's paper proposed a "Garden of Eden" reforestation blitz for Iraq, which, before Saddam, was a regional breadbasket with year-round agricultural production. Within two years, Iraq could replant the missing two-thirds of 30 million pre-Saddam date palms and add citrus trees with vegetables sown in their shade. By treating wastewater and recycling plastic trash into drip-irrigation and rainwater-harvesting equipment, Azzam calculated, they could save enough water to plant a billion trees and have a bounty of exportable fruit and vegetables by 2030.

He imagined tackling the 5 million cubic meters of sewage dumped daily into the Tigris and the Euphrates, and to how to pick Iraq's next national parks. Most audacious of all was his long-term idea for sharing Euphrates water with the Jordan River, via an old pipeline right-of-way for exporting oil to Haifa during British Mandate Palestine. "Removing water from the tensions between Jordan, Palestine, and Israel—even Syria—cannot but help the peace process move forward," Azzam wrote.

"Nature itself gives us the most wondrous example of an interconnected global network," his paper concluded. The president moved Azzam into his Baghdad guesthouse to shape it into a formal Iraqi strategy for the future. The Mesopotamian Revitalization Project was formally posted in late 2021. The following year were national elections. As usual, mayhem ensued, followed by a year of paralysis that only entrenched the Iranian militias more. Eventually, a new president emerged, and the Mesopotamian Revitalization Project Azzam compiled ended up buried in the presidential archives.

As they head back to Chibayish, Abu Haider pauses for a line of buffalo scrambling up a canal bank, water pouring from their ebony flanks. One glance tells him they aren't his; Ma'dan can tell their buffalo apart by their shades of black. Likewise, the buffalo recognize their breeders by their voices and the smell of their clothes.

An afternoon breeze ripples the reeds. The throaty choruses of awakening frogs mask the hiss of geese patrolling the islands' perimeters. "Jassim, is the elevation right here above sea level?" Azzam asks.

Jassim squints at his GPS. "Minus 20 centimeters."

With the world not holding at a 1.5°C temperature rise over preindustrial levels, and probably not 2°C either, "by midcentury, sea level could rise half a meter, so—" Azzam stops mid-sentence, knowing that it means seawater will be all the way to here. The marshes need water, but not that kind.

To the north, Turkey threatens to build yet another dam, and to the south, the Gulf is rising; exploratory oil rigs are circling the marshes; and with chaos in Ukraine causing food riots over cooking-oil shortages, agriculture wants more water. But the good news is that the latest minister of water resources is interested in the marshes. Granted, as an engineer under Saddam, he signed the study for draining them. "Not that he had a choice. He's trying."

"Maybe he'll back the Mesopotamian Revitalization Project," muses Jassim, as the mashouf slips back into the canal in front of Nature Iraq.

"I'll never give up hope," says Azzam as they step ashore. "Its logic is undeniable. The political will doesn't exist now. Things must worsen before people see they must make a politically courageous decision. The crisis will force them to take stock. In the meantime, we keep daring to dream big."

They walk to have tea at Jassim's uncle Fadhel's mudhif, where years earlier Azzam promised the former fishmonger he'd do something about the desiccated marshes. Beyond them, lights blink on the rusting metal bridge the Russians built. The bridge was absurd, Saddam Hussein was absurd, and what petroleum companies do to this world with full knowledge is most absurd of all, with the reckoning coming swiftly to this deepest cradle of civilization, where writing and the wheel were invented and Hammurabi inscribed the first code of laws.

"But Iraq is eternal," says Azzam. "Iraq is the Tigris and Euphrates. It existed before humanity. We are humans with very short lifespans. Iraq and the rivers will be back."

CHAPTER TWO

AGAINST ALL HOPE

i. The Stakes

West of Chibayish, a two-lane highway heads through a desert chalky from shells deposited by a flood of biblical proportions that once covered today's southern Iraq. The road passes Iranian-backed militia checkpoints and the blocky prison complex the Americans built for ISIS until, after 90 kilometers, another blocky silhouette appears on the horizon: the Ziggurat of Ur. To burnish his international image, Saddam Hussein reconstructed the lower third of the tower, which once rose at least a hundred feet. On hot days, its three massive exterior staircases still ooze bitumen mortar. From its top, where ruins of its remaining height lie scattered, can be seen a former US Air Force base, now the airport for the city of Nasiriyah. Just below it is another structure of ancient Ur of the Chaldees that Saddam restored: the home of the patriarch called Abraham by Christians and Jews, and Ibrahim by Muslims.

Archaeologists generally agree that Ur was Abraham's birthplace, and that his house was in the neighborhood of the one Saddam reconstructed. Large, labyrinthine, with myriad arches and alcoves, it suggests that Abraham's father, Terah, was a man of means. If so, the Bible doesn't say why he took Abraham—Abram, before God awarded him an additional syllable—and his brother Nahor, their wives, and Lot, the son of their deceased sibling, from Ur to a settlement called Haran. A chapter later, God famously orders Abram on to Canaan, to father a great nation.

Twenty-five eventful years follow: To save his own skin, Abram tricks two different kings into bedding his wife, Sarai, by claiming that she's really his sister (true—at one point, he reveals that they're half-siblings). His nephew Lot goes to live in Sodom, which God then destroys along with Gomorrah for being immoral, yet spares Lot even after he gets drunk and sleeps with both his daughters. A childless Sarai invites Abram to ravish her enslaved handmaiden, who bears a son, but later—as Sarah—she banishes them after finally conceiving herself, at age 90, when Abraham is 100.

Writing about him nearly 2,000 years later, the Apostle Paul coined a phrase that became enshrined not just in our lexicon but in our consciousness:

> Hoping against hope, he believed that he would become the father of many nations. (Romans 4:18, New Revised Standard Version)

Hope against hope: we all say it, but what does its intrinsic contradiction mean? Why does hope struggle against itself? Etymology isn't helpful: the root of Proto-West Germanic *hopōn*, which became *hopian* in Old English, is lost. But Latin hints at how hope can be opposed to itself—and why that matters to us, especially now.

Latin's *spērāre—esperar,* in today's Spanish and Portuguese—can mean "to hope," "to expect," or "to wait." Which meaning is intended depends on context, but they all flavor each other. You wish and await something to happen, you may even expect it, but until it does, doubt lurks that it may not, or dread that it likely won't. The future is unsure; fate is in God's hands, not ours. (Given the high probability of dashed dreams, early Catholic legend has Hope beheaded along with her sisters, Faith and Charity, in front of their mother, Wisdom, their consolation being martyrdom.)

Before industrial momentum began suffusing Westerners with dreams of molding their own destiny, *hope*'s multi-entendre was also implicit in English. If hopeful expectation is constrained by uncertainty, *hope against hope* makes sense. The first hope yearns for something good to occur; the second concedes it may not. When things get really desperate, the fact that we still hope against all hope means that deep down, like our ancestors we still believe in miracles—even when the odds feel as long as a 90-year-old woman bearing a son named Isaac.

ii. Miracle Quest

What is so doubtful or imperiled today that it begs near-miraculous intercession?

Four things, each big enough to shake the entire planet, so entwined that easing or aggravating any of them affects the other three. Intractable as they might seem, each has spawned unexpected heroes like Azzam Alwash and Jassim Al-Asadi, rising to what will be humanity's most defining moment.

First, the flora and fauna we evolved alongside and depend on for food, shelter, soil, water, climate control, and sheer beauty are rapidly abandoning ship—or rather, being forced to walk the plank.

If that maritime metaphor feels forced, consider its long history. In the Bible, God commands Noah; in the Qu'ran, Allah orders Nūh; in the Epic of Gilgamesh, which preceded both, Ea tells Utnapishtim; and in the Akkadian version, an unseen voice warns Atrahasis the same thing: that in order to save humanity, he must build a boat to save the animals, too (some of these legends also include seeds).

When deities concur with scientists, we should listen. Worldwide, we've lost more than two-thirds of our total animal wildlife population in the last 50 years. That's bad, because nothing happens in isolation. A species' behavior may seem unremarkable—pecking a hole in a tree, for instance—until it turns out that dozens of other species will use that hole or the sap oozing out of it, or the insects attracted to that ooze, or the fungi that feed on it, or the pollinators whose larvae need the fungi. Extract any one, and things can collapse surprisingly swiftly.

It took neither gods nor experts to notice that our windshields are no longer yellow with insect spatter after a long drive, but once entomologists investigated, they found that nearly half of all insects species are declining. Fewer mosquitoes sounds great, until you tally how many birds and bats eat them and how many fish depend on their eggs. Just as no scientist yet knows what detonated or preceded the Big Bang, no one knows how much—or exactly which—wildlife is critical to the survival of our own species. There's only one way to find out, and it's not recommended.

Much of our world is still vibrant. Plants still abound; things still crawl among them; and on seven continents, impassioned people are doing their utmost to save space for them to thrive—often in competition with other people whose depth of knowledge is inversely proportional to their clout.

In a wise world, scientists would make decisions best suited to scientists, and politicians and economists would make decisions appropriate to their expertise, whatever that might be. Unfortunately, that's not what happens.

Some are fighting to lifeboat—that metaphor again—as many as possible into the future before our ark sinks, to repopulate Earth after things die down, literally, as happens with each major extinction. There's a good chance you saw a survivor today. We once thought

Earth lost all its dinosaurs, until scientists discovered that those lumbering reptiles often sported feathers. Birds, it turns out, are small dinosaurs. (Had we gotten to see monsters like New Zealand's 12-foot-tall moa or Madagascar's half-ton elephant bird, we might have drawn the connection sooner, although the resemblance of ostriches to theropods was a clue hiding in plain sight.)

That ostriches, hummingbirds, sparrows, and every duck, dove, and nightingale in between are still alive means their ancestors made it through the Chicxulub asteroid's collision with the Yucatán Peninsula 66 million years ago: an astonishing testament to life's resilience, given that shock waves equivalent to a megaton hydrogen bomb pulsated every six kilometers. Across the planet, incandescent glassy stones rained from 40 miles up, setting the world ablaze, killing nearly all phytoplankton as the oceans sizzled.

Then came cold, as soot from the impact plunged the Earth into a year of dark twilight, during which most birds surely did not survive, even as the forebears of today's miraculously did. Possibly we've already entered another twilight. In the past 50 years, bird numbers in North America alone are down by one-third: 3 billion fewer fill the morning with song. Efforts to save the rest abound, but some of what's causing avian genocide is already out of our control, and what's still within our grasp to fix depends on us remaining around—no longer a sure prospect.

Nothing is, really. In the Age of Uncertainty we're in now, we'd do best to keep every animal, plant, and fungus that we can see—including kingdoms we can't see: bacteria, archaea, amoebas—simply because we can't have a world without them.

Second, our worsening weather finds growing numbers of us beseeching miraculous deliverance from floods, drought, fires, or terrifying winds. By now, even the most bellicose deniers know the climate has come unmoored—they're either so scared that they lie to themselves or they're making so much money that they lie to us.

For decades after scientists realized why atmospheric carbon dioxide is increasing, coal and petroleum merchants tried to confuse us by insisting that the most abundant greenhouse gas is water vapor. That's true, but grossly misleading: evaporation from surface water or leaves doesn't stay airborne long, because it condenses into clouds that fill until they spill rain—and the cycle repeats itself. That started long before industrial times; plants began transpiring a half billion years ago, and oceans appeared well before that.

The last time there was this much CO_2 aloft was 3 million years ago, when seas were 80 to 100 feet higher. Before we revved up our engines, extra atmospheric carbon dioxide came mainly from volcanoes. A particularly spectacular eruption 252 million years ago in Siberia burst through the Carboniferous layer—talk about CO_2 emissions—and lasted more than a million years. That kindled the Permian extinction, Earth's biggest ever, exterminating over nine-tenths of all species. A study by Stanford and the University of Washington suggests that loss of oxygen in hotter seas choked most sea creatures to death. It took millions of years for life to flourish anew, and when it did, it looked a lot different than before. A whole new cast of characters had to evolve, most prominently those large reptiles.

That study and other models project that the amount of carbon dioxide we're expelling could perpetrate another Permian. The warming already underway is accelerating due to positive feedback—only there's nothing positive about it.

It works like this: In a greenhouse, the sun's heat can't escape, because it bangs into a glass ceiling. Instead, it's absorbed by the soil, whose microbes get active and exhale more greenhouse gases, even as more sunlight adds more heat, and so on—so greenhouses must be vented, or the plants wilt. In 2014 came a dramatic example, also in Siberia, of how positive feedback unfolds on Earth—now one big greenhouse, minus the vents. The player in this case was methane, or CH_4, a molecule with one carbon atom surrounded by four hydrogen atoms. Methane doesn't stay in the atmosphere for centuries like CO_2, more like a decade, but while it's airborne it has 86 times the heat-trapping effect.

In small doses, methane isn't toxic: every day, we have several close personal encounters with it. But that July, when Russia's Scientific Research Center of the Arctic discovered three new craters in the Siberian permafrost, it appeared that our planet was also farting. The largest, spotted by a helicopter pilot on the Yamal Peninsula, was 200 feet wide and 100 feet deep—yet no meteors had struck. When methane levels in the craters proved 53,000 times above normal, the scientists realized that with polar regions warming three times faster than the rest of the planet, after two consecutive Arctic summers averaging 4°C above the 1979–2000 baseline, frozen methane was not merely thawing—it was exploding.

By 2020, average Siberian temperatures had risen by another degree Celsius. That spring, geologists at the University of Bonn were watching a satellite map of atmospheric methane over western Siberia. By now, permafrost farts had become routine. But what the Bonn scientists saw in Russia's northernmost Taymyr Peninsula made no sense: two parallel arcs of elevated methane, each around 375 miles long.

By the following March, both had been swallowed by rising methane concentrations all around them. Yet there were nearly no permafrost soils in the region, just outcroppings of limestone bedrock. The Bonn geologists' hypothesis, still unchallenged, is that surface-warming is freeing ancient gas frozen in limestone's cracks and pores. Vastly more methane, they realized, may be locked in bedrock than in permafrost. Now it, too, is being released—positive feedback that, they concluded in the *Proceedings of the National Academy of Sciences,* "may be much more dangerous."

We don't know how to stop what science is calling a phase shift—meaning, having to adapt to a different climate than our species has ever known. The technology for retrieving all the carbon that nature, not needing extra energy to run Earth, had buried away until we exhumed and blasted it skyward in just a couple of centuries—that technology hasn't achieved miracle status yet. It would require somehow building, fueling, and deploying millions of carbon-sucking machines worldwide—using no dirty energy.

But while we're waiting for carbon dioxide removal to become viable on such a scale, stalwart efforts are underway to *slow* the climate's rate of change, to buy ourselves and the miracle workers time to figure out how, if not actually to pull us back from the brink, to smooth our transition to the next phase.

The third and fourth miracles we could sorely use are inseparable.

Like every other animal, we *Homo sapiens* once spent nearly all our waking hours searching for food—searching so far that by the end of the last ice age, we'd reached nearly everywhere but New Zealand.* The archaeological record suggests that in several places independently—first Mesopotamia; followed by the Nile, Indus, and Yangtze deltas; then subtropical Mexico, the Incan highlands, and the Mississippi River valley—we gradually figured out not to eat all the seeds we gathered, but to let some of them grow

* Although every other sizable South Pacific island was colonized by humans long before, we missed New Zealand entirely until AD 1300. Once we finally discovered that landmass, it took only a century to dispatch South Island's 12-foot moas. Their demise cascaded immediately into the extinction of their sole predator, Haast's eagle, the largest known raptor, outweighing Andean condors.

even more for us, conveniently close by. Then we discovered that selecting seeds from the most fruitful plants yielded far more bountiful results than foraging hither and yon. Learning to improve each new seed-bearing generation further by crossing their pollens, we got more to eat and expended less energy.

More didn't necessarily mean better, however, because our diet became less varied. Analysis of ancient human remains reveals that the more we depended on fewer plant species, the worse our nourishment became, leading to maladies like dental cavities and cancers that our paleo ancestors rarely knew. Because we've become adept at fighting infectious diseases, we live longer, but in many ways we're not nearly as healthy as our forager forebears.

What we did with plants—selective breeding—we also figured out how to do with the animals we caught and confined so we could eat them at our leisure. For example, the aurochs (rhymes with "ox"), a megafaunal species we didn't drive into extinction so much as breed out of existence. The wild ancestor of today's cattle, it resembled a lean Texas longhorn but stood nearly a foot taller. We probably captured them with snares made from vines and plant fibers we twisted together. By stringing vines and eventually ropes around tree trunks, we could confine aurochs in enclosures, as well as other quadrupeds such as wild sheep and goats.

From there, it was a small step to shaping branches into fence posts. After a couple hundred generations of corralling the biggest aurochs bulls, with cows selected for meatiness, docility, fast growth, and—compared with wild aurochs' tiny teats—large, easily graspable udders, we got today's fat-marbled, grain-bloated aurochs descendants.

But besides reducing a magnificent Pleistocene creature to lunch meat, breeding for these advantageous traits has downsides. Multiply our 1.5 billion cows by four stomachs apiece, each with such primitive digestion that they must regurgitate and remasticate everything they eat, causing nonstop belching (plus, their stomachs were designed for grass, so the grain we feed them leaves them even gassier), and CH_4 emissions from cattle surpass all the methane we release by extracting and using fossil fuel.

Add what goes into producing that grain, and how much grain we're talking about, and then greenhouse gases really start spewing. Today barely half the world's grain is grown to feed humans. Most of the rest feeds animals that people eat. What remains gets turned into fuel—mainly, the ethanol in American gas tanks, supposedly a planet-saving renewable but really a gift subsidy to agro-industry.

More than two-thirds of American crop calories from grains and soy are fed to animals. After livestock and the ethanol barons evenly split nearly the entire corn harvest, humans are left with a measly tenth—and a third of *that* gets turned into high-fructose corn syrup, which more than doubled diabetes rates since Coca-Cola switched to it in 1980 and other soft drinks soon followed.

To vegans, doctors, and the health-conscious, that's only one reason why stuffing cattle with grain is unconscionable. (Try asking a vegan how much more water it takes to produce a pound of hamburger than a pound of veggie burger.*) But regardless of diet or ethical considerations, what matters is how our meat, grains, soy, and vegetables are grown, because along with industry, transportation, and heating and cooling, agriculture seriously messes with our atmosphere—which brings up the fourth miracle we need, inextricably tangled with number three: feeding ourselves without undermining the source of our bounty.

* A 2023 life cycle assessment by Dutch food sustainability consultancy Blonk, commissioned by Beyond Meat, Inc., concluded that producing a quarter-pound plant-based Beyond Burger requires 34 times less water than what's needed to grow a quarter pound of ground beef—a 97 percent reduction.

Since a few can raise food for many, once we mastered farming, our population grew. To live close to this concentration of food, we became sedentary. Some were cave dwellers for protection and comfort from the elements, but since caves aren't plentiful, we learned to create our own walls. Soon we lived in small clusters of human-made dwellings—then in bigger, city-sized clusters.

But nature limited how fast humanity could grow. Like most species, until 1800 we died nearly as fast as more were born. Half the children didn't survive long enough to have children themselves, so human life expectancy averaged around 40 years. If several good harvests let settlements boom, drought or disease regularly knocked their numbers back. In those pre-vaccine days, epidemics were routine. Even with tribes competing to be fruitful and multiply to outmuscle surrounding tribes, our numbers rose so gradually that our population graph was nearly a flat line.

In 1796, British surgeon Edward Jenner discovered a vaccine for smallpox, which had killed millions every year. The new century added treatments for rabies, anthrax, diphtheria, and tetanus. In 1815, 500 generations beyond the Mesopotamian cradle, human population reached 1 billion. In 1863, pasteurized milk began saving innumerable lives, followed by germ discoveries that led to disinfectants and antiseptics. Lifespans lengthened. Infant mortality plunged. By 1900, we'd added a half billion more.

Then, in 1913, two German chemists, Fritz Haber and Carl Bosch, discovered how to suck unlimited nitrogen from the atmosphere, and our graph rocketed skyward. Nitrogen, which makes up 78 percent of our air, is inert in our lungs, but in soil it's an essential nutrient. Previously, the amount of available N_2 was limited to what manure and a few nitrogen-fixers like beans, clover, and other legumes could add. With synthetic fertilizer, we could grow far more plants than nature ever could.

No other invention has changed the world as radically as the Haber-Bosch process. Without it, half of us wouldn't be here. For literally changing the face of the planet by making vastly more land farmable through chemistry, Haber and Bosch each won a Nobel Prize. Despite two genocidal world wars, by midcentury all that extra food pushed our numbers past 3 billion.

In the 1960s, when agronomists successfully crossbred corn, wheat, and rice to grow shorter stalks, diverting the energy saved into producing up to 10 times more grain, we skyrocketed again. This "Green Revolution" came just in time, because South Asia, breeding faster than fertilizer could keep up, was on the brink of famine. Green Revolution founder Norman Borlaug also received a Nobel—the Peace Prize, for saving more lives than anyone in history. Yet on accepting his award, he warned that when people don't die of famine, they live to produce more people who need more food, ad infinitum—except Earth isn't infinite.

For the rest of his life, Borlaug urged population management—to no avail, due to a problem of perception. Growing up, we assume the number of people we see is normal—but in nature, for an already large population of big organisms to quadruple in a century is grossly *abnormal.*

It seems counterintuitive that more food could be a bad thing (or for that matter, more people). But too much of a good thing is simply that: too much.

With jet-fueled technology exponentially magnifying our influence, there are now far more humans than this planet can comfortably carry—8 billion and counting—without going seriously out of whack. Barring global calamity, by midcentury there will be 2 billion more of us to feed, on dwindling land and water supplies. Equally dire, the energy that Haber-Bosch requires to produce synthetic nitrogen ammonia from its feedstock, methane, generates vast amounts of greenhouse gases. In fields, soil microbes then convert fertilizer to nitrous oxide, N_2O, which traps *300 times* more heat than CO_2.

With half the habitable planet either growing or grazing food for us, we're pushing our atmosphere to a boil and other species off the Earth—not only animals: In the tropics, trees topple to make way for more cows and soybeans. In overheated temperate zones, they simply burn.

Miracle number four, bringing our population down to a manageable size before nature does it for us—COVID-19 was barely a preview—is both the easiest and the hardest.

Easiest, because draconian edicts like China's 1980 one-child policy aren't necessary. In any country—rich, poor, Catholic, Muslim, Buddhist—educated girls with access to family planning solve the problem, because they know how many children they can responsibly care for while using their skills to help support them. Worldwide, girls reaching high school average two children or fewer. Since two children replace their parents, population doesn't grow; with one or none, population shrinks.

Hardest, because after fertility dips below replacement rate, it takes two to three generations for population to subside, because so many in their childbearing years are already alive.

The father of capitalism, Adam Smith, envisioned growth stabilizing in a mature economy, not a pyramid scheme depending on more consumers being born indefinitely. Another reason economists like big populations is the more people, the lower the wages, as poor masses compete for miserable salaries. When fewer workers are born, they become more valuable, so wages rise—and they become happier and more productive, which should more than satisfy CEOs, who are already downsizing their labor force through robotics anyway.

With most humans now living in small urban spaces, children are no longer economic assets in the fields, so families are shrinking. Still, 100 million people predicted in Lagos by 2100 defies any notion of sustainability. To feed everyone, yet somehow save enough terrain for wildlife lest mass extinction crash our ecosystem even before runaway climate, is a challenge seeking miracles.

As long as life goes on, it's hard to conceive of it ever stopping, or our civilization ending, or our planet becoming unlivable. But the usual litany of denial—human ingenuity will solve these things; technology will get us out of the mess it got us into—is belied by cyclone bombs and wet-bulb temperatures that caution us otherwise. When the only dinosaurs that survived the last major extinction are now disappearing, that should give us pause.

Amid all, stubborn, visionary men and women refuse to give up on finding us a way through this shaky century, despite what they know. Some will throw Hail Mary passes that underscore the desperation creeping into our thoughts and vernacular, dystopian science fiction being the defining literary expression of our times. Proposals that a generation ago would have defied credulity—launching clouds of satellites to partially block the sun, or fleets of airplanes perpetually spraying sulfur into the stratosphere to do the same—since the 2015 Paris Climate COP* are now seriously considered in annual climate meetings.

More schemes, like hanging thousands of miles of underwater curtains to shield Greenland's glaciers from warming seas, or blanketing them with reflective silicon pellets, or replacing lost Arctic

* Conference of the Parties.

ice with rafts of floating mirrors to bounce sunlight back into space, will likely fail hilariously—that is, if we weren't talking about a real planet.

But our hopes persist that others will amazingly succeed, and in undreamed-of ways pull us through, and even let us thrive anew.

CHAPTER THREE

FOOD FROM THIN AIR

No doubt about the evidence
things sure have gone awry,
what we know now would
make your sweet hair curl.

Prime ministers and presidents
they don't know what to try,
they're just hoping that maybe
Molly can save the world.

—BRITISH BLUES BAND BASKERVILLE WILLY

i. The Math

To hear Molly Jahn laugh—a huge, mirthful cackle that unexpectedly bursts from this slender woman with loose blond hair, dark blue eyes aflutter behind big glasses—you'd almost think she's talking about something that's funny.

It isn't. It's the Second Law of Thermodynamics: entropy; order crumbling into chaos. It tells us we're inevitably doomed, she says, unless we do something drastically different from anything we've done for the past 10,000 years. She's determined to show us how.

Before inspiring that Baskerville Willy song—written by the band's rangy bald lead guitarist, Bill Rutherford, who also holds a chair in biochemistry at Imperial College London—Molly Jahn was celebrated for a cultivar so ubiquitous in gardens worldwide that even if her name doesn't resonate, you've probably tasted her work. After a boring master's program of sequencing genes at MIT, she followed her own genetic legacy: her great-great-grandfather William Saunders was a pioneering Canadian fruit hybridizer, and his son Charles bred Marquis, once North America's predominant strain of wheat.

After earning her doctorate in plant breeding at Cornell in 1988, she was invited to join the faculty. The next year, a colleague showed her a green-ribbed, long yellow gourd he'd found in an Ithaca grocery store. It was a delicata squash, briefly popular in the US in the 1900s, but so susceptible to powdery mildew, the most virulent squash disease, that it needed repeated doses of fungicide to survive.

Extracting some seeds, she decided to cross it with hardy acorn squash. Ten years of selecting and inbreeding later, she had mildew-impervious Cornell's Bush Delicata. In 2002 it won American horticulture's highest honor, the National Garden Bureau All-American Selections Gold Medal, and is now grown on six continents.

In Molly's office hangs a painting of her Bush Delicata. The artist, Sarah Paolucci, included her own hand and brush rendering a squash blossom to honor the tireless hand pollination behind Molly's creation. On her desk is a framed photograph of her baby daughter reaching for a cantaloupe that Molly bred to resist maladies like watermelon mosaic virus and zucchini yellow mosaic virus. Seductively sweet, it's named for the moment captured in the photo: Hannah's Choice.

In 2006 Molly left Cornell to become dean of the University of Wisconsin agriculture school, a position that alternated with frequent government service leaves at national laboratories—and, in 2009–10, a stint as acting deputy undersecretary of agriculture. There her focus was food security risks in the 21st century—and global security itself, should a convergence of climate stress and shortsighted farming practices destabilize harvests. The world's diet depended on four commodity crops: corn, wheat, rice, and soybeans. Each was vulnerable to disease and weather anomalies. A shortage of any, or of the synthetic ammonia-nitrogen fertilizer they required, could catalyze market mayhem and nutritional catastrophe.

In 2011, she was asked to address a conclave of 20 American and 20 British food scientists who'd just been awarded millions by the US National Science Foundation and UK's Biotechnology and Biological Sciences Research Council to enhance nitrogen uptake in crops. Of particular interest to the funders was deducing how the first photosynthesizer, ancient pond scum—cyanobacteria, often misnamed blue-green algae—fuels its own growth directly from airborne nitrogen, in hopes of genetically inserting that secret into corn, wheat, and rice.

A biochemist, Bill Rutherford, renowned in his field for discovering the chlorophyll key that limits growth in stressed plants, had spent years trying to pinpoint how a single cyanobacteria enzyme splits the H_2O molecule. Two and a half billion years ago, the ability to do that changed life on Earth by freeing oxygen into the atmosphere: the most momentous development in biological history.

Splitting water also yields hydrogen, which cyanobacteria combine with nitrogen to form ammonia: plant food. Humans, supposedly a higher life-form, keep trying to emulate what these simple bacteria do, without much success.

The conclave, dubbed the Nitrogen Ideas Lab, was held in a turreted brick Jacobean mansion-turned-hotel 30 miles from Liverpool, with enormous carved fireplaces, frescoed ceilings, and diamond-mullioned windows. For three days Molly snuggled in a Jacobean four-poster canopy bed, working on a review she was writing of British crop science—until someone knocked and asked if she was ready for her keynote.

She'd forgotten that was why they'd invited her. "Can I have a few minutes?"

"No, Dr. Jahn, they're waiting for you now."

It was a classic academic nightmare come true: an audience full of elite molecular biologists, including the star of her Wisconsin faculty, and she felt totally unprepared. Walking slowly, she thought furiously. Having left plant-breeding to a capable protégé at Cornell, Molly Jahn had moved into planet-scale questions that felt more like koans. Earth was a battery that the sun had been charging for billions of years, until one species figured out how to discharge all that stored energy, which was escaping in all directions. Climate change was too many high-energy carbon compounds aloft in the atmosphere. So much energy had been pushed into humans that there were now billions of us, and obesity was epidemic.

To maximize yields, agriculture was hooked on high-energy nitrogen compounds. Like a junkie's ruined body, it constantly craved more chemicals that fouled soils and waters. Annually, global agriculture was using 200 million tons of synthetic fertilizer: nearly four times the weight of the heaviest human-made structure, China's Great Wall.

In the 21st century, *Homo sapiens* were painted into a Faustian corner. Chemistry had produced far more food than nature could, leading to more people than nature could support. We'd run out of places to practice agriculture without tipping more fragile global ecosystems. Feeding everyone organically might be possible—in a world of 2 billion. Since letting multitudes starve is unthinkable, we keep using more chemistry, tainting more runoff with fertilizer, poisoning more soil with weed and pest killers, and sending more heat-trapping gases skyward—fully knowing that at some point, more becomes too much.

By the time Molly reached the conference room, she'd decided to level with them.

"You've just gotten millions to research nitrogen and plants. But I want you to think about nitrogen and *planet*. You've spent your careers figuring out how to extract more N_2 from the atmosphere and chemically activate it. Excuse me, but it's not clear that the world needs more activated nitrogen."

Heads jerked up. "I'm asking you a favor, for all humanity. Do the math in a planetary context. Add it up. Whatever your brilliant idea is, factor in its potential impact."

Easily a third of them, she could see, were furious at hearing their work deemed superfluous, if not outright dangerous. Another third looked bewildered. But afterward, the rest thanked her, some rueful for being too fixated on their tiny puzzle pieces to consider how they might be altering Earth itself.

Bill Rutherford, needing no convincing, contrived to sit next to her at dinner. A warm conversation quickly heated up to thermodynamics. Recently, Molly told him, Wisconsin's governor had called, saying they had $125 million from the Department of Energy for biofuels. "He wanted to know if the state could grow enough to run its power plants sustainably."

Rutherford groaned. "Stupid governments, thinking biofuels will save the planet. There's no thermodynamic basis for it. You can't stop carbon emissions by replacing petrochemicals with agriculture, one of the biggest petrochemical users and emitters of all."

Even if it weren't, burning corn ethanol produces only 50 percent more energy than what's needed to grow and process it—compared with a 1,100 percent energy gain from mining and refining petroleum into gasoline. To replace equivalent amounts of fossil fuel requires immense amounts of corn. Adding in emissions from tilling forests and grasslands into synthetically fertilized cropland, multiple studies show, ethanol is at least 24 percent *more* carbon-intensive than gasoline.

Nevertheless, nearly half the corn grown today in the US, the biggest producer, is for fuel.

"Growing fuel instead of food. Plowing up habitats. Inciting mass extinction." Molly checked them off on her fingers. "Covering vast swathes of Earth with genetically uniform monocultures. What could possibly go wrong?"

At Imperial College, said Bill, everyone was getting funded to improve photosynthesis in biofuels. "When I mouth off about biofuel at conferences, afterward in the bar people tell me to stop spoiling the party. There is no party. There's a world headed toward the brink."

A year later, when Bill was inducted into Britain's Royal Society, he invited Molly to the ceremony. At one point, they stood with two other honorees: Nobel laureate Steven Chu, who had resigned as President Barack Obama's secretary of energy, his climate efforts largely stymied, and an Oxford don, Dr. Patrik Rorsman, who was fighting a global diabetes pandemic. "We're all battling misplaced energy," Molly sighed.

None of the numbers added up. Two other technologies touted as potential climate game changers—splitting water with solar energy to produce clean hydrogen fuel, and capturing CO_2 to store underground for eternity—would each require, Bill had calculated, triple Great Britain's total renewable energy output.

Globally, the situation had all the symptoms of paradigm failure. "Genuine paradigm failure doesn't happen often," said Molly. "But in the event it does, the same kind of thinking will not solve the problem."

ii. Agency

In 2011, Wisconsin elected a governor who, among other disagreements, was unswayed by Molly's objection to farmers growing corn and soy to burn. Resigning her deanship, a fortuitous conference at the National Renewable Energy Lab in Boulder, Colorado, introduced her to what became a virtual community of kindred spirits who collectively pondered how to get governments to grasp the coming risks. In the US, that meant circumventing the symbiotic coziness of lawmakers whose campaigns depended on the largesse of agrochemical industries, whose vast wealth in turn depended on price supports for commodities: all encouraging overproduction without regard for damages wreaked.

Molly and her new friends worked on finding a vocabulary to define risks and threats. A phrase she used in an article, "multiple breadbasket failure," was noticed by Lloyd's of London, which commissioned her to write a white paper. Titled *Food System Shock: The Insurance Impacts of Acute Disruption to Global Food Supply*, it got the attention of the one agency in the US government with clout over the rest.

The US Department of Defense is the largest organization on Earth, with the most employees and the largest fleet of cars, trucks, tanks, boats, and planes. Accordingly, it's the world's biggest consumer of fossil fuel and emitter of greenhouse gases, and one of the world's biggest users of plastic. But it also has the biggest budget, and Molly was determined to raise its awareness of vulnerabilities in the global food system. It took four years of lobbying the Pentagon and Congress, but the 2018 defense appropriations bill finally required the DOD to report on the security implications of global food disruptions.

Soon, COVID would underscore Molly's warning of how exposed we are to unforeseen disruptions. But just before the pandemic struck, came a phone call that allowed her to confront the transcendental question:

What could actually change things?

Whatever it was would require a great leap in thinking—some paradigm-shifting, completely novel way of creating food.

The call was from a DOD agency called DARPA. "They're willing to invest in very strange new ideas," she told her community.

In 1958, to close the technological gap opened by the Soviet Union's Sputnik, President Dwight Eisenhower created the Advanced Research Projects Agency, ARPA. The *D* for "defense" was later added to focus its mission—then dropped, then readded—but military and civilian priorities so often overlap that the name hardly matters. Born of national urgency, DARPA skirted Defense Department bureaucracy. Its program managers were free to take wild risks, to dream up inventions that might truly change the game if only they existed, and get funding to try to build them.

Thus was born ARPANET, which became the internet and also GUI, the graphical user interface that lets anyone use it by pointing at pictures. DARPA also spawned GPS, without which we'd literally be lost; the computer mouse; weather satellites; voice recognition; infrared night vision; driverless cars; pilotless planes; stealth technology; Siri; drones; and the use of mRNA to teach cells to repel viruses, leading to Moderna's COVID-19 vaccine.

Other DARPA projects have proved less successful: cyborg surveillance insects; plant-powered robots that could recharge by foraging; self-repairing, living construction materials; mechanical beasts of burden to lighten soldiers' loads; and brain-computer interfaces to translate thoughts into action, from driving vehicles to launching weapons. But DARPA's willingness to fail big en route to bold breakthroughs was ideal for Molly Jahn.

Current DARPA gambles include self-healing coastal reefs; turning plastic waste into food; regenerating injured spinal cords; engineering a shelf-stable, universal whole blood substitute; teaching common sense to AI; in-orbit manufacturing; restoring rapid eye movement to sleep-deprived trauma victims; and preventing pandemics. Several, Molly recognized, were technofixes for things that technology broke, but in emergencies you try everything. DARPA was the most forward-looking government entity she knew—a place seeking solutions to problems that hadn't even manifested yet.

When her offer came in early 2020, she'd just returned from a Lloyd's meeting in London. Before her plane landed, a seatmate had inquired about her work. Glumly, she pointed out at Wisconsin, quilted with fields to the horizon. "Showing that this doesn't add up," she said.

But now DARPA's acting director wanted her to tackle food security. Like all its programs, defense was the priority: keeping soldiers fed even if supply chains failed. But once that mission was accomplished, there were 8 billion more possibilities.

In Washington a decade earlier, she'd noticed that Defense Department people were among the quickest to grasp what climate change meant to global security. The DOD might emit more carbon than entire European countries, but its agency DARPA was uniquely charged with anticipating what the future would need—and now it would fund her to do this work.

"It's the only place that invited me to even imagine what that would look like," she told Bill Rutherford. She was asking about his obsession, photosynthesis. Plants perform the miracle of turning sunlight into food, but humans had invented photovoltaic cells nearly 50 times more efficient at converting solar energy. Could we also make food more efficiently?

"The way we grow it hasn't changed since Mesopotamia." That had worked, until humanity crashed into biological limits. Could anything stretch those limits? What about the clever microbes Bill probed that simultaneously photosynthesized and fixed nitrogen?

iii. Plenty

"We're launching a second revolution in how humans feed themselves," Molly declared to the assembled researchers and Department of Defense brass. "Instead of plants for solar collectors, we'll domesticate a vast swath of biology we've overlooked because it's too small to see. Yet below the threshold of human eyesight awaits most of the biological kingdom."

She'd named her project after the mythical horn of plenty. Cornucopia would try to grow food from microbes—bacteria or fungi—far faster than crops grow in fields. Decades earlier, NASA had found microbes that could live off the hydrogen gas produced by electrolyzing—splitting—water molecules. All they needed to add were water, air, and electricity, and maybe yeast.

"Eating microbes is nothing new. We eat billions already: yogurt, cheese, bread, sauerkraut. Our bodies are made of tens of trillions of microorganisms. We just never thought about domesticating them."

Her request for proposals challenged potential contractors—"performers," DARPA calls them—with two scenarios. The first was keeping a platoon of 14 male soldiers in peak physical condition operating for 45 days in an austere environment, providing 3,600 calories each day of nutritionally complete, palatable foods. Whatever grew them had to fit aboard a single Humvee. They would replace the much despised MREs—Meals Ready to Eat—whose bulky packaging, which soldiers routinely fieldstripped to save space, required more calories of energy to produce than the food it contained.

In the second scenario, equipment fitting inside a 20-foot shipping container must sustain 100 rescued refugees of mixed sexes with survival rations—1,500 calories a day—for three weeks. By whatever electrical means the performers concocted, they'd produce hydrogen, carbon, and nitrogen on-site, which a few quarts of dissolved, fast-growing microbial feedstock would then convert to protein, carbohydrates, fats, and fiber. Pressing a button would start capturing airborne CO_2 and pumping out hundreds of kilos of rations. Whatever form they took—pudding, jerky, energy bar, or shake—had to taste good.

But Molly had a much bigger vision than supplying warfighters or disaster relief missions. "What if we had a food source that didn't depend on seasons, soil, rain, oceans, financial transactions, or fragile supply chains optimized for efficiency at the expense of resilience? What if we could produce nutritious food at the point of consumption?"

If so, it could be grown practically anywhere, including in any household. "What if, after all these millennia," she wrote in the journal *Issues in Science and Technology*, "we changed our food paradigm from agri-culture to ubiqui-culture?"

Four performers were selected for a four-year project divided into three phases: First, produce food. Next, add at least six flavors and a variety of textures. Finally, prove it could adequately nourish the humans in each scenario.

One team, from the University of Illinois, proposed to electrochemically combine excess atmospheric CO_2 with airborne nitrogen to form ammonia and acetate, the molecule in vinegar. Together they'd be fed to GRAS—"generally recognized as safe"—microbes, modified to rapidly expand into a rich biomass.

Another, from Harvard Medical School, would bioengineer bacteria to gorge on methanol, an organic molecule with the same carbon, hydrogen, and oxygen atoms as acetate but in a different configuration.

A nonprofit research institute founded by Stanford, SRI International, was collaborating with two California startups on a process combining a high-carbohydrate microalgae with a fast-growing, hydrogen-oxidizing bacteria species that converts CO_2 to protein without photosynthesis—a process first tried by NASA in the 1970s to harvest the carbon dioxide astronauts exhaled. Kiverdi Inc.—whose subsidiary Air Protein Inc. grows chicken and meat substitutes "landlessly"—would capture carbon dioxide from the portable diesel generator each Cornucopia team would be permitted.* Activated nitrogen to feed the high-carbohydrate microalgae would come from Nitricity Inc., which describes its brainchild as "lightning in a bottle"—inspired by what farmers have long known: that crops grow greener wherever lightning strikes, because its intensity ionizes atmospheric nitrogen, which bonds with other atoms to form fertilizer.

The fourth Cornucopia team was led by the Johns Hopkins University Applied Physics Lab, known worldwide as simply APL. Founded during World War II to design fuses that detonated as bombs neared their targets, APL continued as a quasi-separate entity from the university, developing guided missiles in the postwar era and eventually expanding to become the nation's biggest nonprofit defense and space contractor.

On June 8, 2023, five months after Cornucopia launched, Molly Jahn sat with a principal Cornucopia investigator, Collin Timm, on a molded concrete bench in the brushed-chrome-and-glass atrium of APL's new research center in Laurel, Maryland, eating boxed lunches. A biochemical engineer in his late 30s, Timm wore tan jeans, a cobalt jacket, and long sideburns just starting to gray. In a moss green blouse and black silk pants, Molly, now in her early 60s, looked like she'd stopped aging in her 40s.

* Using identical diesel generators would allow comparisons of each team's performance. Eventually, the hope was that DARPA-developed photovoltaics, portable enough for warfighters to carry, would power Cornucopia's field bioreactors.

Timm's team was actually two teams, one in APL's chemistry lab a floor up above them, the other in the biology lab two floors higher. Each was trying to channel excess atmospheric CO_2 into soluble molecules to make glucose that could be fed to food-safe microbes for people to eat. The chemistry team was producing acetate and formic acid, the stinging stuff in nettles, using a reaction that probably had a role in life's origins—which, decades earlier, NASA had also explored for turning simple hydrocarbons into complex edible sugars in space. APL's chemists were accelerating it with a copper foil catalyst, its surface studded with tiny cube-shaped bumps.

In the biology lab, glass bioreactors were bubbling a rose-tinted suspension of *Rhodopseudomonas palustris*—soil microbes from a bog in Woods Hole, Massachusetts, that do everything: photosynthesize, fix nitrogen, emit hydrogen, and eat airborne CO_2. As *R. palustris* also metabolizes any other carbon molecules it encounters—it thrives in wastewater—they were testing to see if feeding it compounds designed by the chemistry lab downstairs would produce even more sugar.

Microbial farming might sound freaky, but that day the air filters in the cavernous APL atrium, abuzz with clutches of Ph.D.s but containing only a few potted palmettos, attested to how much humans rely on engineered nature. Outside, cascading smoke from a firestorm incinerating Nova Scotia's forests, 750 miles north, had pushed New York, Baltimore, and Washington past Delhi and Dhaka for the perverse distinction of having the world's worst air. Drivers groped through yellow-gray grime, squinting at semaphores reduced to red and green smudges. In a twist of pandemic protocols, everyone who'd braved the shocking conditions outdoors to reach APL that day arrived masked until they were safely inside.

Ironically, said Collin, airborne particulates are a feast for microbes like *R. palustri*, which mine micronutrients and key trace minerals from them. "Our soldiers will dine on desert dust," he said. They were even factoring in *airroir*: how dust in different regions bears different nutrients. As the challenge was to give warfighters something delicious while asking them to risk their lives, his teams were currently retooling microorganisms to re-create flavors: "We're aiming for rosemary, vanilla, banana, butter, and chili pepper."

Partners at North Carolina State's chemical engineering department were adding probiotics to APL's microbial confection, plus vitamins A, C, D, E, and K. Another collaborator, Meridian Biotech, which ferments leftover bourbon mash from distilleries into shrimp feed, would texturize their bacterial blends into shakes and energy bars—including for the survival rations, said Collin. "We don't want to feed people nasty gruel in a disaster scenario."

A Johns Hopkins sustainable energy institute had designed a loop system based on high school chemistry to transmute industrial emissions into acetates and other valuable compounds. Supposedly, it used half the energy of other carbon-capture systems: controversial technologies often described as stalling tactics for petroleum companies to emit indefinitely. Thus far, these systems were mainly connected to smokestacks; to build enough machines to suck all humanity's excess CO_2 from the atmosphere, then seal it permanently underground, was likely both unaffordable and physically impossible. But turning captured carbon into food had potential.

"We're scaling their process to provide our microbes CO_2 to turn to sugar," said Timm. *Scaling*: the critical verb for Cornucopia. Whether microbial agriculture could scale to ever really make a difference was the question hovering over all their lab achievements.

"The biggest impediment I see to scaling," said Molly, "is that your bioreactor technology is just an updated cauldron."

"People have been doing bioreactors for a long time," agreed Collin. "Our idea is much smaller microreactors. When you need more, you make more. Need less, you make fewer. Unlike current agriculture's massive energy binge, you operate within the thermodynamic definition of sustainability: live off the energy in your space and time. You're not allowed to invade the past or mortgage the future."

Until the 20th century, most food was eaten near where it was planted, watered, manured, harvested, milled, pastured, and slaughtered. Fossil fuels added many expensive steps: chemically fertilizing, pesticiding, herbiciding, fungiciding, transcontinental and transoceanic shipping, processing, packaging, and marketing.

"In each transition, energy is lost," said Molly. They were walking to APL's auditorium, where she'd be addressing a summit of scientists, military brass, and DOD officials on climate change and national security. "Plants and animals leak a lot of energy on their way to becoming food. If we can move away from them, the difference in energy flow could be orders of magnitude."

Cornucopia was allotted four years to prove whether humans can really farm microbes instead of macrobes—or less, because a DARPA review after 18 months would determine if funding continued. Its progress, however, Molly would now watch from afar. A month earlier, she'd been offered an opportunity—more like a summons. It was to the Office of the Undersecretary of Defense for Research and Engineering, which oversees DARPA and technologies critical to national security, including mission capabilities and missile defense. She would serve as senior advisor to the DARPA director who'd recruited her, under a presidential executive order to advance biotechnology and biomanufacturing.

Her new duties were not limited to food systems. Given all the growing risks—climate, supply chains, cyberattacks—the Department of Defense was looking at not just making food, but making as much as possible as close as possible to where it was needed: a revolutionary switch for the world's biggest energy consumer.

As a DARPA grantee, Molly had already fielded criticism of weaponizing food technology. How did she now feel about becoming a vital cog in DOD machinery? "It's not where I thought I'd end up. But it's the only place I could do this work."

Her decision became even clearer when she ran her promotion by her daughter. "Besides the military, Mom," said Hannah, "won't this also be really good for the rest of humanity?"

"Cue the environment," sighed retired admiral James Stavridis, dangling his N95 face mask. The former NATO supreme allied commander and head of both the US European Command and the US Southern Command, tieless in a civilian blazer, was keynoting the summit. After acknowledging the pall choking North America's eastern seaboard, he projected a grim world tour on three wall-sized screens: a sabotaged Russian gas pipeline in the Baltic; the two biggest CO_2 expellers, the US and China, reneging on their 2015 agreement in Paris; Middle East water woes and wars; the Amazon, lately the smoker's lungs of the Earth; and the Arctic, where global powers were vying over strategic new polar shipping lanes.

"As the ice melts, are we headed to war at the top of the world?" he asked.

The following speakers—admirals, generals, assistant defense secretaries, intelligence officers—all concurred with him that the greatest long-term security threat is climate. Physical rules that governed the planet were morphing before everyone's eyes, heat sensors were melting, Russia's rape of Ukraine had cranked up gas production and stalled environmental momentum, boreal and tropical ecosystems were ablaze—and byzantine social media, the main news source for nearly 4 billion people, instantly muddled any clear messaging on what the world needed. And now came a new danger: artificial intelligence, accessible to anybody.

That afternoon, Molly was on a panel called "Game Changers—Operational Resilience."

Resilience, she explained, comes from simple design principles. "Distributed systems are more resilient. Redundant systems are more resilient. Diverse systems are more resilient. Ten thousand years ago we had the bright idea of intentionally cultivating just a handful of species, although millions out there are edible. Revisiting that design opens new frontiers for how we feed ourselves—for example, with microbes."

Her co-panelists swiveled to look at her, seated in the middle. Nodding, she smiled. "We can grow microbes anywhere," she continued. "We can eat their bodies within three days. This not only creates redundancy, diversity, and distributed capabilities, but eases pressures on the climate by moving beyond Haber-Bosch."

Turning side to side, she made eye contact with each. "To paraphrase Einstein, the thinking that created the problem can't solve it. It means moving some of the most important things we as humans do to smaller scales and figuring out nifty approaches to microreactors and ways to do what biology does, only without all the biology."

Never mind that the food industry profits from huge scales, and that food distribution is one of the world's biggest businesses.

"Our current systems were designed to maximize pressure. The more material we push through them, the richer we get. It doesn't matter what gets wasted, it's about how many times you can crank that crank, move all the stuff through all the systems, not caring about the messes you make, because they're external. If the DARPA project I worked on is adopted at a large scale—which I fully expect, and I don't think it will take lifetimes—the pressure that will come off the planet will be massive."

At the reception, she ran into Tracy Morgan, a systems engineer she hadn't seen in 10 years. They'd met at the National Renewable Energy Lab, and still met online every month as part of a group of 40 scientists and governance designers: Molly's virtual support community. Tracy, who once contracted to the Department of Homeland Security and the intelligence community, was now a vice president at a Silicon Valley firm that was launching a network of satellites to detect wildfires. By passing over every spot on Earth every 20 minutes, they could update firefighters in near-real time. Their image resolution, she explained, high enough to read license plates, could detect a campfire anywhere on the globe and know if it was untended.

Molly explained her own new job. "I get to talk to the big boys now."

If the Cornucopia researchers she funded succeeded, she intended to make certain that growing food from thin air and microbes would be another DARPA invention that jumps from soldiers to everyone. Every kitchen could be a farm. Yes, a huge paradigm shift. "The planet's already going through a paradigm shift. It may as well be one we can control," she figured.

"Just one thing left to do before you retire, Molly," said Tracy. "Save humanity. No pressure."

A shriek of laughter, and Molly recounted biochemist Bill Rutherford's Royal Society induction, when she met Nobel laureate physicist Steven Chu, who compared the world's growth economy to a pyramid scheme. This couldn't go on forever, they'd all agreed.

Later, Bill invited Molly to a blues club in Kent to hear his band, Baskerville Willy. She loved watching an eminent Royal Fellow clad in jeans and a fisherman's cap, flatpicking an amplified six-string acoustic—and then was gobsmacked by their finale:

> So gather round you good folk
> while gathering you can,
> the animals, the birds, the boys and girls;
> give them all a better choice than
> fire or frying pan,

they're all hoping that maybe
Molly can save the world, Good Lord,
hoping that maybe
Molly can save the world.

Come on, Molly:
somebody's got to do it.

CHAPTER FOUR

ARK BUILDERS

i. Biology

The Santa Rita Mountains, a jagged set of silhouetted peaks 25 miles south of Tucson, belong to an archipelago of sky islands rising from the deserts of southern Arizona and New Mexico: the northernmost fragments of Mexico's Sierra Madre Occidental, North America's most biodiverse region. In the US, only in these forested uplifts are found subtropical species such as the giant purple-green Rivoli's hummingbird, the elegant trogon, thick-billed kingbirds, flame-colored tanagers, western twin-spotted rattlesnakes, Chiricahua leopard frogs, ocelots—and jaguars.

While hiking the Santa Ritas in 1982, botanist Ron Coleman discovered an orchid that spends most of its life underground. Blooming only when conditions are right, it sometimes waits seven years before its pinkish stem emerges bearing clusters of white trefoils, each framing a lavender-streaked labellum petal that dies after putting out a seed pod. Leafless, it can't photosynthesize for nutrients, so it steals them from a single species of soil fungus encountered only near the roots of white oaks.

Coleman found just four tiny populations, totaling no more than 200 individuals. In Earth's grand scheme, that would seem insignificant—except to the Tucson-based Center for Biological Diversity, which repeatedly argues in US federal court that no species, plant or animal, is insignificant. What became known as Coleman's coralroot, along with two other endemic Santa Rita plants—the succulent Bartram stonecrop and a tiny perennial aster called beardless chinchweed—are among those species whose existence the Center deems important enough to defend: one long, expensive lawsuit at a time.

By at least one measure, they've been wildly successful. Of more than 900 lawsuits they've filed under one of the world's most powerful environmental laws, the US Endangered Species Act, they've won nearly 90 percent, protecting more than 700 imperiled species ranging from slugs to polar bears.

Since they do this, however, in the teeth of a major global extinction event, winning courtroom battles doesn't guarantee winning the war for survival.

What many scientists now call the Anthropocene, the epoch when human presence grew so prodigious that we've literally become a geologic force, may prove one of Earthly history's shortest if we end up a victim of that war. The version of nature that emerged from the debris of the last major extinction, after an asteroid abruptly ended the 180-million-year reign of big reptiles, included a shrewlike survivor named *Purgatorias*, which evolved into primates and eventually ourselves.

That nature is still the source of everything our lives depend on, from food and water to oxygen. Should enough of it vanish, at one point the bottom would drop out from under us, and it will be too late to do anything.

It may be the biggest question humans have ever asked: How can we stop the great extinction accelerating before our very eyes?

One species at a time is the Center for Biological Diversity's strategy—or die trying. Literally.

In 1985 Todd Schulke left his Iowa farm town, which had to switch to bottled water because its wells were full of pesticides. He eventually found his way to Washington's Evergreen State College, an early hub of environmental activism. There he grew a ponytail and a scraggly red beard, joined the radical eco-defense group Earth First!, and spent his college years scaling tuna-boat masts and unfurling banners protesting dolphin slaughter.

In 1989, en route to Florida, he stopped through New Mexico's Jemez Mountains to hang a banner at an Earth First! protest over proposed steep-slope logging in the Gila National Forest. In Washington, he'd watched Northwest loggers drag tall firs and spruce off treacherous slopes with cables. Steep-slope logging—a damaging, expensive last resort after taking everything easily accessible—had never before been used in the Southwest.

Arriving, Schulke found they already had a banner. But he met a woman, and the aspen glades and mountain meadows of the Gila Wilderness were gorgeous, so he stayed. "Come see this place!" he wrote Peter Galvin, an Earth First! buddy in neighboring Arizona.

Three years earlier, Galvin had returned from a soul-healing backpack trip through Utah's Escalante Canyon to learn that the Chernobyl nuclear reactor had melted down. Instantly, the serenity he'd found in the wilderness was shattered by global environmental disaster. A childhood cancer survivor, Galvin was unnerved by flagrant pollution. Convinced that stupid greed was poisoning his planet, he'd left college to join an ongoing Earth First! delaying campaign until proposed protection for the northern spotted owl under the Endangered Species Act would finally stop timber companies from mowing down the Pacific Northwest.

After three years of civil disobedience arrests—one for yoking his thickly bearded neck to a Forest Service gate with a bicycle U-lock—Galvin was persuaded by a naturalist from Arizona's Prescott College, who'd led his Escalante trip, to finish his studies there. Upon learning that another imperiled spotted owl subspecies lived in southern New Mexico's Gila Wilderness, he chose it for his senior thesis project and wrote Schulke that he'd be coming.

He called US Fish and Wildlife in Albuquerque to request documentation on the bird.

"We just took days complying with a Freedom of Information Act request from some Phoenix doctor about that owl," a ranger told him. "We sent him a whole boxful. You're in Arizona—can I give you his phone number?"

It was an offer the Fish and Wildlife Service would later badly regret. Phoenix was 100 miles south of mile-high, piney Prescott. "Come!" barked Dr. Robin Silver over the phone. "I'll pay your gas."

Silver's single-story brick home, set back behind a line of orange trees, had barred windows and, Galvin noticed, blackout curtains. Other than his wildlife photographs—including one of a large owl with a mournful heart-shaped face, pitch-black eyes, and a brown-spotted abdomen—the prominent furnishings were multiple printers and fax machines. Whenever Silver, a lean, intense, bespectacled emergency room physician in his early 30s, wasn't resuscitating Sun Belt retirees from cardiac arrest, he churned out demand letters and press releases about preserving nature.

He showed Galvin the box of documents. The US Forest Service, he explained, needed to know how many Mexican spotted owls there were to determine if US Fish and Wildlife must list the subspecies under the Endangered Species Act. He offered to pay all of Galvin's field expenses if he kept him apprised of what he learned on his thesis study.

After days spent reading through the entire box, Peter Galvin drove to New Mexico's Gila National Forest. The night he arrived, a guest at Todd Schulke's dinner table turned out to be directing the Forest Service's owl survey. She was also newly pregnant, and scared to tell her boss that she couldn't bounce around the woods anymore.

"Nobody else knows enough about Mexican spotted owls to take over."

"I do," Galvin told her.

The next day, he was hired. "We just had a bump-up on the spotted owl," his boss said. "Hire eight more people fast."

So he hired Todd Schulke, plus another Earth Firster named Cheryl and her boyfriend. The two had met while incarcerated in a warehouse hastily repurposed as a holding pen at White Sands, the first atomic bomb testing range, after more antinuke protesters were arrested at a rally than fit in the local prison. Her jailhouse paramour, Kierán Suckling, was a philosophy student who regularly fled to nature or joined Earth First! actions to avoid writing his Ph.D. dissertation on the relationship between extinction of languages and species.

The pay was terrible, but as Suckling fondly reflected three decades later, "We were itinerant seasonal biologists in our early 20s, living on food stamps, frequent unemployment, Gary Snyder's poetry, and shoplifting." They slept by day, awoke at dusk, drank coffee, then hiked all night up moonlit canyons lined with tall Douglas firs and ponderosa pines that smelled like vanilla, sometimes flushing sleeping elk or bear. Every quarter mile, they'd stop for 15 minutes to listen for the barking hoots of spotted owls.

The Forest Service, they soon deduced, was in such a rush because the tracts they were surveying were poised for timber sales. Naively, they'd assumed that finding Mexican spotted owls would stop planned cuts. Gradually, however, it dawned on them that to avoid yet another war between loggers and tree huggers over yet another spotted owl, the Forest Service was hoping the Mexican subspecies proved abundant enough to *not* qualify as threatened under the Endangered Species Act. They did some research and learned that wherever they found owls, they could legally appeal logging proposals. Soon, the district ranger was not amused to find that his own employees were impeding his old-growth timber sales.

Then Kierán Suckling wandered by mistake into an adjacent canyon not on their survey, where a sale was already underway, and found an owl in an old-growth ponderosa, forcing the Forest Service to suspend cutting. A few days later, however, the sale was again on, with their spotted owl sighting missing from the accompanying survey map. Their boss clearly didn't want to let any rare nocturnal bird interrupt sales.

So they went to the press. The exposé in the *Santa Fe New Mexican* quashed the illegal deal—and got Galvin, Suckling, and Schulke fired.

Robin Silver raced over from Phoenix with his checkbook and a formal petition to declare the Mexican spotted owl a protected species. But the nonprofit he and the unemployed bird-watchers formed, Save the Owls, never had to go court to win the case. Silver's petition was so meticulously researched that US Fish and Wildlife didn't bother contesting it.

It was that easy. Under the Endangered Species Act, every federal agency had a duty to conserve not just imperiled species but also enough habitat for them to recover.

When President Richard Nixon signed it in 1973, they wondered, did he have any idea what he'd unleashed?

To test the power they sensed they'd tapped, their first lawsuit targeted an even bigger adversary than the US Forest Service. This was 1990, and the only remaining Mexican gray wolves existed in captivity. Declaring the species unsavable, the US government was effectively letting it go extinct. But years earlier, an Army base commander of the White Sands Missile Range had acknowledged that wolves could be successfully reintroduced there. The Sierra Club, the Wilderness Society, and the National Audubon Society had lobbied the current commander, who wasn't interested. They were hoping that after he retired, the next commander would be more amenable.

Waiting for your enemy to retire didn't seem like a strategy to Save the Owls. In a move that irritated both the Pentagon and those big environmental NGOs, they filed a lawsuit against the Department of Defense for violating the Endangered Species Act, and sent out press releases.

They were so brash, and rash, that they didn't even have a lawyer. But when the news broke, one called from Santa Fe and offered to join their legal team. "You're it," they told him. They won—and kept winning, even compelling ranchers, those western icons, to stop cattle from grazing along streambeds because they're home to goshawks and southwestern willow flycatchers.

For the first time, courts declared, vulnerable wild creatures were even more valuable than cows.

As the last decade of the 20th century began, it felt like the world was finally coming into environmental consciousness. In 1985, a British geophysicist had discovered a hole in the sky above Antarctica, and within just two years the world's first international treaty to address a global environmental emergency was finalized: the Montreal Protocol to protect the ozone layer.

The following year came NASA climatologist James Hansen's electrifying 1988 Senate testimony that oil, gas, and coal-fired emissions must stop before the seas boil and the planet bakes. Even academia was awakening: a University of California–Santa Cruz biologist, Michael Soulé, had turned a nascent academic field called conservation biology into a crisis discipline that taught young ecologists that their mission was not merely studying nature but safeguarding it.

To prevent species extinctions meant saving the diverse ecosystems that supported them—including our own species; the delusion that technology alone could sustain humanity was akin to being on perpetual life support, hooked to respirators.

Anthropocentrism aside, Soulé insisted that biological diversity had its own intrinsic value. That became the mantra of their nonprofit, which after several name changes settled on the Center for Biological Diversity. Earth First!'s eco-guerrilla tactics had failed to coalesce into revolution—infiltrated by the FBI, its leaders had either left or were behind bars. But even though none was a lawyer, the Center's founders discovered that instead of breaking laws, they could use them as bludgeons.

"You have immense corporations and corrupt government agencies lined up against you, usually backed by Congress. All you need to overcome that juggernaut is to convince one federal judge that this is wrong," marveled Kierán Suckling, who'd abandoned his doctorate to become the Center's executive director.

The Endangered Species Act, Suckling believes, taps a fundamental, spiritual commitment that Western culture has to other species. "Noah's Ark is one of our founding myths, older than Judeo-Christianity, written down at least sixteen hundred years before the Old Testament. People underestimate how powerful that is."

By the year the world first held an Earth Summit to protect the global environment, 1992, their pugnacious little NGO had sued all 11 national forests in New Mexico and Arizona for violation of the Endangered Species Act, on whose clear intent they refused to compromise. At one point, a US Forest Service lawyer grabbed their attorney's throat and began strangling him. Leaping in, Robin Silver punched him; by the time order was restored in the court, any sympathy the government had accrued with the judge was lost. The settlement they quickly reached effectively ended commercial logging in the American Southwest. They didn't stop there.

ii. Geology

August 2016: Todd Schulke hadn't seen fellow Earth First! eco-saboteur Mikal Jakubal in more than 30 years. While Schulke was ascending ship masts, Jakubal was rappelling down dams to paint giant cracks on their faces or perched in tall firs slated for clear-cutting. Now in their graying 50s, they were seated by a campfire near a thicket of red-leafed amaranth on the Río Aros, an undammed wild river that drains the western slope of Mexico's Sierra Madre Occidental into the Río Yaqui, 100 miles south of the US border. They were reconnecting over a two-liter plastic Coke bottle filled with bacanora, a smoky Sonoran mezcal, distilled and purchased at the ranch where the four-wheel-drive Chevy trucks that ferried them and a dozen others for seven hours up a wretched road had paused en route.

Jakubal had long since descended from the trees and settled into the soil as a cannabis breeder in Humboldt County, California. Schulke was still an environmental warrior, overseeing forest protection for the Center for Biological Diversity he'd cofounded.

"Still fighting the good fight, huh?"

"What else should I be doing?"

"Pass me that, por favor," said Randy Serraglio, a skinny dude in a wide-brimmed leather hat with straight gray hair to his shoulders. For nearly two decades, Serraglio had been the Center's campaigns director. It now employed 200, plus dozens of contract lawyers arguing on behalf of animals and plants that otherwise had no say in decisions imposed by an invasive bipedal species that, near the end of the Pleistocene, appeared in the Americas to stay.

Serraglio was currently lead spokesman for the most charismatic species they'd defended since the polar bear. This nine-day rafting trip on the Aros and Yaqui rivers, organized by two transborder nonprofits, Mexico's Naturalia and the US's Northern Jaguar Project, would follow the boundary of an 86-square-mile preserve they owned jointly: the northernmost known jaguar breeding grounds in the Americas.

Except the trip had ended prematurely when an abnormally wicked storm upstream pushed the Aros past flood stage. The first morning, they'd flipped two kayaks and an inflated oar-boat, piloted by a Grand Canyon boatman who had never overturned in 40 years of river running. By the third afternoon, after escaping whirlpools, frothing rapids, and tree trunks sluicing down the river, they managed to beach near an abandoned ranching road. Via satellite phone, they'd contacted drivers to collect them, but that would take two days.

"Once again, weather reminds us we're even in more danger than we thought," snorted Serraglio, relieving Schulke of their last bottle of bacanora.

Jaguars, found only in the Americas, are the world's third-biggest cats after tigers and lions. Older males can top 300 pounds, more than twice the size of the slimmer leopards they resemble, with which they share a prehistoric ancestor. Their jaws kill prey as big as the 10-foot caimans and adult bears they silently pounce on from behind, instantly crunching through their skulls. Unlike any other big American feline—mountain lion, lynx, bobcat—they also roar.

About 15,000 remain, mostly in the tropics. Once, jaguars roamed and bred in the US from California to Louisiana, as far north as the Grand Canyon. Then, after 1915, when the Department of Agriculture started paying bounties on anything nonhuman that ate livestock, in Arizona alone nearly one per year was shot. The last was a female killed in 1963, in a spruce-fir forest above 9,000 feet. Predator controls, it was presumed, effectively had extirpated jaguars in the US.

But in 1996, an Arizona hunting guide treed and photographed a huge feline with a rosette-dappled, golden pelt just over the New Mexico line, 10 miles north of old Mexico. Six months later, a second jaguar was photographed in Arizona, 150 miles to the west. Since the new century, at least five more had been confirmed in the US, distinguished by the patterns of their spots.

The return of these cats thrilled wildlife biologists and captivated the public. Getting the jaguar listed as an endangered species in the US within a year of the first sighting was among the Center's most celebrated successes. The US Fish and Wildlife Service had tried claiming that only jaguar breeding grounds south of the border were crucial to their survival, but three lawsuits later, USFWS was forced to designate 1,200 square miles of southern Arizona and New Mexico—much of it leased grazing land—as critical jaguar habitat and to design a jaguar recovery plan.

Despite assurances that cattle were still permitted there, ranchers didn't welcome a new, powerful predator. Then in 2011, a new jaguar rattled yet another hoary western institution. A young male first spotted in southern Arizona's Whetstone Mountains was soon appearing on camera traps in the nearby Santa Ritas—right near where, every few years, rhizomes of Coleman's coralroot send up blooms through the crushed granite soil.

It was also the site of a proposed open-pit copper mine, which would be America's third-biggest—unless the Center for Biological Diversity and the nearby Tohono O'odham tribe,* which holds the Santa Ritas sacred, could stop it.

The mine's developer, Toronto-based Hudbay Minerals Inc., had been sued for environmental and human rights violations in Manitoba

* Called Papago by the Spanish.

and Guatemala, but in the US it had an antiquated law on its side: the General Mining Act of 1872, signed by President Ulysses S. Grant to attract white settlers west. It had allowed Hudbay to acquire mineral rights to more than 2,000 acres at a price unchanged since it was enacted: $5 an acre. Back then, mining was pick-and-shovel. Hudbay's plan was a pit more than a mile across and a half mile deep. Besides wrecking the stunning view along a designated Arizona scenic highway, the mine, opponents charged, would stack tailings—the rock detritus left behind by digging—hundreds of feet high over four square miles of national forest.

Just eight miles away was Madera Canyon, home to 250 bird species, among the most visited attractions in a state where, according to a study by Arizona Game and Fish and the Tucson Audubon Society, wildlife-watching earns $1.4 billion annually. "Imagine how many tourists and nesting birds they'll get with a mine blasting under lights all night and ore trucks clogging the scenic highway," Randy Serraglio entreated journalists he took there. "The pit will create a cone of depression that'll draw down the water table like a straw. Sixty springs will go dry or be buried."

He'd point to a distant line of cottonwoods bisecting the grassy valley at the base of the Santa Rita's eastern slope. "That's Cienega Creek. It's not just life support for 12 endangered species; it's 20 percent of Tucson's natural groundwater recharge. That straw will suck it right down. Instead of water, Tucson gets mine dust."

Over the next three years, camera traps placed by a University of Arizona contract biologist recorded the jaguar more than a hundred times. The Center for Biological Diversity made the majestic animal a poster boy for endangered species. They held a naming contest at a Tucson middle school whose mascot is the jaguar. The winner was El Jefe, Spanish for "boss," also the nickname of a popular Arizona congressman, Raúl Grijalva, who for years had tried to get the 1872 Mining Act repealed.

In early 2016, at a Night of the Jaguar party the Center threw at a Tucson craft brewery, featuring a beer called Jefeweizen, they showed a 41-second film of El Jefe pacing a Santa Rita trail, to huge cheers. The footage was from the biologist who ran the university's camera traps for US Fish and Wildlife. His repeated pleas to let the public see videos of El Jefe roaming forests and streambeds threatened by the mine—which had just been granted a controversial permit by the US Forest Service—were rebuffed. In frustration, he placed his own field cameras and brought the Center the results.

El Jefe became an internet sensation and appeared on NBC's *Today*, ABC's *Good Morning America*, and CBS's *This Morning*. The biologist was fired.

By then, the Center had sued Fish and Wildlife so many times that the two sides had agreed to a schedule: one new lawsuit to protect one more species per week. Now they sued the agency again over its biological review of the Forest Service's permit, which concluded that Hudbay's mine would not "destroy or adversely modify" enough critical habitat of jaguars or 11 other resident endangered or threatened species to jeopardize their existence. According to a leaked internal memo, a USFWS supervisor with a bachelor's degree had overruled his own Ph.D. biologists' insistence that the Endangered Species Act prohibited *any* alteration to critical habitat, period.

When that opinion was delivered in April 2016, El Jefe hadn't been seen in the Santa Ritas for a year. That was expected: Young male jaguars typically wander 100 miles or more from their breeding grounds to places with plentiful game. There they gain weight and muscle, as seen in camera images of El Jefe over the previous three years, until they grow strong enough to return and challenge older, alpha males for breeding rights. Would the federal judge hearing the lawsuit insist that a jaguar had to be present in order to overrule the USFWS opinion that greenlighted the mine?

They were still awaiting a court date that August when Randy

Serraglio and Todd Schulke traveled to the northern Mexican wildlands where jaguars are born. Although they and the others on their river expedition, mainly donors to the Northern Jaguar Project, mourned their aborted river trip, they still spent two lovely days stuck in paradise, where clouds of yellow butterflies swirled around them to the sublime trills of Sinaloa wrens and blue mockingbirds. At night there was no electric glare on any horizon, no internet or cell service, only the Milky Way's shining pearls overhead and contrails of meteors. The jaguars who left three fresh sets of prints near their campsite proved too stealthy to see, but once before dawn came low growls: the muted roar made by only one animal in the Americas.

Most of the donations to the reserve paid surrounding ranchers to not poison or hunt wild felines to protect their livestock from predation, but instead to allow camera traps. Every jaguar photographed on their land earned them 5,000 pesos; 1,500 per ocelot, 1,000 for a puma, and 500 for a bobcat. Over a decade, the peso equivalent of nearly $170,000 US had been awarded. It was the reserve's biggest, but most effective expense. Otherwise, these carnivores had no safe corridor to their ancestral range across the US border, where they were now reappearing. The hunting ban also meant more wild game like javelina,* which the cats preferred to cows, so livestock losses were minimal.

An extra benefit was that previously, the ranchers themselves made crossings each winter, to earn money working construction on the Dakota Access and Keystone pipelines. Now, they didn't have to.

That fall, with the 2016 US elections nearing, two more jaguar sightings in Arizona—one on a military base, another 60 miles above the border, twice as far north as any jaguar had been seen for a half century—had them believing that the Center's work to repa-

* Collared peccary, a wild boar native to Latin America and the Southwestern US.

triate this fabulous creature to its rightful range was paying off. Moreover, a camera-trap image the Bureau of Land Management released of the second jaguar was cut off at its hindquarters, fueling suspicions that it was female: the one thing needed to reestablish a breeding population in the US.

The jaguars' delighted human allies awaited more rosette-spotted felines. That Donald Trump might actually be elected president and replace the rusty barbed wire separating the US from Mexico with his threatened wall, which would impede the flow of creatures that knew no borders—it just didn't seem possible.

iii. Morbidity

Three years after the impossible occurred, Randy Serraglio and the Center's videographer, Russ McSpadden, hiked into the San Bernardino National Wildlife Refuge on southeastern Arizona's border with Mexico. Picking their way through creosote, prickly pear, and staghorn cholla; leaping creeks that flowed north from Mexico; they came to where earthmovers had scraped a 60-foot-wide swathe of green high desert as far as they could see in either direction. Stacks of 30-foot hollow steel bollards were waiting to be set in 10-foot-deep concrete footings.

The millions of gallons requisitioned from a borderlands rancher's well to pour those footings for hundreds of miles amid a historic Arizona drought violated all reason but no law, because of an executive order waiving all laws that might block construction, including the Endangered Species Act. Center scientists had already tallied 93 species whose natural migrations would be stymied by miles of 30-foot-high bollards spaced just half an inch apart, including jaguars, ocelots, Mexican gray wolves, desert pronghorns, bighorn sheep, and two winged species that don't fly that high: the cactus ferruginous pygmy owl and the Quino checkerspot butterfly.

Along with the Animal Legal Defense Fund and Defenders of Wildlife, the Center for Biological Diversity had unsuccessfully appealed to the US Supreme Court a lower court's decision that Trump had constitutional authority to ignore the Endangered Species Act. Since his election, the Center had added 20,000 alarmed new members, upping their annual budget to $20 million. But without their most powerful weapon, they couldn't stop his wall by showing what it did to the animals—the critically endangered pygmy owl, for one, might actually be extirpated.

Nevertheless, they tried suing again, charging that Trump had violated the National Emergencies Act by pilfering $2.5 billion in non-emergency funds from the military to build a wall he'd once promised that Mexico would pay for. But even if they won, they knew that by the time things ground through the courts, hundreds of miles of wall would have been built. Already they saw it advancing on this wildlife refuge from both directions. To the east, construction had started at the Continental Divide, where one long transborder sky island separates the Atlantic and Pacific watersheds. To the west, a recent protest the Center staged at the border between Organ Pipe Cactus National Monument and Mexico's Reserva de la Biósfera El Pinacate had drawn hundreds but failed to save dozens of namesake organ pipe cacti from Department of Homeland Security bulldozers.

They approached the wall. This was not Trump's fantasized massive concrete barrier, like the ones in Israel he envied. The bollards were hollow, beveled steel slats, stuffed with concrete and topped by slick metal plates intended to deter climbers, but easily scaled with a homemade grappling hook of bent rebar.

"Listen to this," said Russ. He snapped his fingers, sending vibrations ricocheting around the beveled bollards that sounded like bullets whistling by.

"What a joke," hissed Randy. "Anyone with a $100 Home Depot saw and a $15 steel-and-concrete blade can cut these flimsy bollards and squeeze through. If your immigrant-smuggling business is worth millions, $100 won't stop you. Neither will a $2.5 billion wall."

Nonetheless it had done incalculable damage to this exquisite landscape, its yellow cottonwoods on the unravaged Mexican side dazzling in the afternoon light. Wall or not, hardly anyone came through this remote stretch of *la frontera*. A surveillance tower they'd passed driving in, tall enough to scan the entire refuge, detected only five crossers the previous year. All were apprehended with sensing technology that required no wall.

"Wow," breathed Russ, lifting his camera as a full rainbow appeared to the east, one end in each country. They stared at the incongruous beauty shimmering above the hideous wall until it dawned on them why there was a rainbow. Turning around, they saw huge black thunderheads racing toward them. The wind rose, the bollards wailed, and they ran for their truck as a sudden winter deluge drenched the borderland refuge. The previous summer, a flash flood from another unanticipated storm in Organ Pipe had toppled a section of the wall.

As they passed the surveillance tower driving back, lightning struck it, sending blue sparks shuddering up and down its steel frame.

"It's a sign from God," Randy declared. "Nature will surely have her way with this folly."

"What if God's just tossing us bones—a rainbow and a thunderbolt, just to keep us believing we can win?"

"I can feel it."

"When all is lost, I guess we still hope," said Russ.

Two months later, on February 10, 2020, came another shock that smacked of divine intervention. A US district judge not only ruled in favor of jaguars in the Santa Ritas; he upended 147 years of law in the American West by declaring that the 1872 Mining Law didn't allow mine owners to bury 2,500 acres of surrounding national forest land under 2 billion tons of waste tailings piled twice as high as the Statue of Liberty.

When the Center attorney who successfully argued the case texted him the verdict—that by permitting open-pit mining in critical jaguar habitat, USFWS had violated the Endangered Species Act—Randy Serraglio could barely type the press release, he was sobbing so hard. It was so unexpected, and astonishing. Things had looked even bleaker when Donald Trump named a Hudbay Minerals lobbyist Secretary of the Interior. Now, protection for the jaguar would shelter all the other endangered and threatened species in the neighborhood.

But each month brought reminders that the Endangered Species Act, being nearly unique to the US, reached only so far. They could use it to protect Japanese dugongs, nearly extinct relatives of the manatee, because dugongs and three imperiled sea turtles live where the US wanted to extend an Okinawa airbase runway—because even abroad, Center lawyers pointed out, the military must follow US laws, including the ESA. But without such laws worldwide, all the planet's wildlife was now up against a wall.

In late 2022, in Montreal, where three decades earlier world powers signed a world-saving environmental treaty, the world again convened to address another global emergency. The stratospheric ozone layer might be healing, but everything below it was failing. Woodsmoke blanketed entire continents, as forests from Alaska to Australia became infernos. Living species were vanishing far faster than scientists could track.

The 2022 UN Biodiversity Conference, COP15—the biological

analogue to its biannual COP climate sessions—had an equally dismal record: of 20 targets that a 2010 conference had set for 2020, exactly zero had been met. Among them were cutting natural habitat loss and pesticide use by half, slashing fertilizer use to "levels that are not detrimental," and preventing the extinction of threatened species.

To Center senior scientist Tierra Curry, the meeting, which in the end produced a catchy slogan—"30x30": protection for 30 percent of global lands and waters by 2030, but no framework to finance it—felt all but meaningless. For hours, she'd watched hundreds of delegates propose edits to a Word document projected on a screen. When they finally agreed to delete a comma, the room broke into applause. Meanwhile, as they kept debating the meaning of words like *sustainable*, wildlife kept disappearing. Outrageously, nearly every country on Earth was a party to the Convention on Biological Diversity treaty except her own, because US Republicans repeatedly block the two-thirds Senate majority needed for ratification, claiming it puts US sovereignty and commercial interests at risk.*

Tierra, a biologist with long brown hair and a ready smile even when talking about extinction, grew up between two Kentucky coal mines. As a child, she watched their well water turn orange. All the tadpoles and turtles in the creek where she fell in love with nature died. Then both her parents died of cancer. Feeling nearly suicidal at 21, she wandered into the 1997 Million Women March in Philadelphia, where a woman's sign read "Yo soy la Tierra." She, too, was the Earth, she realized. Right there, she decided to change her name and fight like hell for this planet.

After reading about an NGO that halted construction of a Tucson

* The other holdout is the Vatican, to the chagrin of Pope Francis, who named himself for a saint who believed that God made all creatures our brethren.

school because cactus ferruginous pygmy owls were nesting there, she knew she'd found her place. The four white guys who founded the Center for Biological Diversity were still around, but in the new century inspiring women were running programs, like the sainted attorney Kassie Siegel, director of the Center's Climate Law Institute, who got the polar bear listed under the Endangered Species Act. Tierra Curry became a director herself, of the Center's extinction campaign, keeping various snails, slugs, crayfish, and freshwater mussels from being overlooked amid fighting for charismatic species like jaguars. After Randy Serraglio unexpectedly retired, increasingly Tierra was the one to meet with the public.

"Whether or not we're consciously aware," she'd say, "nature is part of our identity. Like little snail stickers on Easter eggs and funny towels embroidered with elephants, we grow up embedded with images of wildlife that underlie our imagination, and our survival, in ways we never contemplate, because we've taken nature for granted. Until now."

iv. Dream

Like a spurned abuser who slashes his ex-lover's face if he can't have her, Hudbay Minerals Inc. had cut a road through its mining claim on the eastern face of the Santa Ritas: a scar just visible enough from the scenic highway to irreparably mar its pristine beauty.

Hudbay may have lost the right to dump its mine's tailings on forest land in federal court, but all along it had a plan B.

Using shell LLCs, it had been quietly buying private land on the mountains' opposite, western face, where there was no national forest. Hudbay still intended to mine its original pit, where 80 percent of the copper lay, but would now transport its tailings to the other side via a slurry tunnel it planned to dig, and chemically smelt the ore in an adjacent sulfuric acid plant. All would be visible to Green Valley, an Arizona retirement community below that had repeatedly been promised that no mining would ever spoil their splendid view.

Amid the Southwest's deepening drought, this plan B would likely use even more water. Tucson and the county protested to no avail. Under the US Constitution, property rights were sacrosanct. Images that Russ McSpadden's drone took atop the Santa Ritas showed the mountains already crosshatched with roads connecting Hudbay's new holdings.

Randy Serraglio knew this activity was partly to impress investors. "That's how these companies work: they make more money mining investors than metal." But then Russ showed him geologists' projections of how the mining would collapse the scenic ridgeline: in effect, mountaintop removal.

Randy finally had run out of energy to challenge encroachment on jaguar habitat. Mining companies, he knew, played the long game. There'd be plenty of years for someone younger to take them on.

With his longtime partner, Louise Misztal, director of Sky Island Alliance, an NGO dedicated to keeping wildlife corridors intact in the Madrean archipelago, Randy Serraglio hikes up into the Whetstones, the uplift 20 miles east of the Santa Ritas where El Jefe first appeared. From the summit he can see the cottonwood-lined ribbon of Cienega Creek, midway between the two ranges, and beyond it, the place the jaguar was last seen—that is, until late 2021, when telltale spots on his hindquarters showed up on a rancher's camera trap in Sonora, Mexico, north of the jaguar preserve. El Jefe had braved the wall and returned to his breeding grounds.

Might he possibly reappear in the Santa Ritas before the big shovels arrive? Perhaps when he's an elder, challenged by younger, more vigorous males like he once was, maybe then he'll again breach Trump's crumbling gauntlet and return to his old hunting grounds, retiring in Arizona like the snowbirds in Green Valley, or like Randy Serraglio.

The Center's battle to strike a livable balance for humanity and wildlife may not be winnable, Randy knows. "With climate change, there's a good chance of losing 50 percent of the planet's species by the century's end. Last I checked, I'm a species. I don't like those 50-50 odds."

But the Center for Biological Diversity, where he toiled so long, also plays the long game: humans or not, its highest mission is to get as many species as it can through this major global extinction event, knowing that each one they lifeboat into the future will help replenish the world when nature revives, as it always does, eventually.

In the meantime, he finds some peace helping Sky Island Alliance restore desert springs, arranging stones into simple, effective structures that can reverse erosion where cattle have denuded the landscape.

He looks across to the Santa Ritas, seemingly just an arm's length away in the unsullied mountain air.

"I have a dream that long after humans catch the last pandemic bus to Valhalla, there are still jaguars. And one walks to a spring in the Santa Rita Mountains that I helped restore with my own hands a thousand years earlier, and drinks. And he looks around, and it's still habitat that works for him. That's my hope."

CHAPTER FIVE

STAR TIME

i. The Perfect Fuel

Hydrogen is such a tease. It's the most abundant element in the universe—three-quarters of all known mass. Like natural gas, it can be burned to produce energy, but with no carbon atoms to form the CO_2 that's causing all the grief overhead, hydrogen just oxidizes into that friendliest of chemical compounds, H_2O.

If we then zap the water in an electrolyzer—like a backward battery: instead of combining chemicals to make electricity, electrolyzers use electricity to pull chemical compounds apart—H_2O separates back into hydrogen and oxygen. We can then harvest the hydrogen, burn it to produce energy, collect its water vapor exhaust, and repeat—the cycle is so elegant it almost seems like perpetual motion. And it's clean as dew.

Even better than burning are refillable batteries called fuel cells: electrolyzers in reverse. Squirt in hydrogen, and, aided by a catalyst, it combines with airborne oxygen to produce electricity. No combustion necessary.

There are catches, of course; otherwise, we'd already do this instead of using filthy fossils. For one, pure hydrogen is rare: on Earth, it's nearly always bound to something else. In water, its bond with oxygen is so tight that it takes considerable electricity to extricate it. Prying hydrogen from hydrocarbons is easier and cheaper but a lot messier: it's done by blasting natural gas with steam, which burns even more carbon.

Using solar or wind energy to power electrolysis is called green hydrogen—but although touted as a clean pathway to decarbonizing, producing renewable electricity to make hydrogen fuel, which is then used to make electricity again—isn't very efficient.

The other big catch is that hydrogen is the most abundant element because it's the simplest: 99.9 percent of it is just one positively charged proton orbited by one negatively charged electron. That also makes hydrogen the lightest, so it takes a lot of it to produce as much energy as denser, dirtier fuels—and, being so light, it needs to be compressed or frozen to fuel vehicles, requiring even more energy. So far, despite many ardent proponents, the hydrogen age has remained perpetually five years down the road.

Speaking of which . . .

ii. The Perfect Fuel, Redux

In 2015, Dennis Whyte hadn't yet given up, but he was getting close. Until then, his entire life had been a crescendo, from ruddy-faced Saskatchewan farm boy to head of a high-energy physics laboratory at MIT. How could this be a dead end?

Growing up on the prairie, on winter nights he'd watch the undulating iridescence of the aurora borealis. It was caused by charged particles ejected from the sun, then deflected poleward by Earth's magnetic field, where they slammed into the atmosphere, producing flashes. He couldn't yet know how much charged particles would mean to his future, but already he loved science. During second grade, his electrician father taught him Ohm's law, which governs the interacting forces in an electrical circuit. He was hooked.

The University of Saskatchewan had a program in plasma physics. Plasma, such as the ionized gases in the northern lights, is the fourth state of matter: heat first turns solids to liquids, then liquids to gas. Keep heating gas, and its electrons strip away, leaving plasma. Although rare on this rocky planet—except when lightning bolts ionize the air, or in the minute quantities in neon lights and plasma TVs—throughout the universe, plasma is far more abundant than the other three states of matter combined.* It is the stuff of stars.

In the 1980s, it was also the stuff of experimentation, and an undergraduate professor recommended Dennis Whyte to the Université du Québec, which housed Canada's first program to see if humans could trap, and tap, starlight's practically limitless power right here on Earth.

Now he was really hooked.

The power of stars arguably derives from a miracle, because we don't yet have any other explanation for it—even geniuses like Albert Einstein and Stephen Hawking seemed to beg the question. Nearly 14 billion years ago, everything we know (and much we don't know) burst forth from a point infinitesimally small and almost infinitely dense, from which this universe is still expanding. What created that point may remain one for the mystics. Hindus believe it keeps happening over and over: Brahma awakens, exhales out the universe, and eventually inhales it back again, ad infinitum. But Brahma is born from a lotus that springs from the navel of Vishnu, who somehow exists *before* creation.

Although physicists still argue the details, most dismiss the question

* That is, known matter. There appears to be six times more "dark matter" than all phases of visible matter together. Because we can't detect it, we know little about it—we assume its existence to explain why planets, stars, and galaxies behave as they do.

of what existed before the Big Bang as meaningless ("Like asking what lies south of the South Pole," said Hawking), because apparently time itself—the interval between events—originated then, too.

Whatever it was, it next took 100 million years for some of the original particles to overcome the sheer momentum of that explosion and start interacting, mutually attracted by a force called gravity (again, not fully understood). Gradually, these particles coalesced, swirling into giant balls of gas—mainly hydrogen, because that's practically all there was; plus some helium, the next lightest element, composed usually of two protons plus two neutrons: particles with no charge.

A tiny fraction of hydrogen atoms also had one or two attached neutrons—more on them later.

The bigger those gas balls got, the more internal pressure the outer layers exerted. When things got so dense and hot at the core—15 million degrees hot—some hydrogen atoms smashed together and stuck. The result of that fusion was more helium—except each helium atom weighed slightly less than the two hydrogens that formed it. A tiny bit of mass had been lost: converted, as Einstein explained in 1905, to a stupendous amount of energy.

That energy is what makes the stars shine. Luckily, they're so big that their surfaces can be 2,000 times cooler than their centers, otherwise stars would burn up all at once. But ever since fusion was discovered in the 1930s, scientists have wondered if we could somehow replicate and harness it. Fusing hydrogen would yield *200 million* times more energy than simply burning it. Unlike nuclear fission, which powers the world's 440 atomic reactors, hydrogen fusion produces no harmful radiation, only neutrons that are captured and added back to the reaction. Instead of radioactive wastes with long, lethal half-lives, fusion's by-product is helium, the most stable element—and a year's worth from a fusion plant wouldn't supply a party balloon business.

But really? A star on Earth? In 1952, during the Cold War, the US tested a brute version of how to achieve such a reaction at Enewetak Atoll in the Marshall Islands, by detonating a bulky nuclear-fission device coupled with a packet containing super-chilled liquid deuterium and tritium—hydrogen isotopes with one and two attached neutrons, respectively, making them heavier, slower, and 100 times more likely to collide. The resulting explosion, which left a crater more than a mile wide, was far more powerful than they had calculated. Two years later, at neighboring Bikini Atoll, a solid-state version using far less tritium proved unexpectedly twice as destructive: the Marshall Islands are still trying to recover. How could we possibly tame such deadly potency?

Dennis Whyte's graduate studies began in a laboratory just outside Montreal belonging to the electric utility Hydro-Québec, where he was shown a device built to replicate that stellar process on an Earthly scale. It was a doughnut-shaped hollow chamber, big enough for a lanky physicist like him to stand inside, based on a design conceived in 1950 by future Nobel Peace Prize laureate Andrei Sakharov, who also developed hydrogen bombs for the Soviet Union. Its name, tokamak, was a Russian acronym meaning "ring-shaped chamber with magnetic coils."

The idea is straightforward: Fill the doughnut with hydrogen gas, then heat it until it turns to electrically charged plasma. In this ionic state, plasma would be held in place by magnets positioned around the tokamak. To achieve fusion on Earth without the immense pressure of a star's interior, scientists calculated, would require temperatures nearly 10 times hotter than our sun's center—around 100 million degrees Celsius. So, the trick would be to suspend the hot plasma so perfectly in a surrounding magnetic field that it wouldn't touch inner surfaces of the chamber that would instantly cool it, stopping the fusion reaction.

The good part about that was safety: in a failure, a fusion power plant wouldn't melt down—just the opposite. The bad part was that gaseous plasma wasn't very cooperative: any slight irregularity in the chamber walls could cause destabilizing turbulence. But the concept was so tantalizing that by the mid-1980s, 75 universities and governmental institutes around the world had tokamaks. If anyone could get fusion—the most energy-dense reaction in the universe—to work, the deuterium in a liter of seawater could power one person's electricity needs for a year: effectively, a limitless resource.

Besides turbulence, there were two other big obstacles. The magnets surrounding the plasma needed to be really powerful—meaning really big. In 1986, 35 nations representing half the world's population—including the US, China, India, Japan, the entire European Union, South Korea, and Russia—agreed to jointly build the International Thermonuclear Experimental Reactor: a $40 billion giant tokamak in southern France. Standing 100 feet tall on a 180-acre site, ITER (Latin for "journey") is equipped with 18 magnets weighing 360 tons apiece, made from the best superconductors then available. If it works, ITER will produce 500 megawatts of electricity—but not before 2035, if then. It's still under construction.

The second obstacle is the biggest: many tokamaks have briefly achieved fusion, but doing so always took more energy than they produced. In 2022, instead of using a tokamak, the US's National Ignition Facility at California's Lawrence Livermore National Laboratory aimed 192 laser beams in an array the size of three football fields at a hollow gold pellet the size of a pea, filled with deuterium and tritium frozen nearly to absolute zero. For less than 100 trillionths of a second, the lasers ignited a burst of fusion energy 1.5 times greater than the jolt that had produced it.

It was a scientific breakthrough, but not a clean-energy breakthrough. NIF's main role is verifying the viability of aging nuclear warheads; studying fusion's high-density requirements sharpens weapons scientists' assessment models. There's no power application for NIF's huge laser array, and its frozen targets proved inordinately delicate—the first five attempts to repeat the success failed. But for a week, its scientists were besieged with questions of how soon before we could hook our wires into a limitless, carbonless, safe fusion reaction. Increasingly aware of atmospheric havoc overhead, people yearned for clean energy that could fuel everything we demand without turning our planet into another furnace like Venus.

Although several experiments around the world were trying variations on the laser technique, most physicists, Dennis Whyte included, considered the simpler tokamak's design far more likely to provide that. Among other reasons, its plasma's density was one ten-billionth of the gold-capsuled laser fuel—far easier and less energy-intensive to produce.

After earning his doctorate in 1992, Whyte worked on an ITER prototype at San Diego's National Fusion Facility, taught at the University of Wisconsin, and in 2006 was hired by MIT. By then, he understood how huge the stakes were, and how life-changing commercial-scale fusion energy could be—if it could be sustained, and also be affordable.

MIT had been trying since 1969. The red-brick buildings of the Plasma Science and Fusion Center in Cambridge, Massachusetts, where Whyte came to work, had originally housed the National Biscuit Company. PSFC's sixth tokamak, Alcator C-Mod, built in 1991, was housed in Nabisco's old Oreo cookie factory. Just as copper wire wrapped around a nail and connected to a battery turns it into an electromagnet, C-Mod's magnets were coiled with copper. Before it was finally decommissioned, C-Mod's magnetic fields, 160,000 times stronger than Earth's, set the world record for the highest plasma pressure in a tokamak.

As Ohm's law describes, however, metals like copper have internal resistance, so it could run for only four seconds before overheating—and needed more energy to ignite its fusion reactions than what came out of it. Like the now 160 similar tokamaks around the world, C-Mod was an interesting science experiment that mainly reinforced the joke that fusion energy was 20 years away, and always would be.

Each year, Whyte had challenged Ph.D. students in his fusion design classes to conjure something just as compact as C-Mod, one-800th the scale of ITER, that could achieve and sustain fusion—with an energy gain. But in 2013, as he neared 50, he increasingly had doubts. He'd devoted his career to the fusion dream, but unless something radically changed, he feared it wouldn't happen in his lifetime.

The US Department of Energy decided to scale back on fusion. It informed MIT that funding for Alcator C-Mod would end in 2016.

"We shouldn't count on fusion to make a difference before the end of the century," Whyte overheard a prominent colleague say. MIT attracted some of the planet's brightest 23-year-olds. He saw fewer and fewer interested in fusion.

"It's uncertain what they can accomplish," he told his wife. "By the time ITER finally comes around, they'd be nearing retirement."

Either he would quit fusion and do something else, Whyte decided, or try something different to get them there faster. There was a new generation of ceramic "high-temperature superconductors," not available when ITER's huge magnets were being wrapped in metallic superconducting cable, which has to be chilled to 4 kelvin above absolute zero (−452.47°F) to have zero resistance to current.

Discovered in 1986 accidentally in a Swiss lab, the new ceramic superconductors still needed to be cooled to 20K (–423.7°F). But with far smaller power requirements, their output was so much greater that a year later its discoverers won a Nobel Prize.

The potential applications were limitless, but because ceramic is so brittle, coiling it around electromagnets wasn't feasible. But one day, Whyte ran into a research engineer in the hallway holding a fistful of what resembled unspooled tape from a VCR cassette. "What's all that?" he asked.

"Superconducting tape, new stuff." The filmy strips were coated with ceramic crystals of rare-earth barium copper oxide. "It's called ReBCO."

ReBCO's rare-earth component, yttrium, is actually 400 times more common than silver. Could superconducting tape, Whyte immediately wondered, be wound like copper wire to make much smaller, but far more powerful magnets?

He assigned his 2013 fusion design class to see. Just one doubling of the strength of a magnetic field surrounding hot plasma could multiply fusion's power density sixteenfold. His brilliant 23-year-olds came up with an eye-opening design they called Vulcan. It produced five peer-reviewed papers—but whether layers of wound ReBCO tape could stand the stress of the current needed to hold plasma suspended while being superheated to ignite a fusion reaction was unknown.

For two years, his classes refined Vulcan. By 2015, with improved consistency of ReBCO's quality and supply, he challenged his students—11 males and one female, including an Argentine, a Russian, and a Korean—to outdo what 35 nations had been attempting for nearly 30 years.

"Let's see if ReBCO lets us build a 500-megawatt tokamak—the same as ITER, only way smaller."

If superconducting tape could let them make a fusion reactor to fit the footprint of a decommissioned coal-fired plant, he told them, it could plug right into existing power lines. To then make enough carbon-free energy to stop pushing Earth's climate past the edge, its components would have to be mass-producible, so any competent contractor could assemble and service them.

They'd meet in a windowless room in a former Nabisco cracker factory, surrounded by blackboards. Divided into teams, they studied how thin-tape electromagnets could be made robust, and how to capture neutrons expelled from fusion reactions to use their heat for turning a turbine—and also to breed more tritium for the plasma: crucial, because natural tritium is exceedingly rare. As ReBCO-wrapped magnets would be so much smaller, shrinking the dimensions of one component rippled through everything else. One team's innovations fed another's, and parts of the design started to link together. As excitement spread through PSFC, members of earlier classes, now postdocs or faculty members, pitched in.

Having been trapped in academic and federal funding cycles for so long, Whyte realized that he'd lost sight of what he really wanted to do. His students, some with doctoral dissertations due, were putting in 50-hour weeks on this, reminding him of why he'd dreamed of fusion in first place.

And then, at the semester's end, out popped their design. Just over 10 feet in diameter, it actually looked like a prototype power plant. While ITER had massive shielding, their tokamak would be wrapped in a compact blanket containing a molten salt mixture of lithium fluoride and beryllium fluoride to absorb the heat of fusion's escaping neutrons—and, through reactions with the lithium, also use them to breed more tritium.

The blanket's heat would be tapped for electricity—except one-fifth of the heat energy would remain in the plasma, meaning the reaction was now heating itself and was self-sustaining: producing more energy than what was needed to ignite it. Net fusion energy.

The ReBCO magnets, although 40 times smaller than ITER's, could deliver a magnetic field strength of 23 tesla (a hospital MRI machine typically operates at 1.5 tesla). That was more than enough to achieve a fusion reaction, yet it would require 2,000 times less electricity than its copper-clad C-Mod predecessor. Everything was designed for easy maintenance or parts replacement without having to dismantle the entire reactor.

Most important, the calculated energy output was more than 13 times the input.

Whyte looked it over for the thousandth time. He was pretty sure they hadn't broken any laws of physics. He calculated the cost per watt and was astonished. Suddenly their goal wasn't just building a much smaller ITER. It was being commercially competitive.

Stunned, he told his wife, "This can actually work."

They called it ARC, for Affordable, Robust, Compact fusion reactor design. "Buildable in a decade," Dennis Whyte predicted. The peer-reviewed journal article his twelve students published in *Fusion Engineering and Design* estimated it would cost around $5 billion. In 2015, that wasn't much more than the cost of a comparably sized coal-fired plant, and one-eighth ITER's price tag.

That May, Whyte gave a keynote about ARC at a fusion engineering symposium in Austin, Texas. Four of his students attended. When he described their plan for a workable reactor by 2025, in just 10 years, conferees were astounded—everyone else was talking decades. Afterward, the MIT contingent went to lunch at Stubb's Bar-B-Q. Two things were clear: First, with the climate eroding and the Intergovernmental Panel on Climate Change warning that yet-uninvented technologies were needed to keep temperatures from soaring into dreaded realms, they had to do this. Second, since the DOE had pulled its funding, how could they?

On a napkin, Whyte started listing what they'd need to do and what each step might cost. Over ribs, they crafted a proposal to spin off a startup to raise venture capital to finance a SPARC, a Soon-as -Possible ARC demo fusion reactor to show that this could really happen. Then, a commercial-scale ARC.

Forming a company would free them from those ponderous academic and government funding cycles, but they were plasma physicists, most still in their 20s, without business backgrounds. Nevertheless, Whyte and a senior faculty colleague agreed to join them, so in 2018 Commonwealth Fusion Systems, CFS, was born. Three of his former students would run the company, and three would remain at MIT's Plasma Science and Fusion Center, which, in a profit-sharing agreement, would be CFS's research arm.

They opened shop up the street, in an MIT startup incubator. Design in hand, they entered the 21st-century techno-feudal system, where most of the world's money is held by a few princes, some dark, some illuminated. Persuade a prince or two that your idea could make them even richer—or maybe do some good—and they bless you with seed funding. Keep convincing them, and serious venture capital—sometimes billions—comes in successive waves known as Series A through D, after which, a company is on its own.

CFS gained the attention of climate-concerned princes like Bill Gates, George Soros, and Jeff Bezos. But they weren't the only ones competing for fusion funds: PayPal prince Peter Thiel was backing Helion, a company working on an hourglass-shaped reactor that would squeeze plasma from two directions into a bottleneck where a laser would hit it. Hedging his bets, Bezos was also grubstaking General Fusion, a Canadian-British firm whose design would inject magnetized plasma into a spinning drum of liquid metal, where steam pistons would implode the plasma until it was so dense that it ignited.

It was a race to see who could make commercial-scale fusion first: one of them, or the Chinese, Japanese, or Koreans. During the Atoms for Peace era of 1950s, with programs around the world building tokamaks or laser systems, all agreed that fusion was so complicated they should share data. Everything was open-source. PSFC published dozens of papers detailing their breakthroughs. As much as possible, the CFS team wanted to keep it that way for credibility, to not operate in a black box. But a company also needed a certain amount of intellectual property to satisfy investors.

They may have been young, but because of their partnership with MIT and its more than a hundred experienced fusion scientists, they had a running start.

By 2018 they'd attracted more than a dozen major venture capital firms, plus Google, a big university's endowment, a pension plan, and two energy giants: Italy's Eni and Norway's Equinor. By the end of 2021, Commonwealth Fusion Systems had raised more than $2 billion and was breaking ground on 47 acres outside Boston for a commercial fusion energy campus, to build SPARC by 2025—and then, by 2030, commercial-scale, mass-producible ARC.

iii. Magnetism

"Fusion is the only energy that doesn't need resources," said CFS's CEO Bob Mumgaard a few weeks before the company's first big test in September 2021. "Other than tiny bits of deuterium you buy for two bucks a bottle, and a bit of lithium that gets reused like a catalyst for breeding tritium, you're not beholden to Mother Nature for fuel. You don't have to dig it up, or wait for the wind or sunshine, or use a lot of land. One small footprint replaces 300 wind turbines."

It's also safe, he added. "When people hear 'a hundred million degrees,' they imagine lava. But fusion's like a candle in the wind: one breath of air into the chamber shuts the whole thing down."

Like Dennis Whyte, Mumgaard—a coauthor of SPARC's predecessor Vulcan design—grew up on the prairie. At the University of Nebraska, he studied how to engineer faster hard drives, but solid-state firmware ended that, so he looked for something else. High-energy particle colliders, dark-matter telescopes—so many neat things to do, but he settled on fusion. In 2008, he wasn't looking to change the world. "I wanted something cool to talk about at parties."

For his master's degree, he designed a way to take measurements inside fusion reactors that is now used worldwide, including by one of CFS's rivals, Korea's KSTAR. By his Ph.D., with a runaway storm looming over Earth's future, fusion was no longer just cool science, but a mission.

"You go from thinking what you could do, to what *should* you do. Everyone's working weekends, late nights. We know we need this fast. We can't wait till 2050."

Or even until 2030, the way temperatures were rising. "At every meeting," echoed chief science officer Brandon Sorbom as test day approached, "we look at what's limiting our schedule. Every week we reevaluate to see what we can do quicker."

They were in PSFC's West Cell laboratory, the cavernous former Oreo factory that previously housed Alcator C-Mod. Lithe and athletic, with a new beard, Sorbom studied electrical engineering at Loyola Marymount University in Los Angeles and was a varsity rower until he tore his hip labrum during his junior year. Bedbound for a month, he picked up a book about fusion. His scholarship allowed a fifth year, so he became a red-shirted physicist. Inspired by a hobbyist website for "fusioneers," during his second senior year, he built a tabletop fusion reactor that actually worked.*

* And so can you: https://fusor.net.

Nevertheless, MIT rejected his grad school application. With a thousand dollars, a credit card with zero percent first-year interest, and no winter clothes except for sweatshirts, Sorbom moved to Cambridge anyway and got hired at PSFC as a lab technician. After toiling in the ranks for a few years, he was elevated to the doctoral program and became the lead author of the 2015 fusion design class's breakthrough paper.

Although gaining and actually sustaining net energy is perpetually called fusion's yet-unreached "holy grail," Mumgaard, Sorbom, Whyte, and 200 CFS colleagues were now confident they could do it—if their magnets held. For three years, straight through the pandemic, they'd worked in the West Cell, furiously solving problems like how to solder thin-film ReBCO tape together into a structure strong enough to withstand 40,000 amps passing through it: enough to power a small town.

The completed SPARC would have 18 magnets encircling its plasma chamber, but for this test they'd built just one. It was composed of 16 layers, each a D-shaped,* 10-foot-high steel disk grooved like an LP. On one side, the grooves held tight spirals of ReBCO film, 270 kilometers in all—the distance from Boston to Albany. "Yet all that ReBCO holds just a sprinkling of rare earth," said Sorbom. "That's the magic of superconductors: a tiny bit of material can carry so much current. By comparison, a wind turbine's rare-earth neodymium magnets weigh tons."

On each disk's flip side, the grooves channeled liquid helium to cool the superconductor for zero resistance. (The design dates to history's first high-field magnet, built at MIT in the 1930s, which used copper conductors and water for coolant.) Each layer was built on an

* As in most tokamaks, a cross section of SPARC's doughnut-shaped plasma chamber will be D-shaped, not round, to concentrate peak intensity at the D's corners, permitting optimal control over plasma's variable electromagnetic field.

automated assembly line. "The idea," said Mumgaard, "is to make 100,000 magnets a year someday. This can't be a scientific curiosity. This needs to be an energy source."

Although COVID-19 had waned, an outbreak could foil everything, so they maintained coronavirus protocols, moving computer terminals outside beneath a tent to avoid crowding within. Others worked virtually. Mumgaard, now in his late 30s, had bet his career on this moment. His usual impish grin was hidden behind a white surgical mask, but his eyes beamed confidence.

For a month, dozens worked eight-hour, continuous shifts. Some operated the electromagnetic coil, encased in stainless steel in the middle of the room, which over a week had to be gradually supercooled from room temperature of 298 kelvin down to 20 K before slowly ramping up to full magnetic strength. Others constantly compared real-time data to redundant models. As the temperature dropped, the internal connections, welds, and valves contracted at different rates, so they watched for leaks. Every 15 hours, they'd pause, slowly charge the coil, run simultaneous computer simulations, match up the data, then exhale in relief as everything did what they were expecting.

Support staff ferried in meals. Whyte, Mumgaard, and Sorbom were pulling 14-hour days, collapsing, then awakening to look at their screens. When they were ready for the final test, access to the control room was restricted to the engineers actually running the coil. Outside, the physicists glazed over checklists they'd tortured for months, trying to anticipate every possibility, because no one had ever seen this state before.

On September 2, 2021, the Thursday before Labor Day, they started ramping up by a few kiloamps, stopping frequently as the current slowly climbed to a parking spot at 35 kA, where they rested before the final ascent to their 40 kA target. At each stop, they'd check what the current was revealing, how the cooling characteristics changed, and how the stresses on the ReBCO coil increased as the magnetic field strengthened to record heights.

Two nights later, Sorbom drew the graveyard shift in the mezzanine control room that overlooked the magnet. With board members and investors in town, the evening after the test he'd be answering questions over dinner at Cambridge's Naco Taco. He desperately needed sleep, but he didn't want to miss the final hours as they cranked the amperage toward their goal: a 20-tesla magnetic field—powerful enough to lift 421 Boeing 747s, or to contain a continuous fusion reaction. They'd been aiming for 7:00 a.m. on Sunday, the 5th. He set his alarm an hour early.

At 3:00 a.m. his phone rang. The engineers had made an adjustment and pushed the current. He pulled on the T-shirt they'd designed for the test, showing a poker hand, all face cards, each depicting a different pioneering scientist whose contributions had brought them to this point.* All their chips were in.

Over that went his lucky red plaid flannel shirt. A half hour later he was in the design center, just as the large screen behind Dennis Whyte and Bob Mumgaard showed that they'd reached 40 kiloamps, and the magnetic field had reached 19.56 tesla.

At 4:30 a.m., they were at 19.98 tesla. Things got very quiet. At 5:20 a.m., every redundant on-screen meter read 20 tesla, and nothing had leaked or exploded—except under the tent, where champagne corks were popping.

With an error margin of plus or minus six hours, they hadn't

* André-Marie Ampère, inventor of the solenoid (and the telegraph); Hans Christian Ørsted, who discovered that electric currents create magnetic fields; Michael Faraday, who explained electromagnetic induction; James Clerk Maxwell, who unified electricity, magnetism, and light as different forms of the same phenomenon; and Heike Kamerlingh Onnes, who discovered superconductivity.

known whether the caterers should bring shrimp cocktails or breakfast pastries, so they'd ordered both. No one minded. Washing down a croissant, Whyte realized he'd been so engrossed he forgot to be nervous. Five years earlier, on its final four-second run, C-Mod's copper-conducting magnet had consumed 200 million watts of energy to reach 5.7 tesla. This took 30 watts, less energy by a factor of 10 million, to produce a magnetic field strong enough to sustain a fusion reaction. The joints that transferred current from one layer to the next actually performed better than expected. That was the biggest unknown, because there was only one way to test them: in the magnet itself. They looked spectacular.

After five hours, they ramped down the power. "It's a Kitty Hawk moment," Mumgaard said.

The world had changed overnight on September 5, Dennis Whyte figured: the possible energetic gain had increased by a factor of at least 30. Questions still remained: Could ReBCO tape be protected against repetitive neutron bombardment? How would they remove the fusion reaction's helium ash, and make dozens of other tweaks, as they built their working prototype, SPARC?

These are problems for his design classes to tackle. "Breakthroughs don't happen with a lone genius thinking in a corner," he tells them. "It's usually a half dozen people pushing themselves, using each other's strengths to find a way." His classes have more than doubled in size as more students, especially more women, want to be part of history—and of a future.

Ramping up clean fusion could limitlessly electrolyze green hydrogen, to replace the mostly gray, carbon-intensive hydrogen from methane currently on the market. "An even wilder dream," says Whyte, "would be powering atmospheric carbon capture and storage, to actually pull us back from this climate precipice."

First, a full-scale ARC must not only work but also be affordable and globally replicable.

"Complicated, technologically aggressive solutions," Jonathan Foley, the director of the carbon-removal nonprofit Project Drawdown, has argued, "are years or decades away from practical use." Unlike the already available renewables, increased efficiency, and energy conservation steps that Drawdown advocates, a technofix like fusion, he says, "doesn't require us to change our ways. In fact," he laments, "it usually allows us to consume even more."

But since curbing human consumption might take longer than revolutionizing technology, we probably need to try everything. "It's scarier than we think," says Whyte. "We're trying to replace 82 percent of all the energy we have. It's probably the hardest thing humanity has ever tried to do."

"Over and over," says Bob Mumgaard, "when society wants to do something and has the tools, the technologies, and understanding, it can do it—provided it has some really big levers."

With a lever long enough and a place to stand, Archimedes once declared he could move the world. The world now waits to see if Commonwealth Fusion Systems really has conjured enough leverage by launching a galaxy of tiny stars on Earth to produce boundless, carbon-free energy.

Their place to stand is 35 miles west of Boston, where they're building SPARC and a factory to mass-produce magnets. At a gathering for locals, Brandon Sorbom was explaining how they'd started, scribbling on napkins at an Austin barbecue joint, when he looked up at the half-completed buildings rising over the 47-acre construction site and it struck him.

"It's definitely going to happen," he told them.

CHAPTER SIX

ON THE REEF

i. Coasts

Although human civilization was born in several cradles, our species emerged far from any of them. Back when so much of Earth's moisture was locked in Ice Age glaciers, African forests dwindled, so our *Australopithecine* forebears began adapting to savannas in the rift valley that formed as plate tectonics slowly ripped the continent in two. Along the East African Rift's lakes and ponds, *Australopithecus* evolved into *Homo* and learned to fashion an arsenal of stone, wood, and sinew tools to hunt game that came to drink, and to gather mollusks, turtles, catfish, even crocodiles.

So we were already accustomed to life at the water's edge when our growing bands followed game out of the rift and eventually out of Africa itself. Even so, the first time we saw a shore with endless water to the horizon surely elicited an epic gasp—followed by a gag, as we learned that all that water was undrinkable. But tide pools, filled with delicious crustaceans, marine mollusks, and fish stranded by the surf with each hypnotic wave, proved irresistible.

Ever since, we've been drawn to seacoasts. Today, eight of the world's 10 largest cities are coastal. So are nearly all the world's great financial capitals, among them New York, Miami, Buenos Aires, Amsterdam, Barcelona, Casablanca, Lagos, Cape Town, Tel Aviv, Mumbai, Dubai, Singapore, Sydney, Shanghai, Hong Kong,

Guangzhou, Shenzhen, Jakarta, Tokyo, and the freshwater seaports of Chicago and Toronto.

Two-fifths of the world's population lives within 60 miles of a coast—within 200 miles, nearly two-thirds—a percentage that keeps growing, even as the coastlines themselves are now contracting and threatening those cities. In any future we have on this planet, we'll need to adjust accordingly (an understatement, given all the money coastal cities represent) and learn to live with a whole new set of maps.

Katie Arkema, slim with shoulder-length brown hair, radiating health in her 40s, was born in the Philippines, where her father worked for Del Monte and where 100 million people live within sight of the sea. When she was six, they moved to Duxbury, Massachusetts, whose salt marshes and barrier beach became her playgrounds. Her undergraduate thesis in ecology and evolutionary biology briefly took her from the sea to research poison dart frogs in Costa Rica's rainforests. But after Princeton, she was back, teaching elementary school on Santa Catalina Island, off southern California.

There she became fascinated by surrounding stands of giant kelp, the world's largest seaweed. A brown algae, giant kelp is the global champion of carbon absorption. Converting sunlight and waterborne nutrients, the long triangular leaves alternating up its stalk, each with a gas-filled bladder to keep afloat, can grow two feet per day. In a kelp forest, these bladelike leaves form floating mats that shelter fish, mollusks, sharks, squid, lobsters, and sea squirts.

The aquatic community that giant kelp supports became the subject of her doctoral thesis in marine ecology at the University of California, Santa Barbara, where she spent 2001 through 2008 diving in lush underwater forests. By then, however, she and her husband, who'd just completed his own marine ecology doctorate, were aware that their careers wouldn't be devoted to fascinating underwater mysteries.

Seas now were warming, expanding, and rising. Meltwater from both poles plus the Himalayas, Rockies, Alps, and Andes was raising water levels even higher. Humanity's annual carbon dump into our air and oceans had jumped from 9 billion metric tons in 1960 to 40 billion metric tons annually. There's no meaningful comparison for 40 billion metric tons—we can't picture 6,780 pyramids of Giza. But to marine ecologists, it meant that seas, which absorb 30 percent of our CO_2 emissions, were therefore 30 percent more acidic than before the Industrial Revolution.

To Katie Arkema, our own fate and the rest of nature's were inextricable. Working to save the richest part of the seas, coastal ecosystems, meant including their most influential organism, coastal humans, so she accepted an appointment at a new Stanford University program, the Natural Capital Project.

Its founder was Gretchen Daily, protégé of Paul Ehrlich, best known for his 1968 book, *The Population Bomb*, but already renowned among scientists for a 1964 paper, coauthored with botanist Peter Raven, that coined the term *coevolution*—the biological arms race between flora that evolve chemical defenses to thwart insect predators, who counter by evolving immunity to them, and so on—which helps explain why nature is so biodiverse. Daily, Ehrlich's successor, believed the only chance to keep it that way is to embrace the enemy that is us.

The Natural Capital Project she founded, which today designs resilience plans for governments worldwide, defines nature in terms that even economists can grasp. NatCap frames environmental preservation as a wise investment by calculating how much the services nature provides are worth to civilization—and how much losing them would cost. Adoption of the Natural Capital Project by development banks and entire countries owes not so much to monetizing biology as to naturalizing money. Without nature, economies wouldn't exist. If nature fails, money will be worthless, so keeping ecosystems intact is in everyone's selfish interest.

At NatCap, Katie Arkema hoped to show that maintaining one of Earth's most precarious ecosystems was economically indispensable in the sultry belt from the Antilles to coastal Texas, where hurricanes roll up metal rooftops like old-fashioned sardine-can lids, tumble semitrailers, and snap centuries-old hardwood trunks.

Walking the coasts of Belize, the Bahamas, and Barbados kilometer by kilometer, she assessed mangroves, oyster beds, coral reefs, seagrass, salt marshes, barrier islands, and sand dunes: all natural buffers from storm surges. Unlike costly seawalls and breakwaters, porous reefs and oyster beds diffuse wave energy instead of trying to resist it.

"After a storm, they recover by themselves," she'd tell property owners she met along the way. "Hard infrastructure gets undercut and fails, but natural infrastructure migrates upslope with rising waters, helping coastlines keep pace by retaining sediments."

She and her team developed computer models to predict how much coral, seagrass, kelp, et al., can protect shore communities during hurricanes—and how big a threat their absence poses. To calculate the drag they create on storm waves, they analyzed everything down to the dimensions of individual seagrass blades and the diameter of sand grains. With drones, they measured how much stormwater wetlands sponge up and how much carbon marine vegetation soaks up. Using open-source NatCap software, Arkema layered all that information, plus data from hundreds of scientific studies, onto maps generated by going house to house to learn from local residents which areas they thought should be conserved or developed.

The resulting digital coastlines let them visualize flood risks by simulating storm surges and tidal waves 25 years into the future—and then see, virtually, how adding mangroves, reefs, dunes, or kelp beds might cushion the damage.

In 2015, Katie went to Andros, the Bahamas' biggest, most pristine island, encircled by one of Earth's longest coral reefs. She and residents mapped different scenarios, from maximum conservation—limiting tourism to reef diving, flamingo-watching, and stalking prized bonefish—to maximum development: a dock for cruise ships and a golf course.

Most islanders opposed the latter, but bonefishing, some pointed out, could benefit from an airstrip big enough that tourists could fly straight from Miami instead of having to change to smaller aircraft in Nassau. The scenario NatCap developed—"Sustainable Prosperity"—would permit minimally invasive tourism; control exploitation of the island's spiny lobster, sponge, and conch fisheries; and keep the protective reef and mangroves in place.

It was a poster-worthy project—until October 2016, when Hurricane Matthew smashed Andros Island with 140-mile-per-hour winds. Through tears, Katie saw the flattened remains of a charming fishing village, strewn with caskets washed up from a seaside cemetery. It was shocking but not surprising: their model had flagged how Andros's offshore shallows were exposed to storm surges, especially where mangroves had been replaced by a shoreline road and seawall.

Because so much of the island's natural barriers were still in place, it could have been much worse, she kept telling herself. She was heartened that people on Andros still opted for NatCap's nature-based, green infrastructure as their storm shield. Nevertheless, the Bahamian government and the Inter-American Development Bank were already discussing loans for conventionally "hardened" shorelines on other islands, meaning more inflexible seawalls. With future storms sure to be more severe and frequent, eventually the Bahamas would show which approach was wisest.

ii. Peninsula

The 1,000-kilometer, Z-shaped Mesoamerican Reef is the world's second longest, surpassed only by Australia's Great Barrier Reef. Beginning along the north shore of Mexico's Yucatán Peninsula, it dips south across the border with Belize, traverses the length of that county's coast, then turns sharply eastward along Guatemala's short coastline and continues along Honduras's broad northern coast. With more than 60 different hard corals, the world's second-largest manatee population, more than 500 fish species, two endangered crocodiles, three dolphins (bottlenose, spotted, and rough-toothed), and spiny lobster, queen conch, and shrimp, it's revered by divers and fishermen as one of Earth's richest marine habitats.

It's also among the most vulnerable to coral bleaching, rising seas, and the pressures of 2 million people living along it.

In 2018, Germany gave the World Wildlife Fund a €5 million grant to help the four Mesoamerican Reef nations brace themselves for what nature would surely throw at their coasts in the coming years. WWF brought in two partners: NASA and the Natural Capital Project.

The following July, they gathered in Mérida, a steamy Yucatán city 25 miles inland from the Gulf of Mexico, with pastel colonial architecture, cobbled streets, and a ficus-shaded plaza framed by arched facades and a colonnaded portico. Costas Listas—Smart Coasts—would be a workshop for Mexican coastal refuge managers, staff and scientists from government agencies and local environmental NGOs, and, arriving via an eight-hour bus ride, their Belize counterparts. An identical session in Honduras, which would include Guatemalans, would follow. In each country, site visits were planned to see the challenges facing them.

"It's so depressing," declared WWF conservation officer Abby

Hehmeyer over a patio dinner of a local specialty, panuchos: barbecued shark, pork, tomato, and avocado atop black-bean-stuffed tortillas. A wildlife biologist still in her 20s, this was not the world she had signed up for.

She'd arrived from Washington, DC, with Ryan Bartlett, WWF's climate resilience and risk management director. Following the workshops, Ryan, angular and bearded, his hairline receding, would be getting married—a blissful life passage but for the pall his work cast over the future.

"We climate scientists," he said, pulling on a Yucatán pilsner, "are forced to deal with a grieving process."

Nodding across the table were two data modelers from NASA's Goddard Institute for Space Studies at Columbia University: Manishka De Mel, a short, black-haired Sri Lankan, and tall, fair-haired Meridel Phillips. They provided the numbers that WWF and the NatCap crew would show the Mesoamerican Reef people.

Katie Arkema, they'd heard, had arrived but was resting. Months earlier, riding to work from her home in northeast Seattle, her bicycle hit a patch of black ice. She'd landed so hard that her helmet cracked. The injury had left her sensitive to light and struggling with migraines. She'd taken medical leave and canceled numerous conferences, but she was determined not to miss this. It was her first international trip since the accident, and the long flight with a complicated plane change in Mexico City had been taxing.

Dessert appeared: stewed nance, a cherry-sized golden Caribbean fruit. Bartlett was still brooding over his cerveza. "You still hear talk about stopping or reversing climate change. It's not possible. Physical science says we cannot get there. We need to deeply think about how to survive and still thrive on a much warmer planet."

Abby Hehmeyer twisted her long chestnut hair. Manishka and Meridel were still nodding.

"When there's no room left on Earth, the balance between ecosystems and humanity gets very complicated," said Ryan, thumping down his empty.

"There's no hope," GISS analyst Stacie Wolny mumbled to her NatCap colleague Jess Silver the next morning.

"Uh-huh," replied Jess, as they loaded thumb drives with the open-source NatCap software that would let users compare different preservation scenarios.

Jess, her red hair cropped close in front, was a technical analyst of ecosystem services: things nature provides that humans too often take for granted. The Mesoamerican Reef's ecosystem includes corals, rainforests, soils, mangroves, rivers, coastal dunes, seagrass meadows, and all the creatures that make a living from them, humans included. All interact with each other in myriad ways that NatCap's free software, InVEST*—Integrated Valuation of Ecosystem Services and Tradeoffs—was designed to reveal, by showing what happens if any go missing.

Stacie, whose gray-streaked straight hair reached the middle of her back, created the layered InVEST maps that give a regional overview, then zoom anywhere into granular detail. As they chatted with arrivals about how the NatCap software would help them integrate climate change with coastal management, Stacie's pride in the tool's capabilities belied her despair over the big picture.

For two days, 70 participants would sit around nine round tables in an air-conditioned, windowless hotel conference room, sealed off from Mérida's heat and humidity, strategizing a compromise between their species and everything else to keep coastal life viable. As each group introduced itself, some highlighted successes, such as protecting jaguar and manatee habitats. Others mentioned disquieting

* https://naturalcapitalproject.stanford.edu/software/invest.

changes, such as reeking piles of sargassum that the ocean was heaping on beaches. Mats of the brown algae now stretched from Africa through the formerly transparent Caribbean to the Gulf of Mexico, fouling fishing nets and engines, smelling like rotten eggs, repelling tourists and nesting sea turtles alike. They were caused by unprecedented synthetic nitrogen runoff gushing down the Amazon, Mississippi, Congo, and Orinoco rivers, meeting millions of tons of phosphorus-rich dust blowing off the Sahara. Stoked by heat waves, together they'd fertilized a continent-sized algae bloom.

The attendees also dealt with developers uprooting mangroves for more seaside dwellings, and dredgers pulverizing endangered corals: adversaries these scientists and wildlife managers had known for years. It was the unknown that now made them nervous. In climate summits, *mitigación* had morphed to *adaptación*. Humanity was losing the battle against rising temperatures, and now must adapt accordingly, Ryan Bartlett's welcoming remarks implied.

A man in a T-shirt that read "Pronatura" raised his hand. "Who's paying for this adaptation?" He looked around the room. "We all need money. Maybe some of the good ideas we're about to hear will work in the US, but this is Mexico."

"Climate is what you expect," Manishka De Mel told the room, quoting science fiction author Robert Heinlein, "but weather is what you get."

Pues, así era—well, that used to be true, someone hissed, evoking titters. No one knew what to expect anymore.

Actually, Manishka continued, they had some idea. "When we do projections, we think of climate as an average over 30 years. We already see increases in land and sea surface temperatures, the number of hot days, and the number of storms. Most relevant for you are extreme heat events. They're increasing not only in intensity, but also frequency."

"Part of the challenge," added Meridel Phillips, "is there's a range of possible changes." She projected a graph with a line that began in the lower left and broadened into a multicolored fan swooping upward to the right. It represented the four different scenarios, or RCPs, that the Intergovernmental Panel on Climate Change has posed for the 21st century. RCP stands for a name only a bureaucrat could love: Representative Concentration Pathways. Although *concentration* refers to greenhouse gases in the atmosphere, the familiar parts per million of CO_2—280 preindustrial, more than 400 since 2015—aren't the whole story, because carbon-trapping methane and nitrous oxide also count. Rather esoterically, RCPs are expressed in wattage the Earth receives per square meter.

Averages, Meridel explained, can be misleading. "So our models show high and low estimates: RCP 8.5 and RCP 4.5."

At the tables, the Mexican and Belizean scientists exchanged bewildered glances. The high estimate, they knew, is the worst-case, business-as-usual scenario, in which emissions keep rising relentlessly. But these modelers, they realized, weren't even using the IPCC's lowest scenario, RCP 2.6, representing the fastest transition to clean energy, but instead, midrange RCP 4.5, which assumes that global emissions won't stop rising until 2040.

Just a year earlier, the IPCC had warned that humanity must cut emissions in half by 2030 to limit global warming to 1.5°C. Were NASA's modelers acknowledging that this was no longer possible?

Manishka pointed to where the high estimate's line turned sharply upward. "When you come to midcentury, you see these really diverging. So you want to mitigate as much as possible by 2050. Reducing emissions to the midrange RCP will really make a difference."

She paused, sensing confusion in the room. "I know—it's hard to communicate about the long term. But you can see the difference really begins to widen midcentury. By 2100, the results are quite different. So it depends on the future we choose. We still have a little time to reduce emissions. But as we go along, change will need to be more drastic to be effective."

At "we still have a little time," several participants visibly slumped in relief.

She switched slides to a bell-shaped curve, its vertical axis showing days in a year and its horizontal axis showing temperature. "This curve shows that, as you'd expect, most days are average, some cooler and some hotter, and that a few days are very cold or very hot."

She then superimposed another bell curve, overlapping the first, but moved slightly along the horizontal axis toward higher temperatures. "The challenge is that just a little warming shifts that whole curve to the right." The gap between the leading edges of each bell represented, she explained, many more very hot days per year; already, the number over 36.5°C (97.7°F) had doubled since 1995.

"By the end of the 2020s, Yucatán can expect six more hot days per year. By midcentury, a month's worth."

There was also sea-surface temperature. By 2100, the number of days over 30°C (86°F)—the threshold for coral bleaching—could increase from the current 140 per year to 342: practically the entire year. Plus, temperatures would still be rising.

"You'll see heat extremes that didn't really exist before," Manishka said. "It's impossible to imagine. That's why it's so important to reduce emissions."

"You see why we collaborate with NASA-Columbia," Ryan Bartlett said to the subdued room in flat, fluent Spanish, after Meridel had concluded with rainfall models that projected overall regional drying, despite more and stronger storms than ever.

"Their data confirm that it's best to prepare for the worst rather than the best case."

He asked each table to select one variable, then list all the impacts it had on their region and how they might become more resilient. The exercise, intending to clarify, only proved how complicated things were. Biologists from Ría Lagartos, a nearby nature coastal reserve, saw no way to stop beach sands from becoming hotter—meaning more sea turtle hatchlings would be female, because with many reptiles and fish, gender is cued by incubation temperature. Green and loggerhead turtles were already arriving at beaches much earlier, so maybe they instinctively knew to compensate, but that could skew seasonal feeding patterns.

An adjacent table was equally uncertain what to do about tapirs and jaguars, driven out of drying habitats in search of forage and game, crossing highways into urban areas. The next table worried that more storms meant fewer fishing days and also fewer tourists. A suggestion of adding more indoor activities evoked derision: Who travels to the seashore to stay indoors?

More flooding also meant more mosquito-driven diseases like dengue, and saline contamination of the aquifer. One table surprised Manishka by predicting more forest fires—she wasn't aware that parts of tropical Belize had already burned, including stands of mangroves growing right from the water.

"We're dealing with changes that none of us can combat," said a woman from a Mexican blue carbon* NGO. "All we can do is tell our fishermen to build stronger boats."

During a working lunch of soft tacos, a discussion about disappearing mangroves overheated.

"The US should pay for their restoration to atone for what its carbon-fired capitalism has ignited," demanded a university professor,

* Carbon captured by oceans and coastal ecosystems.

standing at a rear table, arms spread. Whether he was implicating the workshop leaders—excepting Manishka, all Americans—was unclear, but that moment, a pallid Katie Arkema, wearing a loose-fitting cotton blouse and pants, appeared in the doorway. She smiled. The tensions calmed.

Her voice, hesitant at first, gradually strengthened as she talked. "You've heard Manishka and Meridel. We know that we need to adapt to climate change. But how? That's the big question."

Although her team was running the models, finding the answers would be a bottom-up effort, coming from people who live and work in their region, she said. "This afternoon, we'll work on how the future might look if we're actually able to adapt."

For a half hour she showed how NatCap's InVEST software proves that protecting ecosystems protects coastal communities.

"It's impossible to quantify all the benefits the nature provides, and not very useful: we'd get too much information and might not tell you which places matter most." InVEST, she said, simply measured risk by combining exposure with vulnerability.

"Exposure is wherever waves and storm surges likely lead to flooding and erosion. Vulnerability is the consequences of that exposure. For that, we need to know where are people most at risk. For example, young children unable to evacuate on their own or poor, elderly people."

Those demographics, and data about economic vulnerability, would come from the attendees themselves, she told them, just as she'd gathered information door-to-door in the Bahamas. They would add things like how hard or soft or high a shoreline is, habitat types, wind and wave exposure—"we don't actually need to model storm surges: it would take months, and yield too much detail. Instead, the distance from shoreline to the edge of the continental shelf works quite well. Where it's steep, water stays down, like in bathtubs. But where it's wide and shallow, storm surges have nowhere else to go but ashore."

She projected a map of the entire Mesoamerican Reef coastline, showing communities, commercial zones, maritime installations, tourism sites, and protected areas, studded with blue and red dots representing lowest and highest risk. Their data sources ranged from satellite images and sensor-equipped buoys to Twitter posts and Instagram images, which reveal places tourists favor.

She showed a second map. "We ran the model again, without all the protective ecosystems as a buffer anymore." Nearly all the commercial zones were now red. "Imagine those businesses lost."

She flipped between the two maps. "Compare all the places that switched from blue to red. When you lose just your officially protected refuges, more than double the number of people are at risk. Remove the rest of the coral reefs, forests, seagrasses, and mangroves, and it's three times the number."

To be really useful, she concluded, the model should be run three, four, even five times, each incorporating more data from locals who know their environment. "Tomorrow we'll give you maps to draw on, to make this much more specific."

She paused as a thunderclap flickered the fluorescent lighting. "But first"—she shouted over rain pelting on the roof—"despite everything negative you heard earlier, we want to shift our mindset and try to be positive." Several lifted their gaze from laptop screens, amused.

"We tend to think the future will be worse. Get rid of that. We want innovative ideas to come from this, to make this future a more pleasant place for our kids and grandkids."

More thunder, and all the lights went out. "Time for a break," announced Jess.

Two days later, tall, gray-bearded Christian Mario Appendini and curly-headed Alec Torres-Freyermuth, coastal engineers at Yucatán's branch of UNAM, the National Autonomous University of Mexico, skipped the training session for using the NatCap software that Stacie had distributed in thumb drives and headed north for the Gulf in a university Subaru SUV. Leaving Mérida, they crossed a ridge about three miles wide that eased into a broad basin just 15 feet deep, a descent so gradual it's visible only in computer-enhanced imaging taken from space. Hundreds of miles long, semicircular, the ridge was first noticed in the 1940s by a Petróleos Mexicanos geologist looking for oil deposits, who assumed it was a volcanic formation. Thirty years later, another PEMEX geologist deduced from samples of shocked quartz and glass tektites he'd gathered that it was actually the rim of an impact crater.

At the time the meteor struck there, 66 million years ago, Yucatán was beneath the sea. The collision spewed rocks into outer space and filled the planet's atmosphere with fire and ash for months, igniting most of Earth's forests, killing nearly everything on land except some burrowing creatures and a few small feathered dinosaurs whose beaks had evolved to eat seeds, which they foraged among the dead trees. As the atmosphere cleared and photosynthesis resumed, ocean zooplankton resumed depositing their shells, which precipitated to the seafloor for eons, forming thousands of feet of limestone, while their bodies, along with dead phytoplankton and algae, were compressed into immense deposits of oil.

Around 34 million years ago—halfway between that cosmic crash and today—two smaller meteors struck. One hit Siberia; the other's crater became Chesapeake Bay. The Earth had already cooled since the days of the dinosaurs, its atmospheric CO_2 greatly reduced by enormous blooms of fast-growing aquatic ferns that sequestered vast quantities of carbon when they died and sank to the ocean's depths. But those two meteors' ashes blocked enough sunlight to turn Antarctica's beech forests to tundra. As reflective snows begat more cooling, a glacial cap formed. And as more of Earth's moisture was locked in polar ice, the seas lowered until the Yucatán and Florida peninsulas emerged.

Being a former sea bottom, Yucatán is pancake flat. Lacking hills, it has no rivers; rains simply percolate through the porous limestone, dissolving scores of caverns whose roofs have collapsed, forming cenotes, the peninsula's popular swimming sinkholes. During the conference, Christian told NASA's modelers that they might be overstating bleaching of the Mesoamerican Reef, because Yucatán's limestone is riddled with subterranean streams that satellites can't detect. As they reach the sea, their cooler runoff likely explains the many pockets of still-living corals.

"That's a perfect example of why we're here," Manishka replied, "to strengthen the models with local knowledge and observation."

During the second day's sessions, the NatCap and WWF crews had given each table maps of their respective coastal zones. They had participants write conservation priorities on blue sticky notes. On red notes, they listed everything already in place, natural or human-made, that might help them live with sea level rise. On orange, they noted the benefits if their knowledge was applied: improved chances of preserving fishing stocks, dunes, mangroves, and homes.

Patterns emerged. People who had earlier stared at a doomed future were now engrossed in pairing desired outcomes with mechanisms to make them happen, and seeing possible convergences. At one point Jess asked everyone to pretend that a time machine had transported them to the year 2060, where they were journalists writing headlines the day after a big storm hit their coast.

"The headline needs to mention sea level rise, and needs to be positive." Everyone snickered, and some headlines were sardonic: "Sand Dune David Defeats Hurricane Goliath," "Flooding Fills Cenotes: More Water for Swimming!"

Her next request, to add nature to their headlines, produced "The Nature of Water Is to Enter Our Houses!"

It was a corny exercise, but Christian and Alec found themselves impressed that the Americans' efforts weren't just earnest but effective. As participants superimposed sticky notes on their maps, they started imagining adaptive strategies and scenarios, not more impending disasters.

But as he and Alec reached the coast, the sobering reality of their own task lay before them. Stretching to each horizon were private beach houses and businesses, many built atop dunes, fronted with concrete seawalls or, where once mangroves grew, stacks of long, sand-filled polypropylene bags. Most homes' beachfronts were framed by breakwaters of either steel or rocks piled between wooden stakes, to keep the northeast current from sweeping away the sand. But each breakwater cut off the natural sand flow to their neighbors on the west, who in turn did the same to the next neighbor, resulting in a coastline resembling the edge of a scallop shell.

Earlier they'd heard Katie Arkema describe the Netherlands' Sand Motor, a 128-acre sand peninsula created in 2011 near The Hague. Instead of having to dredge sand annually, a single massive dredging had extracted 21.5 million cubic meters of offshore North Sea sand, dumped it in place, and let nature take its course, replenishing beaches and dunes to either side for at least two decades. "You can do that in Europe," said Christian, whose mother was from Denmark, and who spent a decade working for a Danish hydraulic agency before returning to Mexico. "In Holland, beaches are for the public."

"Here it's unmanageable, eroding private frontage," agreed Alec, who did his doctoral work in Spain. "Regenerating dunes and planting mangroves? It would cost a fortune, and who would benefit? Condo owners who think trees make garbage because they drop leaves? I sincerely don't see it."

Christian and Alec's university's campus is in Sisal, a town that before the invention of nylon shipped henequen fiber for making rope all over the world. Thirty-five klicks to the east, however, Sisal's pier was now dwarfed by the world's longest wharf: Progreso, which juts 6.5 kilometers into the Gulf of Mexico, to where the continental shelf drops deep enough to dock cargo ships. Although there are two berths for cruise ships, among its main purposes is importing refined hydrocarbons in return for Mexican crude oil exports.

Because more than half its length is earthen causeway, to these coastal engineers it's also the world's longest breakwater, playing havoc with natural currents and sand deposition, and another reminder of their planet's most powerful, dirtiest industry, to whose energy their species is terminally addicted.

So frustrating, because with its constant northeast breeze, Yucatán's flat continental shelf would be matchless for offshore wind generation. Another missed opportunity; a Chinese solar project here that replaced a mangrove forest with a chain-link-enclosed compound was never connected to the grid, and ultimately abandoned.

Just before Progreso, they grabbed a seaside lunch in Chicxulub, where a broad white awning shaded outdoor booths with racks of beachwear. Next to a taquería was a small, neglected plaza, where unmowed grass and a henequen agave partly obscured a poured-concrete plaque shaped like a giant bone.

Christian posed for a picture. Alec read aloud the hand-incised Spanish inscription: "Sixty-five million years ago, an enormous asteroid 10 kilometers wide crashed into this spot, detonating a global cataclysm, finishing off 60 percent of everything alive, most spectacularly the dinosaurs, allowing the emergence of mammals, thus beginning the story of humanity."*

Returning to their SUV, they headed west to Chelém, a coastal town where, due to Progreso's massive wharf, more sand had washed away than currents could replenish. Their beaches gone, houses were now fronted with masonry seawalls or stacks of sand-filled geotextile tubes. Some had already succumbed to the surf. A brown pelican perched on a new seawall already significantly undermined by waves.

"They build new ones every year. The old ones keep falling," said Christian.

"Everyone says the beach used to be 50 meters out there," said Alec, pointing.

"With another half-meter sea level rise by midcentury," said Christian, "we won't need another asteroid here."

Before the visiting scientists left for the next workshop in Honduras, the Mexicans took them up the coast to Yucatán's Ría Lagarto UNESCO Biosphere Reserve, an inlet lagoon filled with crocodiles, tiger herons, white ibis, and pink flamingos. A park ranger named Jaicy Maldonado, sporting silver turtle earrings beneath a Comisión Nacional de Áreas Naturales Protegidas ball cap, led them up the beach where sea turtles nest. Green turtles lay up to 140 ping-pong-ball-shaped eggs, she explained. Hawksbills, up to 200.

"How many nests here?" Katie asked.

"About 8,000. We monitor a quarter of them," Jaicy said, "by following mother turtles' incoming tracks, then placing a numbered stake at every nest we check."

---

* Generally accepted figures today are 66 million years and 75 percent of all species.

Nearly all the eggs hatch, she added as they continued up the hot strand, a few cloud wisps floating in the cerulean sky. "Of those, 90 percent make it to sea without being snatched by a crab or seagull."

"That's a lot," said Abby Hehmeyer.

"Yes, but just one in 1,000 reaches maturity. The sea has plenty more predators."

Ten meters from the lapping sea, the light brown sand was pocked with thousands of faint depressions. Somehow, Jaicy knew which were still active. They stopped at one where baby sea turtles had already clawed their way to the surface and instinctively aimed for the surf. Jaicy was soon up to her armpit, excavating and then counting every broken eggshell: 113 in all. "It's the only way to get an accurate estimate of reproduction."

A lot of work, but necessary to justify funding for a nature preserve. With so little wildlife left in the world, micromanagement of populations like these had become routine for conservation biologists.

"¡Miren—está viva!" She held up a live, wriggling baby hawksbill, half the size of her palm. "We often find some alive, from eggs buried so far beneath the others they can't fight their way up."

Immediately Katie and Abby were up to their armpits in adjacent nests and soon extracted five more, which they lovingly cradled while Jaicy fetched a white plastic pail with sand and water. "We take them to a dark spot in our office. They go to sleep. We liberate them at night, when it's safest."

After a sweltering hour of turtle census, Jaicy jumped into the foamy surf, fully clothed. A colleague named Rosario appeared, arms loaded with plastic bottles she collects daily along the tide line. "It's a lost cause," she said, "but we do it anyway."

Abby was holding the bucket with the baby turtles. "Can you tell the males and females apart?"

"Too small."

"Then how do you know what the sex ratio is?"

"We put temperature probes in select nests," said Jaicy. "Above 31°C (88.8°F), they'll be female. Below 27.7°C (81.9°F), male." She frowned. "Lately we're getting 80 percent female. In some nests, 100 percent."

"Is just taking the sand's temperature accurate?"

"At nests where we log temperatures, we also take blood and skin samples to confirm."

She picked a hatchling from the bucket and regarded it fondly. "We have to stay optimistic."

iii. The Depths

Two days later, Katie Arkema and colleagues stood on a windswept sand bar in Omoa, Honduras, near the mouth of the Río Motagua, the border with Guatemala. Gustavo Cabrera, a thickset man whose monogrammed polo shirt identified him as the director of the Cuerpo de Conservación Omoa, explained they were in the transboundary biodiversity corridor that the Smart Coasts project was trying to protect.

On the long, bumpy ride from Ría Lagarto back to Mérida after too much hot sun, her throbbing head glued to a pillow, Katie had considered skipping the Honduran workshops. But she wanted to meet people who will use their program to prioritize what to save.

"You can't just tell them to 'Download the software. Here's a list of instructions,'" she'd argue with NatCap programmers who believed if they made the tools well enough, they shouldn't need to spend time teaching people how to generate solutions. Katie disagreed. "There are so many technical barriers. It takes engagement."

It also took witnessing what conservation was up against. "People hunt jaguars in this corridor," Cabrera was saying. "They kill the mothers for their pelts and sell their cubs to zoos. They steal our endemic yellow-headed parrot chicks for caged pets." Looking at the ground, he shook his head. "We forgive them. They need money."

Evidence of why was all around them: the remains of houses destroyed by the advancing sea. Fragments of concrete walls listed in the sand as the surf pounded them. "Twenty-five years ago, the shore was 800 meters from here. There were sand dunes, sea grape trees. People grew watermelon and rice in the wetland behind us. The soils were good. But they're all under salt water now."

When a Parque Nacional Cuyamel-Omoa was proposed in 2006, they'd started monitoring red mangroves. Each year, they noticed the farthest mangroves they'd marked the year before were now standing in the water. During the past decade, wealthy people had built new beach houses five kilometers up the coast, each with breakwaters. Now, little sand reached here.

Those homeowners weren't the kind of wealthy you could approach—you were stopped at an armed checkpoint before driving past their enclave. Cabrera's conservation corps had warned this hamlet at the river's mouth, Barra del Motagua, that within 10 years, they'd be underwater, but no one paid attention. By 2013, when Cuyamel-Omoa was recognized by the Ramsar Convention on Wetlands of International Importance, the global treaty for wetland conservation, they'd already moved power lines six times. In winter, constant 40-mile-an-hour winds piled the remaining sand from the shrinking beach up to their windows.

Cabrera himself lived in this village. "We're four kilometers inland. We should be safe. But it's so flat here, who knows?"*

He led them along a raised dike that skirted former rice fields, now

* A year later, in 2020, heavy rains hitting the Honduran coast submerged 80 percent of Cuyamel.

strewn with Styrofoam, plastic packing bands, nylon rope, tangled fishnets, deflated soccer balls, toothbrushes, unpaired flip-flops, and a puzzling number of foam insoles. At a clearing, about 20 people stood in a circle. Of 150 families, Cabrera said, half had left to live with relatives or to sneak into the US or Spain. Most who remained were packed into the school or the church. Little else was still standing. The mayor, as officials do, somehow had found the money to build himself a house on pylons they called "Noah's Ark."

Behind them rose the Río Motagua's source: the evergreen Central American highlands, with 150-foot forest canopies of buttressed kapok, salmwood, white olives, gumbo-limbo, crabwood, mahogany, and yellow Caribbean pine. But the skirts—"That's another problem," said Cabrera. "We've lost so much wildlife habitat to African palm."

Thousands of hectares where migratory birds once wintered had been converted to a crop devouring the tropics from here to Indonesia. According to the World Wildlife Fund, palm oil is in nearly half of all packaged supermarket products, including pizza, ice cream, doughnuts, chocolate, shampoo, deodorant, toothpaste, lipstick, soaps, detergents, and makeup.

"We always had farms here. Then people you can't refuse offer to buy your land. Narcos tore up our fields and put in palms. So people take their cattle higher and clear pastures, which harms the aquifer, already low because of less rainfall. Rainforest trees with shallow roots are drying, and dying. Like the sea grapes when the beach sands got so hot and dry."

The villagers were begging the government to relocate them before they drowned. The government offered them land, but said the town had to come up with $415,000.

"We need to get out from behind our computers and see places like this," Ryan Bartlett said to Katie Arkema. A woman approached him. "Mr. American visitor, I'd like you to help us buy the land. We'd really thank you. We need it, because 250 people here risk losing our lives if we have another Hurricane Mitch. The US gives Honduras $40 million a year, supposedly to eradicate poverty. It can afford another $415,000, no?"

"Gustavo," Ryan asked Cabrera on the way back, "what can be done?"

"Short term, relocate them. Long term, replace dunes, plant mangroves, clean up the garbage. Carefully approach palm growers to persuade them not to use chemicals. Since palms like sunlight, maybe let people grow crops like tomatoes in between the rows."

He pointed out the bus window at the land the government wanted to sell them.

"But it's also right by the beach."

"Yes, but higher. It should last 20 more years."

"They'll have to move every generation?"

"At least they'll know the warning signs, start adapting."

"Gustavo, how do you not go crazy?"

He laughed mirthlessly. "Do you know the story about the hummingbird? Once upon a time in the rainforest, there was a terrible fire, like in the Amazon right now. All the animals were running, desperate. But the hummingbird, that tiny little creature, flew to the river, took some water in his beak, came back, and spit it on the flames. The rest of the animals said, 'What the hell?'

"'Hey,' an anteater yelled at him. 'Are you crazy? How do you think you're going to put out the fire with that teensy bit of water?'

"'If the fire doesn't go out, it's not my problem.' The hummingbird replied. 'But I'm doing my part.'"

They slowed for the checkpoint. Along the shore, security lights winked around the narco mansions.

"So that's what I do. If the planet destroys itself, at least I know I did all I could."

CHAPTER SEVEN

GOING DUTCH

i. New Amsterdam

The brick building at 112th and Broadway where Manishka De Mel and Meridel Phillips worked was better known for its ground-floor corner restaurant—Tom's, whose facade appeared in the 1990s TV sitcom *Seinfeld*—than as where James Hansen ran NASA's Goddard Institute for Space Studies until 2013. It remains a den of climatologists, including Kate Marvel, whose job was to ponder what to do with the data NASA synthesizes from 33 global models.

While Manishka and Meridel were on the Mesoamerican Reef, Marvel was pondering the Paleocene-Eocene Thermal Maximum, a large-scale release of carbon dioxide about 55 million years ago, probably through plate-tectonic activity: "It's the best historical analogue for what we're going through right now."

On her screen, jagged red lines on a graph traced two eons' worth of temperature gradients. "What's fascinating about the PETM is despite really pronounced warming, there's no geologic record of a mass extinction. This was just 10 million years after the dinosaurs died. Maybe the species that survived could basically survive anything?"

The pigeons at her windowsill, descended from those survivors and oblivious to the fumes and tumult rising from Broadway, might be evidence of that. It's a bit encouraging that there wasn't a major die-off during the PETM, even though the atmosphere was stuffed with an extra 2 trillion tons of CO_2—since, so far, humans have only emitted an extra 1.5 trillion tons (albeit in just slightly more than two centuries; the PETM happened over 15,000 years). Not that Kate Marvel was looking for encouragement. In her regular essays for *Scientific American*, she'd made it clear that she doesn't have any hope.

"I have something better: certainty," she summed up in a tweet. "We know exactly what's causing climate change. We can absolutely 1) avoid the worst and 2) build a better world in the process."

Whenever she heard people lament that it would take a miracle to escape this existential mess we've made for ourselves, she'd retort:

"I believe in miracles. I live on one."

Just down the hall was the office of NASA's climate impacts director, Cynthia Rosenzweig. In 1969, she and a boyfriend dropped out of Stanford to rent an olive grove in Tuscany. After three years, they got a farm in the Hudson Valley, where they raised corn and pickling cucumbers. A community-college technical degree to learn tractor repair and apple pruning eventually led her to ag school, then to graduate studies at Rutgers in crops, soils, and environmental science.

In the early 1980s, during her master's, she took a job at Goddard. James Hansen then was modeling what would occur if atmospheric CO_2 doubled. Since Rosenzweig was an agronomist, he asked her what would happen to food. Would more CO_2 mean more plant growth? The simulation she ran led to her first publication, which suggested that Mexican wheat crops would suffer heat stress, but production would increase in the US and Canada—although they might need to switch to more heat-tolerant varieties. As the models sharpened over the next decade, she kept at the question. Soon, it grew clear that positives from extra carbon dioxide were canceled out by negatives, such as needing more fertilizer to maintain high yields.

Expanding her inquiry worldwide, Rosenzweig became a pioneering expert on how climate impacts food systems and was lead author of three Intergovernmental Panel on Climate Change global assessments. But climate affects everything: In 2008, she and NASA Goddard geologist Vivien Gornitz published a warning that a hurricane plus climate-driven sea level rise could strike New York City. In 2010, now cochairing the New York City Panel on Climate Change, she and CUNY–Hunter College geographer William Solecki published a climate-risk management plan for the city's subways, tunnels, bridges, electric grid, and water supply. In 2011, they followed with projections of what, given sea level rise, might happen to New York in a hundred-year storm.

In 2012, Superstorm Sandy confirmed all her predictions. It wasn't the strongest hurricane to hit New York—80-mile-per-hour winds, Category 1—or the wettest. But nurtured by an overheated, moisture-laden atmosphere, it was surely the biggest, extending 1,150 miles—triple the size of Katrina. Arriving at the full moon's highest tides, it slammed into an unprecedented confluence of a high-pressure system to the north—a lobe of the jet stream, adrift because its anchor, Arctic ice, was losing its grip—and an eastward low-pressure system, steering it straight for New York and New Jersey. A 14-foot storm surge flooded lower Manhattan, several subway lines, and two tunnels, and drowned or damaged 650,000 homes.

A decade later, she was warning, "Sea levels in New York City are rising at almost twice the global average rate. It's gone to warp speed." Nowhere was safe: with Meridel and Manishka, she'd coauthored a paper titled "Catastrophic bleaching risks to Mesoamerican coral reefs in recent climate change projections."

"Yet I'm definitely one of the people with hope." For a woman charged with the weight of the world, Cynthia Rosenzweig, still girlish in her 70s, much of her hair still blond, is upbeat even when discussing how ice caps and glaciers melt slowly—until they tip and rapidly accelerate.

"Remember, one of the two words in climate change is *change*. There can also be tipping points in response. Suddenly people are working on everything from batteries to alternative meat. As soon as cities learned about climate change, they began making informal agreements, pledging together to reduce emissions. It raises the question of which governmental level is really best to tackle climate change."

The Dutch founded New York, née Nieuw-Amsterdam, in 1626, conning the Lenni-Lenape out of the verdant, creek-lined island they called Manahatta for 60 guilders. In 2013, the Dutch returned, in the person of 46-year-old Henk Ovink: tall, thin, head shaven, with intense blue-green eyes and large hands that punctuate every point. When Superstorm Sandy hit a year earlier, Housing and Urban Development secretary Shaun Donovan was in Europe with his family. On learning he'd be overseeing the federal government's response from President Obama, he made an impromptu detour to the Netherlands to see how that country had managed to keep from drowning for the past 800 years. Ovink, then Dutch director of spatial planning and water management, took him to tour the gigantic storm-surge gates that protect Europe's biggest port, Rotterdam; The Hague's beach-restoring Sand Motor; and towns whose water boards, each headed by a locally elected *dijkgraaf*—"dike count"—effectively form a second national government.

No one in the Netherlands argues about taxes that water boards levy, because ever since windmills began turning scoop-wheel pumps to reclaim seabed, creating the dike-enclosed polders where two-thirds of them now live, water management essentially defines the Dutch. As the exposed peat soil dries, polders shrink, then sink, forming bowls sometimes six meters below sea level. With survival at stake, over centuries cooperation became cultural behavior.

Dikes to hold back the North Sea are only half the Dutch story. The country is western Europe's drain, where its two biggest rivers, the Rhine and the Meuse, meet the smaller Scheldt in a great Dutch delta. In deltas, rivers meander and spread. This delta spreads over two-thirds of the country. One of the Netherlands' biggest cooperative decisions came in the mid-1990s after repeated river floods displaced a quarter-million evacuees. Everyone was saved, but water boards agreed that with noticeably higher European precipitation, channeling rivers with dikes had reached its limits.

"Evacuation," says Henk Ovink, "wasn't how we wanted to live with water." The 2006 program he directed, Room for the River, changed centuries of Dutch history by re-landscaping 38 Rhine and Meuse tributaries to allow the rivers to spill into their former floodplains during times of high flow, taking pressure off dikes rather than raising them continually.

"The river is a living thing," he'd say. "You figure out what it needs."

Room for the River, which pays generously to relocate homeowners and indemnifies farmers for lost harvests when their fields become flood catchment basins, cost €2.7 billion. "In damages and evacuations avoided, not to mention human lives," says Ovink, "it has saved many times that amount."

Hours after Donovan's US embassy van had dropped him off following their week together, it occurred to Ovink that taking his ideas to a stage as big the US could change how the world addresses rising seas and growing storms. Apologizing for being forward, he emailed the HUD secretary and confessed he wanted to work with him.

"You're being just forward enough," Donovan replied. "When can you start?"

Dutch water boards operate beyond politics, but Henk soon grasped that nothing remotely similar existed in the region mauled by Sandy. New York City, New York state, New Jersey, and the federal government that had hired him seethed with conflict, among each other and internally. At a post-Sandy resilience conference at New York University organized by New York's scientific community, he was asked to speak. His topic was intergovernmental collaboration, so key to Dutch survival.

But it turned out no scientists from New Jersey had been invited. "We had France, China, Japan, the Netherlands, but not the neighbors."

After acquainting himself with the waterlogged Rockaways, Coney Island, Hoboken, Staten Island, and lower Manhattan—where defenses that had failed were already being rebuilt—he organized a competition called Rebuild by Design. From 148 proposals, 10 finalists were chosen. Each was instructed to spend three months interviewing residents across economic and cultural spectrums, then, based on what they learned, to submit five solutions.

Seven winners qualified for nearly $1 billion in HUD grants. They included Living Breakwaters, a project creating artificial oyster reefs off Staten Island to absorb waves and restore water-filtering shellfish beds; and the BIG U: redesigned coastal reinforcements around lower Manhattan to protect people and parks, and to cut fossil-fuel consumption by restricting FDR Drive to public transit.

The product of much public input, that last part of the BIG U—a play on the acronym of its designer, the international Bjarke Ingels Group, which has an office in Brooklyn—was approved under New York City mayor Michael Bloomberg but subsequently quashed without discussion by Mayor Bill de Blasio. That sparked a lower Manhattan class war, and accusations that by preserving FDR automobile traffic, de Blasio was a petroleum-industry toady. Eventually, all sides agreed that something had to happen before the next superstorm, so work on a modified BIG U plan commenced. Other winners—Hoboken, the Meadowlands, Long Island, and, after acknowledging that Connecticut was also affected, Bridgeport—would build combinations of quays and parks to turn their hard urban shells into sponges for rain collection and flood defense.

In 2015 the Netherlands named Henk Ovink its special envoy for international water affairs, a cabinet-level position to raise global water awareness and promote its most valuable intellectual property export, Dutch water technology. Since Hurricane Katrina in 2005, engineers from the premier Dutch water think-tank, Deltares, had been regularly visiting New Orleans. As sea level rise showed no more sign of abating than fossil fuel usage—well into the 2020s, carbon emissions had yet to decline—Dutch companies and government agencies had technology-transfer contracts in Indonesia, the Philippines, Vietnam, Thailand, Colombia, Chile, Argentina, Uruguay, Brazil, Peru, Ecuador, Paraguay, Mexico, Iraq, the UAE, China, South Africa, South Korea, Egypt, India, and Bangladesh.

The whole world was turning to the Dutch.

ii. Below Sea Level

In black jeans, black sneakers, and a blue poplin jacket, Henk Ovink bikes from his solar-powered home in the province of Overijssel to the old city of Kampen, about 50 miles northeast of Amsterdam. He points out where houses on the River IJssel, the northernmost branch of the Rhine delta, have metal flood-protection doors, and where citizens place temporary embankments by hand along their postcard riverfront when water is high.

"Every year, they practice. Otherwise they'd have to build a wall next to the river. You don't need a wall 365 days a year. Perhaps you need it once every five years. Why build a wall for once every five years?"

At the twin brick towers that flank the Koornmarktspoort gate, built in the 1400s when Kampen was a walled fortress, the bike path turns south into cornfields. At the top of a new levee, he dismounts. Below, the waterway has gotten complicated. He's arrived at the Kampen Bypass, one of the largest Room for the River projects. Eventually, the historic town's handmade defenses wouldn't withstand the rising rainfalls that annually pummel Italy, Switzerland, Germany, and France, swelling the Rhine and Meuse and pouring into the flat Netherlands. To lower flows before spring floods reached Kampen, an entire new river branch was dug to divert part of the IJssel by means of a flip gate.

The three-and-a-half-mile diversion, the Reevediep, connects to lakes that drain into what was once the Zuiderzee, a saltwater inlet of the North Sea until Dutch ingenuity turned it into a freshwater reservoir, the IJsselmeer. (Over centuries, freshening water and saline soil became routine here, aided by ample rainfall and an ancient trick of sowing reeds with hollow roots for fast aeration, making polders ready for agriculture within a decade.)

Henk loves the Kampen Bypass, which showcases so many of Room for the River's objectives. Wherever homes were displaced, people were compensated handsomely. Where dikes or levees were moved to give the river room, exposing fields to flooding, farmers plant knowing they risk losing their crops, but also knowing those losses will be covered. Where Dutch Railways had planned a land route for a new line linking Amsterdam to the Netherlands' northeast, it agreed instead to build a costly bridge over the Reevediep.

All these arrangements were forged during meetings where water engineers and national and local governments did everything to accommodate everyone in the water's way—meaning everyone. Locks, never cheap, were even approved for recreational boaters to access the new waterway, which, depending on current flow, might be lower or higher than the IJssel. Nature lovers got a reed marsh for endangered bitterns and reed warblers, and soon otters and beavers were spotted among stands of bell-shaped fritillary lilies, their seeds borne in by high river flows. Oak, poplar, birch, and willows were planted along cycling paths built atop the Reevediep's dikes.

"It's a whole new floodable infrastructure, a public park where you can bike, watch nature, or fish," effuses Henk. They've made this concept palatable throughout the country, he says, "by always maintaining quality living as well as safe living. If people objected to something, the planning department gave them money to come up with an alternative." For example, he says, at a polder east of Rotterdam whose fields were designated for overcapacity stormwater storage, farmers refused to move. "So we raised the ground underneath their farmhouses by building mounds, called 'terps' in the old days. So now they can stay when their lands flood."

A moist breeze cuts through the July morning. Henk shades his eyes to look eastward, where, behind a flock of barnacle geese, clouds are darkening. The land is so flat you can see to where the horizon bleeds into Germany, which doesn't have a program like Room for the River.

"In 2016, when we chaired the European Union, big floods caused casualties in France, Germany, and Austria, but nothing in the Netherlands, even though that same water ended up in our rivers." Traveling the world as the Dutch water envoy, Henk encounters other countries even messier than their European neighbors.

"In Louisiana, the Mississippi is a canal, not a river anymore. The coastline loses a football field an hour. All the Asian rivers are full of dams, like the Mekong, and deltas are dying because there's no sedimentation flow. Between that and groundwater pumping, delta cities like Shanghai are sinking, and salt water is intruding."

Henk insists he's not a trade representative for Dutch water technology. "At the UN's 2015 Disaster Risk Summit, we formed the Delta Coalition with scientists from across the world. Why? If you look at the flood-vulnerability map of the world, you see South Asia, parts of the Americas, and the Netherlands. We are first to drown. If we don't work with the world to change course to a 1.5°C increase ceiling and implement resiliency measures, we're lost. So, this is what we have to offer. What we did 1,000 years ago, the world needs to know and learn from."

Henk gets back on his bike to beat the rain. "There's no Dutch plane flying around the world with solutions that fit everywhere. This problem is bleak, but bleakness is an opportunity to try working on it. Each time it works, there's an opportunity to go from local to regional, then to a systems-level course change. We have to try. There's no other way."

Every culture has a creation tale based on murky, embellished, distant truths. The Netherlands has a destruction tale, no embellishment necessary. Over the centuries, great lobes of land that comprise its deltaic coastline, separated by long estuaries reaching inland, have shifted or even vanished as rivers from the east changed course, stranding or washing away former ports. From the west, the North Sea has clawed through dunes and obliterated dikes, turning polders back into inlets. In 1216, 1287, 1375, 1404, and 1421, each storm hitting the Dutch coastal province of Zeeland, which borders Belgium, killed thousands.

What today is the Netherlands then consisted of various duchies, lordships, kingdoms, and minor empires, regularly at war with each other. With no friendly place to flee, flood survivors had no choice but to rebuild and erect more barriers and windmills to keep pumping. Then, in 1530, a huge chunk of Zeeland simply disappeared in the St. Felix Flood, which drowned 18 villages and 100,000 people.

Along with invading water, there were also occupying Spanish, but later that century alliances formed to repel them, which led to a Dutch republic and eventually today's Kingdom of the Netherlands.* As various water boards cooperated, dikes connected terps and merged into networks. Rivers were channeled and defenses strengthened. In the 19th century, a deep canal to the sea, the Nieuwe Waterweg, was dug through the silting Rhine–Meuse mouth at Rotterdam, which soon became Europe's busiest port.

In the 20th century, walls went up across the Zuiderzee, creating lakes to hold the sea farther away from Amsterdam. The Netherlands felt safer than ever—until the Nazis invaded.

Rotterdam's dikes survived the German bombardment that leveled much of its maritime infrastructure. After liberation in 1944, the port's business resumed. Any illusions that postwar prosperity meant security, however, were dashed at 3:00 a.m. on February 1, 1953, when a storm surge in the North Sea breached a dike 25 miles south in the 11th-century village of Ouwerkerk. Seventy years later, it's why Dutch children must pass a swimming test with their clothes on.

---

* The name Netherlands—literally, "low countries"—originally referred to the rival polities that formed a republic in the 16th century, which then included what later became Belgium and Luxembourg. Today, all three nations are collectively referred to as the Low Countries, although the Netherlands (Nederland) maintains that name. Partly to avoid confusion, but sometimes causing more, it's also called Holland, often even by the Dutch, although technically that term refers to the coastal provinces North and South Holland, which between them contain every major Dutch city except Utrecht—which, at a comparatively lofty elevation of 17 feet, may someday be the last still above water.

"Everyone was sleeping. Suddenly they awakened, freezing cold, and realized they were lying in water."

Eric van der Weegen stands at the visitor center coffee bar on an artificial island in the middle of Oosterscheldekering, the Netherlands' biggest line of defense against the sea. Eyes closed, ruddy face tilted skyward, he channels that moment. "You don't know what's happening, but instinctively you run upstairs. From the windows you see the flood coming fast. You see houses being torn apart."

It was the fatal collision of the century's worst gale—a nor'wester reaching 12 on the 12-point Beaufort Wind Scale—with a spring tide just past full moon. It swept the North Sea ashore in all directions, sinking nine ships and swamping coastal Scotland, England, and Belgium, but the Netherlands worst of all. A storm surge 18 feet above sea level roared up the shoals so forcefully there was no low tide that day, blowing a hole in Ouwerkerk's dike that by the next high tide was 640 feet wide. Villagers were on their rooftops, surrounded by a sheet of water unbroken to the Thames, watching it rise until the houses beneath them collapsed.

The dead numbered 1,836. Seventy-two thousand more were evacuated. Ten percent of Dutch farmland lay underwater. More than 100,000 farm animals drowned. For months, salt water kept gushing into Ouwerkerk's polders, threatening to return the southern Netherlands to the sea from which it had been retrieved over centuries.

Ships laden with rocks, sandbags, and gravel were steered into the breach, uselessly. Finally, the Dutch water agency, Rijkswaterstaat, had surplus World War II caissons, built to construct artificial harbors for the Normandy invasion, towed across from England. Each was a hollow reinforced-concrete box, 62 meters long, 18 wide, and 20 high. It took four to finally close the gap on November 6, eight months after the storm.

They're still in place today, together housing the Watersnoodmuseum, one of several Dutch flood-themed museums where families take children not just to educate but to scare them. Another, 12 miles away where the inlet reaches the sea, is at Oosterscheldekering, the barrier where Eric van der Weegen is drinking coffee with Zeeland photographer Rem van den Bosch. In a nearby surround-screen immersion theater, a virtual simulation of the 1953 storm has real howling winds and an all-too-believable girl stranded across a raging dike breach from where her parents are trying to free a truck, screaming, "Mama! Papa!" They holler back for her to get inside, but as she does, more of the bank succumbs to the torrent, collapsing their house's brick facade and dousing the lights.

By now, children brought to watch are screaming, too. By the time the hysterical girl reappears, clutching a lantern at the edge of a second-story room now missing its front wall, her parents even more unreachable, most are sobbing in their mothers' arms.

Topped by a highway, the 9-kilometer Oosterscheldekering—Eastern Shoal Barrier—is the longest and most expensive of 13 flood-control structures that the Netherlands decided must be built following that disaster. The Delta Works, as they're collectively known—six dams, three locks, and four movable storm surge barriers—were completed in 1997, more than 40 years later. Together with the Zuiderzee walls and an inflatable rubber bladder dam at the mouth of the River IJssel, they are considered among history's most ambitious civil engineering feats. The Oosterscheldekering is a dam for slightly more than half its length. The rest, owing to public demands to keep shellfishing and the vast estuary ecosystem it encloses alive—now Oosterschelde National Park, the Netherlands' largest—is a line of 62 sluice gates, each 138 feet wide, suspended above the water to keep the tidal zone refreshed but ready to drop into place when needed.

The question Eric and Rem are discussing is whether they're high enough.

"People say the water is getting higher than we think," says Eric. "Unless we take measures about the environment, I don't think it's possible to stop. It's like seeing a car coming at you at 250 miles an hour, but you're still standing there, doing nothing about it."

The year that Eric became the Delta Works communications director, 2015, the Netherlands hosted a stage of the Tour de France. Cyclists passed several Delta Works en route to the finish line inside the Oosterscheldekering visitor center's two-story atrium. There, a floor-to-ceiling color photo by Rem van den Bosch of a Dutch girl in a yellow dress and bonnet, holding a sheaf of wheat stalks, still hangs between two yellow pillars, each with bold black lettering reading, respectively, "Water you thinking?" and "Bring in the Dutch!"

"When we built this there was another state of mind. People believed we must take care that this never happens again. But now we're talking about one to two meters. Zeeland—"

"—will be gone," Rem finishes for him, his fingers worrying his thick grizzled hair. "There's one main highway out of here. If a car breaks down in a tunnel, everybody will be stranded. In Middelburg"—Zeeland's capital—"where we're building a hospital it's like a bathtub: 5.5 meters below sea level. In central Holland, where it's 5 meters below sea level, they're planning 200,000 new houses. A woman I know wrote the housing policy that says roofs must have windows so people can flee. There's a reason they commissioned her to write that. When it floods, you have a mortgage that's literally underwater."

Rem's latest exhibition, currently displayed at the Watersnoodmuseum and about to travel to Belgium, is titled "Here Comes the Flood." He is known for portraying Dutch women—both fair-skinned, their families here for centuries, and richly hued immigrants—in unexpected settings, wearing striking, boldly colored folk dresses he designs himself. Lately, he's been posing them in water—or atop ladders, climbing out of harm's way.

"I'm in a difficult situation because I work for the government," Eric tells him, hooking his sunglasses into his white shirt's open collar, "but if we have to start closing these sluice gates 50 times a year, the environment will start dying. If we wait until the water is two meters higher to look for another solution, politicians will then say it's cheaper just to close it."

Rem doesn't dispute that. "My wife wonders if we'll get to grow old in our house. My children ask if they'll lose their future because we're next to the water." They live 18 miles to the south, in a brick home with skylights and solar panels on the Westerschelde estuary, which Delta Works didn't close because it's the shipping route to Antwerp. "At some point, building higher seawalls isn't possible. In my opinion the only solution is living with the tide."

"How?"

"To give the rivers room to move, they built high mounds for people and their cattle to be safe. They should do that here."

Eric looks dubious. It would be very expensive.

"Maybe tourists would climb a hill 50 meters high to look over the estuary. There could be a coffee shop there," says Rem. Eric sees he's not joking. "Something will have to happen. If hundreds of thousands have to leave their homes, it will be anarchy."

Once in a promo film, Eric, wearing the same blue serge suit as now, called Oosterscheldekering the eighth wonder of the world. "Together with the other Delta Works, it ensures that more than 9 million people keep their feet dry."

"Only we don't have a new plan for what's coming," he says a decade later. "People have been driving cars for a hundred years or so. That's all it's taken to destroy a world millions of years old."

"In the perspective of 24 hours," says Rem, "it's like one second."

"What will this world look like in another hundred years?" wonders Eric. "I wish it weren't true, but sometimes it takes a disaster to clean things up. This time, we're the dinosaurs."

Deltares's eight-story rectilinear Delft headquarters is not Vermeer's turreted Delft; it adjoins the Delft University of Technology, whose glossy angular architecture suggests a taut future, not an immortal Golden Age. When the research institute's senior climate expert and former science director, Jaap Kwadijk, received his geology Ph.D. on December 10, 1993, his dissertation topic, climate change impacts on the Rhine, seemed a distant prospect—until December 20 brought the river's all-time flood.

"Then, in 1995, another all-time high—250,000 people had to evacuate. So you're a prophet and end up science director at Deltares," Kwadijk says, chuckling.

Henk Ovink calls Kwadijk very scary, but he laughs easily, even though he's known that his country was in terrible trouble ever since James Hansen's seismic 1988 global forecast paper.

"We were prepared for one-in-10,000-year floods. But . . ." he scratches his shiny head. "First, Katrina in 2005. Then, in 2006, Al Gore's *An Inconvenient Truth*, with a slide showing two-thirds of the Netherlands underwater. It wasn't so much that the Dutch believed him, but the government feared others would and stop investing here."

In 2007, the government formed an expert commission. Kwadijk's group had to estimate the maximum sea level rise they could plausibly expect in the next hundred years. They were put in a hotel and not allowed to leave until they came up with an answer. "After three days, we said, well, 1.3 meters."

Assuming that were true, the commission next asked the Deltares engineers, Could the Netherlands defend itself? "Yes, we said, for about €1 billion per year. Parliament unanimously agreed, and that's

Henk Ovink's climate program. It's already survived a number of different administrations, from left to right."

Kwadijk's second-floor office has bare white walls, a picture window that overlooks a patch of lawn and covered parking, and a broad white fir desk, its top empty except for hand sanitizer and his coffee cup. Over the past decade, a series of freak events started making him nervous, like the record 2020 Canadian heat wave that killed 500, unpredicted in any climate scenario. Then, in 2021, it hit home. "The Meuse hit an all-time high—in summer!"

Europe generally floods in winter, when ice dams break up. No one expected the July rain cloud, half as big as Germany, that sent a monsoon-strength deluge down narrow river valleys, their engorged Meuse tributaries sweeping away German and Belgian villages, killing 240.

The floodwaters next poured into the Netherlands' eastern hills—the highest, and usually safest, part of the kingdom. In the storybook village of Valkenburg, the country's top tourist destination after Amsterdam, the Geul River rose eight meters above its ancient stone channel that winds through the town's tiled streets, swamping picturesque footbridges, leaving water stains high up interior marl columns of the old pre-Reformation church, whose priest had to be rescued from a second-story parsonage window in an excavator's bucket.

But no one died. Because they've invested so much in defenses, no Dutch has perished in a flood since 1953 ("Delta Works cost us $5 billion, over 40 years," says Henk Ovink. "Hurricane Katrina alone cost the US $125 billion. Do the math.") But now every year, sometimes every month, comes a new stunner, like the vertiginous disintegration rates of Greenland's ice cap or new signs of weakening of the Atlantic Meridional Overturning Circulation, the planet's temperature regulator, which flows by them.

After a paper appeared in the journal *Nature* showing how precarious the West Antarctica ice shelf had become, one of Kwadijk's colleagues called for a daylong brainstorming event they dubbed a hackathon. Deltares geologists, ecologists, engineers, and physical geographers gathered in a room to contemplate what a two-meter sea level rise would mean for the Netherlands. What about four meters? Six? Ten? Twenty?

"What was interesting was that from zero to two meters, the engineers were talking. From two to six, everybody got nervous. And beyond six meters, the geologists were talking."

The geologists reckoned it was unclear what parts of existing deltas might remain, as it would be a very different coastline. The engineers reckoned there would always be something they could do. All agreed that if nothing happened fast, the Netherlands would become a very strange country. But how to get policymakers to grasp that?

"If you show them figures, it doesn't really help. We had to make it simple." So they brought in a cartoonist. "Cartoons, anyone can understand."

Kwadijk and colleagues went to Amsterdam to explain to the government what was happening in Antarctica. "'Look,' we said, 'this really limits how we might be able to protect us. This might be beyond the limits.'"

Their cartoons showed five possible scenarios for the Dutch delta, given recent developments: 1) a towering fortress wall along the coast, with enormous pumps to lift Europe's rivers over it; 2) deep polders surrounded by equally high walls, connected by roads resembling roller coasters; 3) cities that were half lagoons; 4) key civic buildings perched atop terps; 5) entire cities migrated to artificial islands.

All looked fanciful, except they weren't.

"The administration didn't have an answer." He smirks. "They don't like it if they don't have an answer."

Everyone knew why they didn't.

"It's difficult for us working here to understand how the same ministry that tries to protect us against floods isn't eager to ban fossil fuels. Simply, the first is a technical issue, while an energy transition is a political issue."

He sighs deeply. "There are extremely capable water engineers in this building. I think we could cope with one-meter rise per century. But it's not so much the amount of sea level rise, it's the speed. That will be hardest to handle, and it will continue. In the end, a lot of people will have to move. If there's sufficient time for that, it's a less-worse case than if we don't."

A colleague recently asked what he was telling his children.

"Get the best education you can," he said, "and leave before you become a refugee, because refugees aren't wanted."

iii. Reality

Every autumn just before storm season, an annual test of one of the world's biggest machines has become, for residents of Rotterdam, a celebration of Dutch mastery: their below-sea-level city never floods. Their glee worries the hell out of Rijkswaterstaat's Flood Defenses Program director Richard Jorissen.

The machine—Maeslantkering, the final Delta Work, completed in 1997—is a double set of curved steel walls that are actually pontoons, each 72 feet high and 777 feet long, welded to 6,800-ton arms whose white lattice trusses are often likened to two halves of the Eiffel Tower lying on their sides. Controlled by a computer that scans the weather every 10 minutes, should a storm surge three or more meters over sea level be predicted—it's only happened three times: in 2007, 2018, and 2023—the largest moving parts ever built float together from their respective banks, swiveling on ball joints

10 meters in diameter. It takes two hours for them to meet, forming a graceful V midstream on the Nieuwe Waterweg. Their ballast chambers then fill, sinking the wall to the bottom to create a watertight seal. When it's safe to reopen, the computer signals 30 pumps to empty them and float them back.

"On the one hand," says Jorissen, an amiable, broad-shouldered man in a blue short-sleeved shirt, "we're probably the best protected delta of the world. On the other, we're also one of the most vulnerable, because high protection leads to a kind of social disconnect."

"People assume there's a solution for this thing," agrees Jaap de Heer. "They don't remember anymore."

De Heer, who has a fringe of white hair and a trim goatee and wears a navy blazer, is the director of the Netherlands' biggest water technology transfer ever, Bangladesh Delta Plan 2100. The two old friends sit in the Maeslantkering cafeteria, which overlooks one of the barrier's gargantuan arms, its tubular crossbeams six feet in diameter. The Maeslantkering was designed to withstand storm surges up to 5.5 meters, factoring in a possible future half-meter sea level rise. The Oosterscheldekering, 45 kilometers south, completed a decade earlier, contemplated just over half that. Whether both will have to be raised—and how—is a frequent discussion.

Since 1977, Maeslantkering has closed on average once every decade. "With one meter," Jorissen says, "we'd close every year. Probably more."

Whenever the immense gates shut, the force of the water radiates to the ball joints. "With higher loads and more frequent closings, the chance of structural failure increases," says Jorissen. "Same with Oosterscheldekering: bigger waves, more fatigue. In winter, its gates could close every 14 days, with each spring tide. That exceeds structural performance criteria. I think with a bit of tuning we're still relatively comfortable to 2050. But then we'll have to rethink these things."

The rethinking is already underway. Jorissen still hopes they won't see 1.5-meter increases this century, so they can keep tweaking their semi-closed, semi-natural system, using short-term hard measures when necessary—like how they helped ring New Orleans with dikes and barriers to buy time while allowing protective coastal marshes to come back. But they've begun considering long-term options.

The most drastic, already quietly being mentioned, is retreat. Since two-thirds of the 17.8 million Dutch population live in the two-thirds of the country either below sea level or prone to river flooding, moving that many to higher ground feels unfathomable; apart from Monaco and Vatican City, theirs is already the most densely populated country in Europe.

"If we had to evacuate the western Netherlands—which is included in our safety standard plan—I'm not sure we could survive it as a nation."

Alternatives once thought far-fetched are getting serious attention. "Like what has been proposed in Jakarta," says de Heer. "Outer seawall and a lake"—meaning build a wall out into the sea and store river discharge behind it.

"Here, too," says Jorissen. "I saw it again on Twitter a couple of days ago. It's not just some engineer's wet dream. This guy wants a real discussion on what scale preventative measures can be justified. He proposes building two sets of dams, one between Brittany and Cornwall and the other between Norway, Shetland, and Scotland."

Versions of the map that oceanographer Sjoerd Groeskamp of the Royal Netherlands Institute for Sea Research and meteorologist Joakim Kjellsson of Germany's GEOMARCH published in the *Bulletin of the American Meteorological Society* had proliferated on the internet. The €250 to €500 billion cost for a Northern European Enclosure Dam—NEED—the researchers wrote, would equal about 0.1 percent

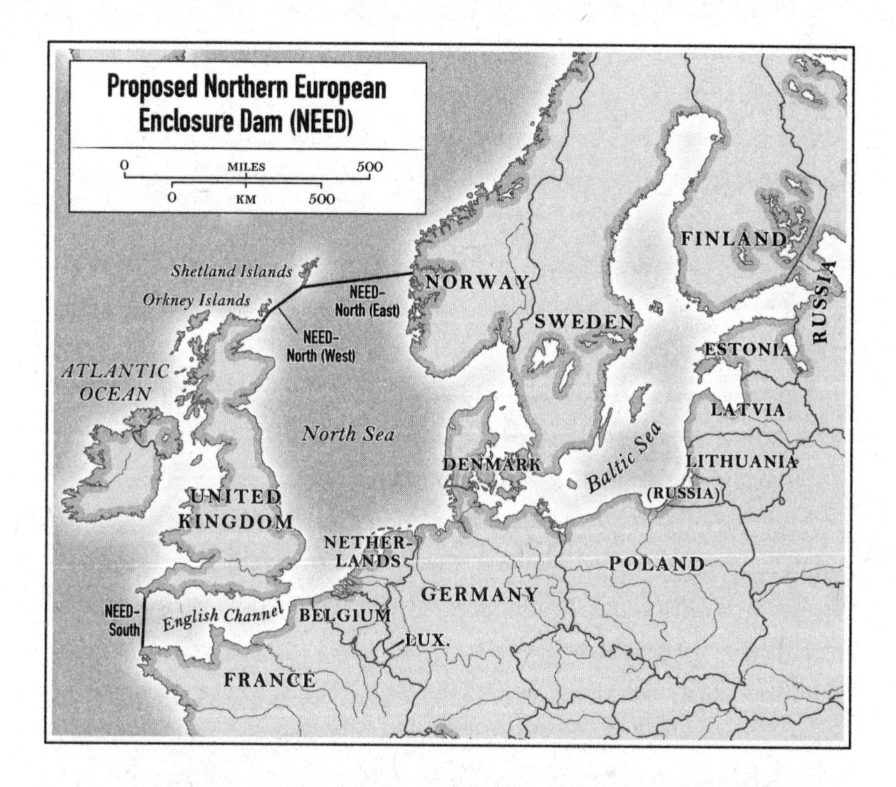

of the annual gross domestic product of all the impacted countries, over 20 years. Given the protections such a dam would provide, they argued, it would prove cost-effective, notwithstanding losses to fishing as the North Sea becomes a freshwater lake and disruptions to shipping—plus the constant pumping of Rhine and Meuse inflows over the dam. Their plan's acronym, they emphasize, underscores that it's a warning, should mitigation fail.

But is what Bangladesh contemplates any less radical? It essentially wants to clone what the Netherlands has done, only nearly four times bigger, in a delta that drains a vast chunk of Asia and is exposed to enormous cyclones, with monsoon-driven rivers like the Ganges and the Brahmaputra that dwarf the Rhine, pounding down from the Himalayas with Everest-sized loads of sediment. "A

1959 Himalayan earthquake is still delivering sand," Jaap de Heer says, marveling.

For 17 years, he's practically commuted there. He's grown bald and frost-bearded on this project, and hopes to see it through before he's through himself.

"I find Bangladesh's delta plan inspiring," says Jorissen. "Beyond flood protection and adaptation, it's linked to their national sustainable development goals."

But to afford its multibillion-dollar Delta Plan 2100, Bangladesh must become a middle-class country. "Or upper-middle," says de Heer. "The challenge is where to get the money."

And therein lies the conundrum. Beyond Maeslantkering's latticed arm, the view from the window reveals why the Netherlands can afford the world's biggest plumbing project and its tidy, perfect country, filled with healthy people riding bikes atop green dikes—a country where everyone got paid during COVID, where the government indemnified Valkenburg's flooded businesses, where houses displaced by rivers are replaced with better houses, where nutrition is so good that the Dutch are the world's tallest people (a fact often credited to its dairy production, so prodigious that two-thirds is exported; after the US, the tiny, efficient Netherlands is the world's second-biggest agricultural exporter, easily capable of feeding all of Europe).

Yet agriculture barely accounts for 1.5 percent of the Dutch GDP. On the south bank of the Nieuwe Waterweg rises a forest of crude oil and petroleum derivative storage tanks, plus petrochemical refineries and western Europe's largest coal terminal. This is Europoort Rotterdam, the continent's biggest seaport, where income from fossil fuel imports and exports, including from the country's North Sea natural gas fields, nearly triples the value of all other goods passing through here.

The Netherlands' wealth sluices through Rotterdam on an oil

slick.* Dutch prosperity is not merely levitating above the reality of the planet; it depends on the very substances causing the country's rivers and seas to rise to levels modern humans have never seen, engulfing its future.

iv. Against Long Odds

On November 1, 2015, Marjan Minnesma, her ash-brown hair secured in a bun with an orange scrunchie, slipped into the gear that Patagonia had sent her: a red-orange zipper fleece turtleneck, water-resistant hiking pants, and a red quilted parka. Grabbing her new Patagonia backpack and hugging her husband and three children, she climbed into her hybrid Opel. Moments later she was atop the ring dike built in 1611 to create Beemster, a 27-square-mile North Holland polder, looking down at their brick farmhouse, six meters below sea level. Heading south, in 20 minutes she passed through Amsterdam and in another half hour arrived in Utrecht, where her team was waiting to accompany her on a hike. To Paris.

It was crisp and sunny, the trees in euphoric color, perfect for walking. If they averaged 22 kilometers a day, they would arrive in four weeks, by the beginning of COP21, the UN Climate Change Conference. Minnesma's team, also outfitted by Patagonia, was small; to avoid layers of committees, she limits Urgenda, the organization she directs, to 15. But dozens more awaited to accompany them, and the throng grew to hundreds as word spread through the networks and more joined, some for a day, some for the duration.

On nights they'd needed lodging for 200 people, someone always provided. Once they slept in a castle, another time in a convent. A

* Also through Amsterdam, the world's largest gasoline port.

Dutch public television crew filed live daily stories. Tall, blue-eyed, and articulate, at 49 Minnesma still had prominent dimples that made her face both disarming and resolute, which the camera doted on.

They crossed the Rhine and Meuse, entering Belgium near Antwerp and continuing through Brussels into northern France. Restaurants there were closed for the season, but sustenance kept appearing. A village mayor fêted them in a schoolhouse with a banquet of local wines and cheeses.

By the afternoon they arrived in Paris, on Saturday, November 28, more than 2,000 of all ages had joined, including mothers pushing prams. COP21 was beginning, but Minnesma never bothered attending. She also skipped the side meetings that NGOs hold all week at global climate summits.

"Blah, blah, blah. Someone makes a report, and nothing happens." Marjan Minnesma had walked 600 kilometers to Paris to inspire people to *do* something, to take to the streets and let the world know that climate action was no elitist eco-fad but an exponential, exploding populist movement that can't be ignored any more than a cyclone.

She didn't try to see anyone, because everyone came to see her, as they've been doing since 2013, when she pulled off what everybody thought was impossible. She still hears that.

"I thought it was possible."

She stayed only long enough to dine with a prime minister. At 11:00 p.m. she retrieved her Opel, which during the walk had ferried their luggage. Her son's birthday was the next day. She arrived home at 6:00 a.m.

Urgenda's office is four classrooms in a former brick elementary school in Zaandam, across the North Sea Channel from Amsterdam.

Minnesma grew up 10 kilometers away, on land where her family raised sheep and cows. In school she studied math, biology, chemistry, physics, business administration, philosophy, and international law, earning three bachelor's and three master's degrees. One of her dissertations was on Royal Dutch Shell. She wrote another after attending COP1 in Berlin in 1995, the first climate conference.

Her first job, with the Netherlands Institute for Energy and Environment, took her to Albania and Estonia, where EU money was bringing green energy projects to eastern Europe. She stayed to lead a Greenpeace climate change campaign and became acquainted with the world's leading climate scientists. The best ones, she could see, were the most worried. But excepting James Hansen, most scientists were afraid to state that publicly, lest they be dismissed as activists, their credibility undermined.

"If you don't shout out," she'd tell them, "we have a serious problem."

She was named head of an institute at Erasmus University Rotterdam devoted to sustainability transitions. Soon she told her codirector, Jan Rotmans, "We have 40 Ph.D. students who will produce books that at most 10 people will read. What are we really changing?"

Rotmans agreed. They brainstormed an action plan for every year up to 2050, spotlighting projects showing that sustainability is more than new lightbulbs, home insulation, or switching to plant protein. What they called their Urgent Agenda for 2050 ended up on the front page of the leading Dutch newspaper, *NRC*, generating thousands of emails—and thus was born Urgenda.

She left her university job to run it full-time. In 2010, she started wondering why there weren't more solar panels. Why was everybody waiting? The simple answer was that despite government subsidies to help homes become energy-neutral, only the wealthy or businesses could afford an €80,000 to €100,000 outlay for solar.

"That much isn't necessary," she argued. Government experts disagreed. So she emailed the Urgenda membership, saying that if they bought panels jointly, they could bring the price down themselves. Who'd be willing, she asked, to finance €35,000—what they generally spent on energy in 15 years—if they knew they'd never have another energy bill? "You could pay it back over 15 years, for what you now spend on energy per month. It's a solution with no additional cost."

When the banks wouldn't loan her money, enough members were willing to advance a portion that she flew to China herself and bought 50,000 cheap solar panels, plus connecting hardware. Today, Urgenda makes houses energy-neutral for €35,000. "We give them around 20 solar panels per house, heat pumps, an induction stove, and an electric boiler." Today, 2 million of the Netherlands' 7 million houses are solarized.

Urgenda imported the country's first electric vehicles, from Norway. It preempted a sluggish national tree-planting initiative through an email campaign encouraging people to find saplings in parks or woodlands destined to be culled and bring them to tree hubs for others to transplant. "More than 80 percent survive," says Minnesma. "We've saved 1.5 million trees that otherwise would be dead."

Then, in 2013 Urgenda made environmental history—not just for the Netherlands, but for the world. Two years earlier, Minnesma had read a book by Dutch attorney Roger Cox, *Revolution Justified*, which argued that if governments failed to safeguard their constituents and future generations against catastrophic climate change, the only remaining democratic recourse was to sue them. The Dutch government was a signatory to statements by both the EU and the United Nations Framework Convention on Climate Change that industrial countries should reduce greenhouse gases by between 25 and 40 percent by 2020, but it had done nothing to remotely meet that goal. A 2012 letter from Urgenda demanding a 40 percent emissions reduction elicited a letter in response, agreeing that climate change was serious—but if the Netherlands tried to be a frontrunner, it would jeopardize its economy and make environmental measures unaffordable.

So Urgenda hired Roger Cox, and they sued. Coining a term, *crowd-pleading*, she amassed both contributions and 886 co-plaintiffs, the youngest five years old, hundreds of whom accompanied them to deliver the summons in person, suing their government in civil court on the grounds that failure to address climate change was endangering its citizens' lives.

After more than a year of filing motions, the case was heard in The Hague in April 2015. The government argued that the Netherlands was too small to impact the global climate. Cox replied that by that argument, no small nation had an obligation to do its part, but the Dutch government had repeatedly signed accords stating the contrary for industrialized nations, regardless of size. Two months later, the judges agreed: at minimum, the government was obliged to reduce national emissions by 25 percent by 2020.

The verdict electrified environmentalists worldwide. The government appealed on 29 grounds and, assuming it would be overturned, did nothing. In 2018, however, the appeals court not only rejected all 29 but added that the Netherlands was violating a legal obligation under the European Convention on Human Rights to protect its citizens.

The government appealed again, to its own supreme court—which in 2019 upheld both lower courts' rulings, forcing the Netherlands to reduce greenhouse gas emissions by 25 percent by the following year. It was unprecedented anywhere: a climate case successfully adjudicated against a government under its own laws, and environmentalists worldwide rejoiced. But winning would be meaningless if the government couldn't comply, and Urgenda was ready, having prepared in advance a list of 54 actions that together would mitigate 17 megatons of CO_2 to fulfill the court order. The measures ranged from more rooftop solar to energy-neutral rental properties, reducing speed limits, reducing dairy production by 30 percent (at no loss of profits), compulsory LED lighting in greenhouses, and extinguishing office and business lights after-hours.

The government adopted 30 of Urgenda's recommendations, but avoided angering dairy farmers,* making up the breach by shuttering three coal-fired plants. In 2022 Marjan Minnesma was awarded the Goldman Environmental Prize. The Urgenda court ruling, said the Goldman judges, "has major implications for governments around the world, many of which are far behind on their climate promises. Inspired by Marjan's work, activists in Belgium, France, Ireland, Germany, Britain, Switzerland, Norway, and New Zealand are making similar stands, now with a model for victory."

Urgenda formed a climate litigation network to assist these legal challenges worldwide. Among the first was in its own country. Using their same language (and same lawyers), in 2018 Milieudefensie—Friends of the Earth Netherlands—filed suit in the same district court in The Hague against the country's biggest, most valuable company, Royal Dutch Shell.

"International climate policy is focused on nations. That excludes the biggest world's polluters," said Milieudefensie director Donald Pols. "The 25 biggest multinationals—especially oil multinationals—literally produce half. That's 25 CEOs—who can fit into a city bus—responsible for half of all emissions. They are not regulated, because they are multinationals. Multinationals outside policy frameworks can avoid paying taxes and CO_2 reduction."

The civil suit that 17,000 co-claimants brought for Shell's failure to protect people from harm, thus violating their human rights, demanded the company cut its worldwide CO_2 emissions by half. (Orig-

* As the Netherlands has since been compelled by the EU to cut nitrogen emissions in half by 2030, it's had to reconsider.

inally they'd sued on behalf of "current and future generations of the world's population," but the court ruled that the entire world lacked standing in a Dutch class action and limited the co-plaintiffs to vulnerable northern Dutch coastal residents.)

Despite Urgenda's resounding victory, major foundations shrank from funding Milieudefensie's lawsuit, Pols said. A tall, bald, bearded South African in his 50s, he had come to the Netherlands to study and was happily trapped by marriage. "One foundation actually said that Shell is so powerful in decision-making bodies, it wasn't in their interest to make an enemy of them. They also didn't believe we would win."

Instead, they raised more than a half million euros purely on individual contributions. "*Milledefensie v. Shell* is a misnomer. This was truly the people versus Shell."

On May 26, 2021, the court found that Royal Dutch Shell was putting people's lives at risk and ordered the company to reduce its worldwide CO_2 emissions by 45 percent from its 2019 levels* by 2030.

"It was," reported *The New York Times*, "the first time a court ordered a private company to, in effect, change its business practice on climate grounds."

"Shell emits 10 times as much CO_2 as the whole Dutch economy," said an exultant Pols. "If we halve that, we have cut five times the amount of Dutch emissions. That's the solution needed for a global problem. It brings hope back into the discussion. That's even more important than beating Shell."

But victory isn't simple. When Pols told someone at the Ministry of Finance that among the easiest ways to cut emissions is covering every house with solar, he was informed that 14 percent of the

* In 2019, Shell emitted 1.65 billion metric tons of CO_2.

ministry's income comes from household energy taxes. "So we're not going to do that. Sorry."

Shell, of course, appealed the decision—although Milieudefensie wasn't too worried, as it would be facing the same appellate and supreme courts that heard the Urgenda case.

"I'm convinced the court verdict will be confirmed," said Pols, at their brick headquarters near Amsterdam's Rijksmuseum. "With concrete examples of climate change taking place right now, the science only gets clearer. Cutting emissions has been accepted by shareholders of ExxonMobil and Chevron. All this supports maybe even a more ambitious verdict."

Shell didn't wait to find out. In November 2021, after 131 years, it dropped *Royal Dutch* from its name and moved its headquarters to London.

CHAPTER EIGHT

A LINE RUNS THROUGH IT

Beating Shell in April 2021 was Milieudefensie's second win over them that year. The first, in February, came 15 years after it joined four Nigerian farmers and fishermen in suing the petroleum giant over thousands of oil spills from rusted pipelines in their Niger Delta villages. They filed suit in The Hague because, said Milieudefensie's Donald Pols, "Shell is a state within a state in Nigeria. They can do whatever they want without any consequences."

The judges denied Shell's argument that the case belonged in a Nigerian court. They ordered it to clean up the horror, install a leak alert system, and negotiate compensation, resulting in a €15 million settlement.

Unfortunately, 15 years is a long time for justice. By then, Milieudefensie's four co-plaintiffs were dead. The Niger Delta, among the most fertile spots in Africa, is now also one of Earth's most polluted places. Human life expectancy there is 45 years.

The Shell–Niger Delta case reveals three hard realities:

1) Litigation routinely takes years, especially when suing a corporation with its own phalanx of lawyers. The longer they delay proceedings with motions and appeals, the longer they keep profiting by polluting, and the sooner plaintiffs grow

cash-strapped and exhausted. There's also the risk of drawing a judge who is biased, or bribed.

2) Although congresses or parliaments can act faster than judiciaries by passing laws, on climate issues they're largely paralyzed because so many legislators are beholden to energy companies, electric utilities, or their surrogates, who fund their campaigns, lobby them relentlessly, stuff their pockets, hire their relatives, etc.

3) While elected assemblies dither and courts grind on, damage deepens. Once oil spills, methane leaks, or CO_2 is emitted, they aren't easily recaptured.

i. Rice

There's also a fourth reality: the other side does not play fair.

Minnesota is called the Land of 10,000 Lakes, but there are really at least 13,000, plus innumerable ponds and marshes. During the past 100,000 years, mile-high glaciers regularly slid back and forth here, gouging holes that filled with meltwater when they retreated. By the 20th century, ice sheets were due back anytime, until *Homo sapiens* started fiddling with the planet's thermostat. Until that fever eventually breaks—maybe when someday there's no more fuel to mine, or no one left to mine it—except during what remains of winter, Minnesota will remain ice-free.

Seen from above, the state resembles an immense sponge. From its lakes flow ribbons of water that nearly all drain into a single winding channel that first emerges from a wishbone-shaped northern lake as a clear trickle. After wandering north through sodden lands redolent of peat, it then loops south, laying down oxbows and broadening as it gathers water from nearly half the North American continent by the time it reaches the sea. Only South America's Amazon and Africa's Congo and Nile have watersheds larger than the Mississippi's.

The land surrounding its headwaters was once forested with white pine, red oak, spruce, juniper, birch, aspen, basswood, ash, tamarack, and sugar maple. Its inhabitants were bear, caribou, moose, lynx, wolves, mink, muskrat, otters, beavers, ermines, raccoons, badgers, passerine birds, turkeys, woodpeckers, and raptors, including eagles. The water held trout, whitefish, pike, bass, walleye, sturgeon, muskellunge, hard- and soft-shell turtles, frogs, salamanders, assorted ducks, coots, grebes, loons, herons, pelicans, and trumpeter swans. Blackberries, blueberries, chokecherries, and currants grew on shorelines, and cranberries in the nearby bogs. The only native grain on the continent (excepting maize, native to southern Mexico) grew here from a five-foot-high grass that emerged from shallow lakes and river bottoms.

The first people here, called Dakota, arrived well after the Pleistocene ended, but it's unclear from where, because they call Mni Sóta Makoce* their birthplace. They harvested this grain, but were more skilled at hunting. Eventually, another group arrived from the east, and there were clashes. Then came white-skinned people. Wars with both, and better hunting prospects, drove most Dakota west to the buffalo plains.

The ones from the east remain today. They call themselves Ojibwe, a nation of the Anishinaabeg, meaning "original people," one of the Algonquian-speaking tribes from North America's North Atlantic seaboard. Most accounts agree that around the beginning of the last millennium, the Anishinaabeg received a fiery vision to move west. It came from the first of seven prophets who would lead them on a

* Land of sky-colored water.

path of islands resembling turtles.* That took them along the St. Lawrence River, following signs of sacred cowrie shells. The Second Fire prophecy guided them to seas of sweet water: the Great Lakes. The Third Fire, which came after they briefly lost their way, told them that they would know their final destination by food growing from water.

Their journey—by birch-bark canoes, by dogsleds, by foot—took hundreds of years. They paused awhile where the Ottawa River joins the St. Lawrence near today's Montreal, then kept on. At Niagara Falls, some stayed—today, the Mississaugas, settling north of Lake Ontario. Others continued along Lake Erie to the Detroit River. At one point two prophets appeared bearing the Fourth Fire. They predicted the coming of white-skinned people, warning they could be friend or foe, and it might be difficult to tell the two apart.

Around the same time, an Italian explorer employed by Spain sailed into the Caribbean.

Some Anishinaabeg wandered toward Lake Michigan and became today's Potawatomi. Others, ancestors of today's Odawa, remained at Lake Huron, while the rest continued, following the northern and southern shores of Lake Superior. Finally, near the western tip of the lake they called Gichigami, they found an island—Madeline Island today—whose profile resembled a tortoise shell. From its bay grew grain, as the vision had promised.

More than anything, this water-sprung grass they named *manoomin* defines the essence of the Ojibwe: the Anishinaabeg who completed the journey. "It is the first solid food a baby eats," says Ojibwe elder Winona LaDuke, "and the last we taste before departing for the spirit world."

* Turtle Island was also their name for the North American continent. These would be turtle islands within a turtle island, echoing Algonquian cosmology of the world created from dirt heaped atop a turtle's shell.

It's gathered in late summer in shallow canoes, one Ojibwe poling through what resembles a wheat field springing from the water, while a seated companion uses one cedar stick to pull stalks of ripe grain over the boat and another to knock off the seedheads—a two-stroke rhythm often likened to a heartbeat—to fill the canoe. Much later, travelers who had seen Chinese harvesting paddies would call it wild rice. Lakes and rivers of northern Minnesota especially are replete with stands of manoomin. During the season it's common to see 30 or 40 canoes ricing, including on the chain of lakes along the channel they call Misi-ziibi.

By the early 1600s, they had met white-skinned French trappers near Sault Ste. Marie. Although the Fifth Fire had warned of a looming great struggle, this was an amicable encounter, the Anishinaabeg trading pelts in exchange for muskets, steel kettles, and knives, and French traders taking Anishinaabe wives. But in the 1800s, after Ojibwe were long settled into Minnesota and Wisconsin, came other white-skinned men with different dealings in mind.

Their government was near Turtle Island's Atlantic coast, not in distant Europe. They had come wanting land, and were well armed. A series of treaties ensued, which reserved to the "Chippewa," as the whites pronounced Ojibwe, land that was solely theirs in exchange for vast tracts that together would form new, manifestly destined American territories and states. The treaties recognized "Indian" tribes as sovereign nations, dealing as equals with the whites' government, whose constitution deemed treaties supreme law of the land.

Because the treaties specified that they retained rights to hunt, fish, and, most important, gather manoomin on ceded lands, the Ojibwe signed, the concept of land ownership baffling to them anyway. The Earth—in their stories, the Mother who gave birth to all—simply wasn't ownable. Her bounty was to share, which they believed the treaties codified in white man's language.

But following the Treaty of 1855, loggers—some today enshrined as Minnesota's founders—arrived, and the Ojibwe learned what else they meant. Entire forests vanished; millions of pines turned into houses, barns, and churches as the US advanced toward the Pacific.

"At one point," recounts Ojibwe scholar Anton Treuer, "nine of the ten largest timber mills in the world were in Minnesota." A loophole created by Congress permitted sales of treaty land by "competent Indians"—meaning racially mixed. A calipers-wielding University of Minnesota anthropologist named Albert Jenks claimed that nose measurements and hair samplings showed that 90 percent of Ojibwe on the northwestern White Earth Reservation had white blood, qualifying them to sell. Bribery and coercion ensued; protests over the resulting land swindles ended at rifle-point, or in a noose, as new military forts enforced white treaty interpretations.

As the great timbering era peaked, cleared land became white farms. With game scarcer, the Ojibwe became farmers themselves, raising corn and squash on reservations that were mere slivers of their former territories. But their staple remained manoomin. Each year, bands from northern Minnesota's seven Anishinaabe reservations harvested and parched it, danced to separate the hulls, then winnowed and stored tons of it.

The Sixth Fire had warned that their grandchildren would be turned against them. That came to pass in boarding schools that cut their hair and forbade the Ojibwe language and teachings. Returning broken-spirited, many turned to drink. They lived on the reservations, isolated from white towns burgeoning around them, especially after, says White Earth historian Robert Shimek, "the state decided that GIs returning from World War II needed a vacation playground." Soon, sparkling northern Minnesota lakes were ringed with summer cabins and "No Trespassing" signs.

"They used their laws to harass and intimidate us until people were afraid to rice."

ii. Slick

To the Anishinaabeg, rivers are the veins of Mother Earth, but in the past half century, whites have imposed another sinuous network beneath the land to carry darker fluids. Water quenches fires. These do the opposite.

After the 1960s, rather than ship petroleum by rail, oil companies realized it was more profitable to dig ditches stretching far beyond the horizon, uprooting every tree for 50 feet on either side, to bury pipelines that in a day carried as much as 10,000 oil tankers. Among the longest, Line 3 stretched a thousand miles from a tank farm at the Hardisty, Alberta, rail hub to the Port of Duluth-Superior, near Madeline Island. En route, it crossed two Ojibwe reservations, on rights-of-ways the US Bureau of Indian Affairs awarded without consulting any actual Indians. Near Grand Rapids, Minnesota, it tunneled under the Mississippi River.

In March 1991, a gusher exploded there, coating tree branches just north of town 40 feet high. Rather than investigate when the line's pressure dropped, its owner, a subsidiary of Canada's Interprovincial Pipe Line Company—today, Enbridge Inc.—simply increased the flow, until someone informed the company that their crude oil was flooding the Prairie River, a Mississippi tributary. Luckily, it was still covered with 18 inches of ice. Crews were able to squeegee and vacuum much of it, but not before 1.7 million gallons spilled.

It was the biggest inland oil disaster in US history, and it wasn't Line 3's first: in 1973, another rupture had caused the second worst, in northwestern Minnesota, which barely missed polluting the Red River of the North, whose valley is North Dakota's and Manitoba's breadbasket. At least two dozen more leaks from faulty seams followed—yet instead of closing Line 3, Enbridge was only required to cut its capacity by half.

Stopping the flow entirely was out of the question. Water, vital to life, might be priceless, but petroleum has a price: very high.

In 2002, a quarter million more gallons, this time from Enbridge's Line 4, spilled in Cohasset, Minnesota, a town on the Mississippi; to keep it from entering the river, they had to burn it away. In 2007, another pipeline exploded at Clearbrook Junction, Minnesota, above the Mississippi's headwaters. Then, in July 2010, Enbridge Line 6B burst into the Kalamazoo River, which flows into Lake Michigan, Chicago's drinking supply.

Again, the company had responded to a pressure drop by just increasing the flow—twice—until a frantic Michigan utilities crew phoned. By then, 18 hours had passed. Thirty-five miles of the Kalamazoo River were covered in a million gallons of bitumen that sank to the riverbed, killing everything in the water column, including thousands of fish. The half-billion-dollar cleanup is still the costliest in US history.

A visiting alien might wonder what humans could possibly be thinking, risking irreplaceable water—the Mississippi system is nearly half the drinking supply of the US—by letting oil pipelines anywhere near it. But with petroleum, we don't think at all, unless supplies are cut—and then we *really* don't think, because we panic, like the addicts that we are. What's shocking is that this addiction isn't even 200 years old, and that civilizations like the Sumerian, Egyptian, Incan, Mayan, Roman, and several Chinese dynasties thrived without burning oil.

In 2014, Enbridge announced plans to replace the most corroded section of aging Line 3. Not from fear of liability—for ruining the Kalamazoo River, they were fined $177 million, pocket change to a company whose revenues average $36 billion. With mining heavy bitumen crude from Alberta's tar sands now cost-effective, they planned to double capacity. Exhuming and replacing the old line with 36-inch pipe would mean two years of downtime. Also, that section ran through two Ojibwe reservations, Leech Lake and Red Lake; after years of leaks, both refused to renew Enbridge's lease.

Enbridge proposed a new path that would avoid reservation land by dropping south for 60 miles, then heading east to Duluth-Superior. White Earth Band Ojibwe were appalled. The new route would cross 212 rivers, streams, lakes, and wetlands, including their choicest manoomin and maple sap region, where three sovereign treaties guaranteed their harvesting rights.

It would also tunnel twice under the Mississippi, where it first flows north, then south. Enbridge's own records acknowledged 1,000 leaks or spills since 1996. Another could poison one of Mother Earth's main arteries.

"We've put our bodies on the line. It sucks. We shouldn't have to," said Gina Peltier, a 30-year-old Turtle Mountain Band Chippewa, long braids framing her soft features.

For the next eight days Peltier and 100 others would remain here, until the county sheriff either issued trespassing citations or arrested them, which some chose, refusing to abandon this spot. They were camped on a long timber-mat boardwalk Enbridge had laid to the reedy banks of a stream, so earthmovers wouldn't bog in wetlands. The stream was the Misi-ziibi, still barely 10 feet across, weaving through this grassy valley just east of the White Earth Reservation, seven miles north of its source, Lake Itasca.

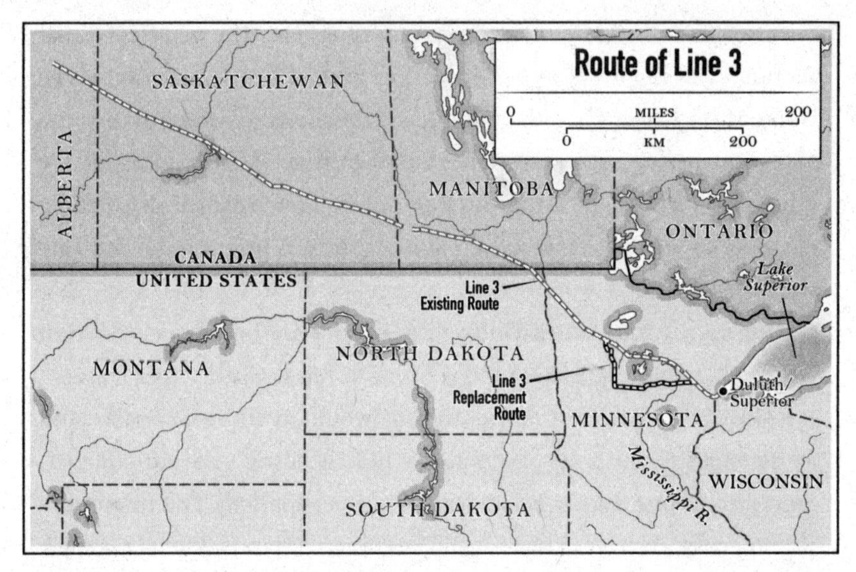
Route of Line 3
0
MILES
200
0
KM
200
SASKATCHEWAN
ALBERTA
MANITOBA
ONTARIO
CANADA
UNITED STATES
Lake Superior
Line 3 Existing Route
MONTANA
NORTH DAKOTA
Line 3 Replacement Route
Duluth/ Superior
MINNESOTA
Mississippi R.
WISCONSIN
SOUTH DAKOTA

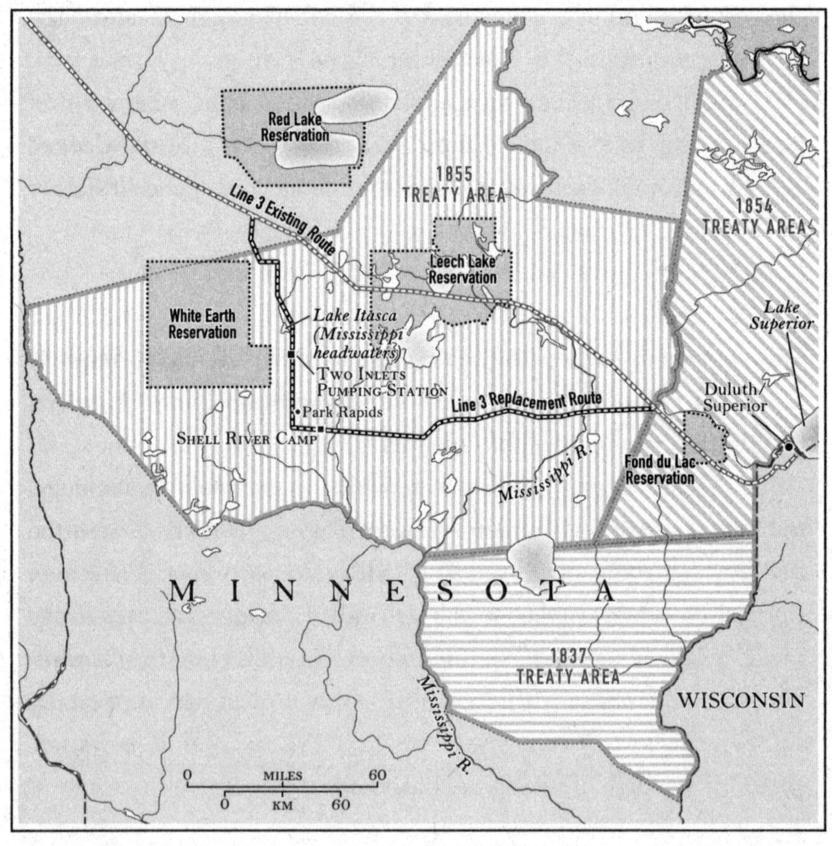
Red Lake Reservation
1855 TREATY AREA
1854 TREATY AREA
Line 3 Existing Route
Leech Lake Reservation
White Earth Reservation
Lake Itasca (Mississippi headwaters)
Two Inlets Pumping Station
Lake Superior
Duluth/ Superior
Line 3 Replacement Route
Park Rapids
Shell River Camp
Fond du Lac Reservation
Mississippi R.
MINNESOTA
1837 TREATY AREA
WISCONSIN
Mississippi R.
0
MILES
60
0
KM
60

This place, Fire Light Camp, was where Line 3 would first cross the Mississippi, unless these women could stop it. On their first night here, bullets were fired over their heads. They stayed, sleeping in tents, using composting latrines, and bathing in the river they guarded.

The day before, June 7, 2021, 2,000 Indigenous rights and environmental activists, many from Minneapolis and St. Paul, four hours away, plus celebrities—Jane Fonda, Rosanna Arquette, Bill McKibben—had demonstrated here. Some stood in the water amid clouds of dragonflies, arms linked. Others performed round dances. Women came to speak from other camps they'd pitched wherever the pipeline would cross rivers.

To Ojibwe, men are firekeepers and women are water protectors. One spirit woman cares for water in the sky, rain, and snow; another, in rivers and lakes; a third, in the oceans. A fourth spirit protects the water in woman herself: her amniotic fluid that bathes babies with life. At Ojibwe ceremonies, a woman prays holding a cup of water.

Sitting under an awning in red canvas camping chairs, these women wore full skirts appliquéd with colored stripes and T-shirts that said "Honor the Earth," "Stop Line 3," "Love Water not Oil," and "Ninga Izhitchige Nibi Onji"—I Will Do It for the Water.

They were there because Alberta tar sand crude rivals coal as the dirtiest of all hydrocarbon fuels. Two other conduits intended to carry it had been stopped: the Dakota Access Pipeline, DAPL, which Enbridge partly owned, after attacks on Standing Rock Sioux and environmental protesters at a Missouri River crossing—using dogs, tear gas, and water hoses in subfreezing weather—scandalized the nation; and Keystone XL, which US president Joe Biden canceled on his first day in office. The carbon from Line 3's daily capacity, nearly a million barrels, would be equal to adding 50 coal-fired power plants. Six years earlier, the world had agreed in Paris to limit global warming to 1.5°C. For that, repeated calculations agreed, tar sands must remain in the ground, or risk incinerating humanity's future.

Ojibwe's Seventh Fire represents a choice: return to a natural existence along a green path, or follow a hardened road that speeds to ruin. Steel pipelines, for instance.

"For the world's fate to hang on a handful of Ojibwe women is insanity," said Peltier, wiping tears.

Yet Biden hadn't halted this one, which he could have with a pen stroke. The governor of Minnesota, Tim Walz, who initially opposed it, inexplicably had reversed course. His lieutenant governor, herself a White Earth Band Ojibwe member and adamantly opposed, had to back off, lest she be excluded from decision-making. In contested rulings, Minnesota's Public Utilities Commission and its Pollution Control Agency had approved it. Most of the latter's environmental advisory group quit in protest.

These Water Protectors weren't surprised; their own tribe was divided. "What Enbridge does," said Leech Lake Band Ojibwe Nancy Beaulieu, pulling off her white irrigation boots, "is tell tribal council members, 'We want to put an easement through.' With money they offer the tribe, they persuade three out of five. All they need is a majority. Then Enbridge says they've got tribal approval."

"It's the colonized thinking we were born into," declared Fond du Lac Band of Lake Superior Chippewa grandmother Debra Topping, her neck wrapped in a wet bandanna against the rising June heat. "You want money to pay for your house, your cars, all your toys. I remember thinking I'd made it big when I could buy paper towels and paper plates. Now I think I'm contributing to genocide." With her porkpie hat, she fanned herself. "Once we could portage to the Mississippi, put in a canoe, and go anywhere in the world. Now we need to buy bus passes and $7 permits to enter Itasca State Park."

"I tell them the Creator didn't tell me I needed a pass," says Beaulieu. At the demonstration, she stood knee-deep in the river and vowed, "We're protecting the sacred for all those not born yet!"

Enbridge, however, was applying muscle and money to deny them. A condition for their permit was that the company create a $750,000 escrow fund to cover costs of protecting Line 3 from protesters. This included forming the Northern Lights Task Force of law enforcement personnel from each county along its route—except with perceived security threats growing as the pipeline advanced, the fund eventually burgeoned to $8.5 million as others around the state lined up at the trough.

In effect, Minnesota public law enforcement became a Canadian corporation's private militia. For this huge protest, sheriffs and deputies came from counties as far south as the Iowa border, billing Enbridge's escrow fund for travel, hotels, meals, and "personal protective gear"—including riot shields, pepper spray, tear gas canisters, and flash-bang grenades. Even the Department of Homeland Security sent a helicopter, presumably from the Canadian border, to repeatedly buzz demonstrators who'd scaled fences to break into the construction site of Enbridge's Twin Inlets pump station, 20 miles away from the Mississippi riverbank protests, attempting to dislodge them with clouds of choking dust.

"Stay! Don't let them weaponize our Mother Earth against us!" someone yelled, although many had already padlocked themselves to earthmovers and infrastructure. At a training on an isolated ranch the previous afternoon, protesters had chosen whether to demonstrate peacefully at the river or risk arrest in this act of civil disobedience. About 200 were here, several handcuffed together inside "sleeping dragons"—lengths of pipe intended to stymie sheriffs with Enbridge-provided bolt cutters from separating them.

"They've known about climate change since the '80s," yelled a 27-year-old woman in a floppy canvas hat. "We should've been getting off fossil fuels before I was born."

"There is no planet B!" hollered the woman cuffed to a tractor axle.

Mined from an ancient seabed in Alberta's Athabasca formation, tar sands crude is sludge. To keep it moving requires multiple pump stations like this one, making Enbridge one of Minnesota's biggest energy consumers. Enbridge ads showing solar panels atop pump stations infuriate protesters. Keystone XL, rejected by Obama in 2015, then reinstated by Trump, was rejected again by Biden because it would mortally worsen a badly damaged climate. The same with DAPL. So why not Line 3, which carries the same filthy hydrocarbon? Did the president, a Democratic politician who horse-traded for decades with Republicans, think he could split the difference with voters by stopping some pipelines but allowing others? With all the carbon in Athabasca crude, by compromising, everyone loses.

The nearly cloudless day was unusually hot for northern Minnesota in early June. Volunteers passed out canned soda and oranges. Two protesters, Quaker women in their 80s, had come in a peace caravan from Massachusetts. "I've lost hope in the power of prayer without action," said the one using canes. "We need to do something transformative, not just the same old rallies and letters to our senators. I came here to learn from Indigenous people's connection to Her, the Earth."

"We're at war," said her companion. "Back in Cambridge, away from the front lines, privileged people think we're still safe. But my sister now has asthma, and my grandkids refuse to have children."

Shouting interrupted them. From orange buses that had pulled up to the construction zone, nearly three dozen sheriffs, deputies, and Minnesota Department of Natural Resources conservation officers marched through a breach in the temporary fence. Wearing helmets, riot shields, and holstered automatic pistols, they formed two lines.

Following brief exchanges of ultimatums and profanities, they moved in with truncheons, handcuffs, and bolt cutters. Zip-tied, screaming protesters were carried to white police vans. As each filled, they were driven six miles to the overwhelmed Hubbard County Jail in Park Rapids, Minnesota. By evening, five area jails had received 186 prisoners, who were strip-searched and given orange V-necked prison suits and bath sandals. The charge was criminal trespassing on critical public service facilities, a gross misdemeanor punishable by up to a year in prison and $3,000 in fines.

Trespassing laws safeguard citizens' privacy on their own property. Under eminent domain, governments can condemn private property for public use, but Enbridge had actually purchased much of Line 3's right-of-way, thereby leveraging Fourth Amendment privacy rights to shield its pipeline project, considered critical public infrastructure, from public scrutiny.

"We are heartbroken," said Katy Grisamore, director of the Hubbard County Jail Ministry, which offers Bible studies to prisoners, as she and her husband, Wayne, ferried an elderly detainee, released on his own recognizance because the prison couldn't provide his medication, back to his rental car that evening. Their own property, a piney tract that her grandfather had bought on the Straight River, a premier trout stream, was in Enbridge's path.

An eminent domain attorney advised them to sell. They were the last on the route to sign off. "Otherwise they would have just taken the land anyway."

Near the pump station entrance, lights on five sheriff's department vehicles were blinking red, orange, and yellow. Over an LRAD—a long-range acoustic device, used earlier to blast protesters with nausea-inducing frequencies—a Hubbard County deputy announced that everyone still locked down was about to be arrested. Among them were the two octogenarian Quaker women.

"We are praying that President Biden pays attention," Grisamore told their passenger. "We're praying for somehow this to stop. We love what Winona is doing. To these protests, we say—"

"Amen," concluded Wayne.

iii. Honor

Winona LaDuke, a cofounder of Honor the Earth, the oldest of the Indigenous groups that organized the protests, had attended the Mississippi gathering and later stopped by the pumping station. Then she returned to her camp 30 miles south of the Mississippi headwaters on the Shell River, a tributary that the pipeline was slated to cross five times. LaDuke, in her early 60s, sturdy with green eyes and a gently aquiline profile, was deeply bronzed from farming corn, beans, squash, heritage potatoes, Jerusalem artichokes, traditional tobacco, and, lately, fiber hemp. Over the years, she'd invested prudently and bought back 800 acres of what the White Earth Ojibwe ceded in the Treaty of 1855, which, she said, Minnesota now ignored and the US violated.

"When my ancestors signed those treaties, the land was full of maples, the rivers full of sturgeon, the prairies full of buffalo. You could get sugar from trees, wild rice from the lakes, drinking water from every creek. It was the world as it's supposed to be. Why would you want to mess with that?"

Her camp was a pine clearing with a few tents and two canvas-covered wigwams. Nearby was a corral for the horses several Ojibwe teens rode on patrol. When the pipeline replacement was first broached, LaDuke and some Anishinaabe companions rode the entire route on horseback to draw attention to what a foreign outfit with a long accident record was proposing: ramming a leaky tube filled with explosive fuel through Minnesota's cherished lake country and hallowed Mississippi headwaters—fuel that wasn't even staying in the state. They would create a few temporary jobs, but mainly they'd bring pipeline assembly crews from the Dakota oil fields, notorious for preying on minor Native girls.

That was seven years ago, and it was finally coming.

Earlier that week, in a tank top that read "Water Is Sacred" and a long, flowered skirt, she asked camera crews and celebrities, "How does a Canadian corporation get to take the Great Lakes hostage and put millions of citizens, tribal and nontribal, at risk? Somebody needs to be the adult in this situation."

Now the crowds were gone, and they were waiting. Construction had begun the previous December, in 2020, mid-pandemic, after Minnesota's agencies signed off, despite a state environmental assessment predicting that resulting social damages, including CO_2 emissions from tar sands, could total $287 billion: a hundred times more than what Enbridge would spend here. Lawsuits had been filed in every possible court, including against the Army Corps of Engineers for issuing a permit under the US Clean Water Act with no federal environmental impact statement—especially since Enbridge was requisitioning 500 million gallons of water to drill beneath rivers.

Winona and her Water Protector sisters erected ceremonial lodges and resistance camps on frozen rivers, blocking equipment, getting arrested. After a pause for spring thaw, the mud had dried and Enbridge's crews were rushing to get pipe into the ground while the courts cases dragged on.

They hadn't yet arrived at her Shell River camp. No machinery, only red-winged blackbirds and hermit thrushes. Few mosquitoes: too dry. Stripping down to a two-piece black bathing suit, LaDuke climbed into an aluminum canoe. Past a clump of cattails, through transparent, drinkable water she could see big freshwater mollusks that give the Shell its name, so plentiful back in the 1900s, before plastic, that there was a button factory nearby.

Catching the current, she glided through flowing stalks of wild rice, still green, lying flat on the surface to soak up sunlight. But soon she was dragging the boat more than paddling. With scant spring rain, Minnesota's rivers and lakes were low. Shell Lake, this river's source, was where she rices. Everyone wondered about the fall harvest.

When Enbridge announced that it had miscalculated and actually needed *5 billion* gallons and the state still didn't balk, multiple Water Protectors including LaDuke joined an unusual plaintiff in suing Minnesota for endangering wild rice's clean water supply and infringing on Ojibwe treaty rights to harvest it. Their co-plaintiff was wild rice itself; earlier, the White Earth Ojibwe had passed a "Rights of Manoomin" resolution, declaring that wild rice had a legal right to exist and flourish—in effect, personhood.

Their inspiration was a growing worldwide "Rights of Nature" movement to grant full legal status to, among others, the Ganges, Himalayan glaciers, salmon in Seattle, and Ecuador's entire ecosystem. Although the tribal court openly sympathized, since the affected waters weren't on reservation lands, prior US Supreme Court rulings held that Indian tribes possess no authority over treaty lands they use but don't actually own.

"In the deepest drought anyone remembers," seethed Winona, "Minnesota's Department of Natural Resources awarded Enbridge 5 billion gallons without even talking to the tribe. The Shell is 75 percent down. Most of their withdrawals come from this watershed."

Over 40 years, Winona LaDuke had engaged in Sisyphean battles for the first people on this continent who'd ended up last in line. It was a war she chose. She was born off-reservation, in Los Angeles, where her White Earth Ojibwe father played Indian extras in films and wrote books based on tribal lore. He met her Jewish artist mother while selling wild rice in New York. When her parents divorced, Winona lived with her mother but went with her father to powwows and visited the White Earth Reservation during summers.

In her first year at Harvard, she heard an American Indian Movement organizer speak about how colonialism, then corporations, systematically crush Native peoples. Her questions prompted him to hire her as a researcher for a 1977 UN conference on Indigenous rights in Geneva, where she briefed AIM leaders on uranium mining on Navajo lands. Only 18, she became the youngest person to address the United Nations.* While completing her degree in Native economic development, she plunged into antinuclear campaigns and tribal fights in New Mexico and North Dakota over mines.

At a rally in the Black Hills, a distant relative, AIM leader Vernon Bellecourt, told her to go work for her people. Her first job, at 23, was as principal of the White Earth Reservation's K–12 school. She began to grasp the squalid history of treaty law that called tribes "sovereign nations" but, through foreclosures and other tricks, purloined their land and resources: nine-tenths of the reservation was held by non-Indians. She went to court and started an Anishinaabe organization to regain that land. She founded the Indigenous Women's Network. She won the Reebok Human Rights Award and used the money to start the White Earth Land Recovery Project, which expanded into recovering language, traditional farming practices, and heirloom seeds.

Time magazine called her one of the 50 most promising leaders under 40. The Thomas Merton Award and Ann Bancroft Award for Women's Leadership followed. *Ms.* magazine named her "Woman of the Year." She proposed a constitutional amendment requiring environmental policies to project their impact on the next seven generations. She wrote books about Native struggles and gave talks all

* Until 16-year-old climate activist Greta Thunberg in 2019.

over the world. Ralph Nader twice selected her to be his Green Party running mate.*

Now she sat in her Shell River camp, in treaty land just south of the White Earth Reservation, awaiting a nightmare to commence.

They came for her in early July. She and six other Water Protectors sat in blue beach chairs in straw sunhats, linked by quarter-inch-gauge chrome-plated chain around their waists. Behind them stood seven horses, the riders bareback, some standing on their mounts, facing the line of sheriffs and deputies in front of the women. The lawmen, in tan and brown uniforms, wore sunglasses, their belts sagging with black holsters for pepper and mace canisters, radios, phones, and handguns.

That week, an Enbridge frac-out had spilled a hundred gallons of drilling fluid into the nearby Willow River. "They're the ones who committed a crime, but you're here to arrest *us*?" Winona asked the sheriffs. That month would bring eight more frac-outs and a pierced artesian aquifer leaking 100,000 gallons a day that Enbridge didn't report. An activist's drone with infrared imaging had spotted it.

Along the banks, drummers played. The chained women sang and swayed as perspiring deputies passed plastic water bottles. More activists arrived in canoes. In the previous two weeks, more than a thousand had been arrested. Now, as the yellow zip ties appeared, there would be seven more.

Facing multiple counts, Winona LaDuke spent three nights in the county jail. After a year, charges against her companions were dismissed. It took another year and more lawyers before hers were also thrown out. She'd been arrested for Line 3 protests in several counties, and each took turns wasting her time before also dropping the

* She insists they didn't cost Al Gore the presidency: "Gore's inept campaign—he couldn't even win Tennessee, his own state—was responsible for that."

charges. It proved hard to claim she was trespassing on lands she'd legally used for years—or, in one case, actually owned.

Line 3 was completed in September 2021, the final section laid east of the White Earth Reservation at Walker Brook. Within days, Enbridge trucks were back there, trying to staunch a new aquifer breach. Within months, a dozen were discovered. Enbridge was fined $11 million and charged with one count of criminal negligence for concealing the first, which lost 73 million gallons of groundwater during the year it took to seal it with a half million gallons of grout.

To Enbridge, $11 million for a little sloppiness as they raced to finish the pipeline before some judge stopped it was a pittance. After another $1,000 fine, the state forgave the criminal charge. During the entire job, just four pipeline workers were arrested for sex-trafficking Native minors—touted as an improvement over the chronic sex-for-drugs violence in North Dakota's oil field "man camps."

Meanwhile, damage to the Mississippi watershed had commenced, with more to come as joints inevitably failed and pipe corroded. Each year, CO_2 from burning Line 3's payload crude would equal the fumes of 38 million cars.

"Canada," said Winona LaDuke, "is a petro-state. Its economic plan is extraction. Three-fourths of the world's mining corporations are Canadian, hauling shit around the world like the world's dirtiest oil that should be in the ground, not in the air. We're all fossil-fuel addicts, and we're letting the dealers write public policy."

She was driving past flat potato fields in an aging Dodge Ram down Minnesota Highway 34, headed toward her farm. "We need to deal with our addiction, and Canada needs a plan B. We want Canada to be a beautiful place where the water's clean and we get some wind energy and we all live good."

She used to live there, on James Bay. Now a US president she campaigned for wouldn't cancel permits the Trump administration had rushed through during its final days to allow Line 3 to start. This was much more than screwing the Indians again; climate collapse is the first existential threat most white people have faced, and they don't know how to grasp that it may take down everybody. There's no word in English for genocide against our entire species. Or in Ojibwe.

"In the time of the Seventh Fire, we reach a fork," she said, pulling up to a red steel industrial building with white garage doors. "The old path is worn and scorched, leading to our destruction. Or we can take a new, unworn green path. If we do, an Eighth Fire can light the way to *mino-bimaadiziwin*—living the good life."

Literally. The sign read "8th Fire Solar." Mounted on the building's south side were four-by-eight-foot modular dark-glass panels: inexpensive solar furnaces that gather thermal heat from the sun and conduct it into the factory where nine Ojibwe men and women manufactured them.

"Free heat that pays for itself quickly," said assembly tech Nick Bellrock, who was in high school when he started here.

"They're just airtight hotboxes," said plant manager Ron Chilton. The base of each was coated with black oxides to absorb solar heat. A fan circulated interior air out through the panels to be warmed by the sun and returned indoors. In winter on the reservation, choosing between heating and eating was all too familiar. In frigid Minnesota, these panels wouldn't heat a house entirely, but they lowered firewood consumption or propane bills considerably. Since 2019, they'd been installing them on surrounding buildings. They were now even getting off-reservation clients.

Winona started 8th Fire Solar, a community development nonprofit, to apply for renewable energy credits that must increase, she believed, if there's a chance for a clean future.

At James Bay, she had wed a Cree chief; together they fought hydroelectric plants on tribal land, but her travels proved a strain. They separated but remained married until his death by heart attack at 51. She brought their son and daughter back to Minnesota, had another son, and adopted three more boys. A few miles from the solar-furnace factory, a houseful of grandsons, both biological and adopted, greeted her at the project that makes her happiest, no matter what else is imploding.

"Our Creator gave us a chance to live this great life, and I'm taking it."

This sign, in front of a market in a converted tractor garage, read "Winona's Hemp: CBD, Tea, Clothes, Food." Like the barn behind it, it was painted with Ojibwe murals and cannabis motifs. She'd purchased this 40-acre farm with proceeds from speaking engagements. There was a garden, fields beyond, and corrals where her grandsons raise ponies and goats.

Across the road, another sign on a steel outbuilding read "Anishinaabe Agricultural Institute." There, her team worked with hemp, the wonder weed with the same botanical name as marijuana, *Cannabis sativa*, but less than 0.3 percent THC, compared with pot's 15 percent—"like poodles and Dobermans," growers say: same species, but very different.

LaDuke was providing her tribe and a locally sourced Park Rapids store with that pain-killing CBD oil, "and our tea from toasted hemp." She batted a fly with her straw hat. "But I want to make canvas—the word comes from *cannabis.* Sailcloth was once hemp canvas. In colonial Virginia, hemp cultivation was actually required. You can use every part of the plant to make fiberboard, plastic, hemp concrete, insulation, oil you can eat or run your car on. It's antimicrobial and antifungal, high in omega-3s, and hemp fiber has six times the tensile strength of steel."

Eighty percent of clothing in the US was once made from hemp.

Henry Ford even made a hemp car—lighter and stronger than steel-bodied autos and requiring a tenth of the energy to build—that could run on hemp-based fuel. But when the 1937 Marijuana Tax Act outlawed *Cannabis* cultivation, he lost his raw material supply.

"Carbohydrate versus hydrocarbon. Guess who won." Hemp, not dope, was the law's real target: pulp-timber industries lobbied for the prohibition because hemp's long, low-lignin fibers are a far cheaper and faster-growing feedstock for paper.

The 2018 US Farm Bill finally legalized hemp again. Outside a lodgepole hoop house where LaDuke was growing it, two women from the Parsons School of Design's Healthy Materials Lab were testing a mixture of her hemp and lime from the namesake white clay of Winona's reservation for making a light, durable building material: hemp-lime. "It's biodegradable and far less energy-intensive than Portland cement."

She had a recipe for making pasta from hemp seed, but local flour mills said the oil gums up their equipment. "Minnesota used to have 11 hemp mills. I want to rebuild local food infrastructure, but we'll need farms producing hemp on the scale of the Hubbard Prairie."

Seven miles to the east, most of the Hubbard Prairie had been converted to giant circles by the nation's biggest potato producer, R. D. Offutt, which had plowed thousands of acres to grow McDonald's french fries, using center-pivot irrigation spray mixed with chemical fertilizer.

"With water stolen from the Shell," hissed Winona. After Enbridge, Offutt was her second-worst bête noire. "Their french-fry processing plants are the area's biggest employers, and they're draining and poisoning the groundwater. We call it 'wiindigoo cannibalism': make everybody dependent on you as you destroy them."

No one here was alive when buffalo filled the Hubbard Prairie, but her grandsons, galloping across snowy potato fields under the full moon, can feel their shaggy ghosts. The last thing Anishinaabeg need to eat are McDonald's french fries, their grandmother tells them. That water belongs to manoomin. Even the Minnesota Department of Natural Resources, the biggest beneficiary of Enbridge's largess, became alarmed about the impact of thirsty potatoes on fishing and put a halt to any further expansion.

Nothing is impossible, Winona LaDuke believes—unless you never try. Line 3 may be a fait accompli, but that doesn't stop a president someday from telling Canada to keep its tar where nature had stored it away from doing harm. A brave Michigan governor, Gretchen Whitmer, had joined local Anishinaabeg in telling Enbridge exactly that regarding aging Line 5, which passes under the most vulnerable point in the entire Great Lakes, the Straits of Mackinac, and which the United Nations had called on the US and Canada to decommission. So there's always hope.

"To bring all that industrial agriculture back to prairie, we'll need to get toxins out of the ground and pull carbon out of the air. The plant that does that the fastest is hemp. So we intend to bring back a hemp economy. The only path to hope," says Winona, "is to hope big."

A red-winged blackbird on a nearby fence post trills a song that survived an asteroid. She picks up a hoe.

"The University of Minnesota's Norman Borlaug founded the first Green Revolution. It's time for ours."

CHAPTER NINE

THE PLAN AND THE SUN

i. Smoke

The oldest brothel in Bangladesh sits atop a muddy dike bordering a drowned polder in the southwestern hamlet of Banishanta, which lies across the Pashur River from Mongla, the country's second-biggest port, 35 miles upstream from the Bay of Bengal. Before shipping waned here in the late 1990s due to aging facilities and silting, a ferry would shuttle sailors all day between Mongla and the rickety docks serving the brothel's line of thatched huts, where hundreds of women charged the equivalent of $3 US per trick. The place was so famous that the only excuse for what the government did in 2022 might be that no official in the capital, Dhaka, realized that anything else was here.

But of course they knew. Every inch of Bangladesh that's above water is occupied or cultivated. With 175 million people in a country the size of Iowa (population 3 million), except for microstates, Bangladesh is the planet's most densely populated nation—more than twice as packed as Europe's most crowded sizable country, the Netherlands.

Besides, one glance at Google Earth reveals the extent of Banishanta's rice paddies. Nevertheless, Bangladesh had recently purchased two Chinese dredgers, and one was parked in front of the village. The idea, the government announced, was to make the Pashur navigable for heavy barge traffic to serve a new economic free zone in Mongla, which would bring jobs to the region. The map their planners allegedly consulted showed only empty expanses behind the town, perfect for dumping river-bottom sand.

Villagers were apoplectic. With so much salinity already pushing upriver from the warming, rising bay, they had to keep switching to more salt-tolerant strains as fast as the Bangladesh Rice Research Institute could breed them. But these new varieties were viable for only a single season, after summer monsoons freshen the watershed. During winter, formerly triple-cropped fields now lay fallow—possibly why planners thought the land was abandoned. Burying them under dredging debris would assure that.

Without rice, how could they live? In October 2022, they planned a protest. They erected a speakers' platform in the town square and invited two of the country's most celebrated activists, Sharif Jamil and Sultana Kamal, cofounders of Bangladesh's foremost environmental nonprofit, BAPA.*

The two arrived together from Dhaka in Jamil's white Toyota SUV. Kamal, an attorney, began defending women's rights even before Bangladesh's 1971 liberation war from Pakistan. Her bespectacled visage is one of the most recognizable faces in the country. Jamil, a burly man in his 50s, his thick hair and beard more salt than pepper, coordinates Bangladesh's branch of the international Waterkeeper Alliance and is the designated riverkeeper of Dhaka's heavily polluted Buriganga. Although Bangladesh's rivers are so crucial they have the same legal status as people, defending them has brought so many death threats that he no longer appears on TV talk shows.

Six months earlier, their drive would have taken eight hours, including a long ferry crossing of the Padma, as the Ganges is called in Bangladesh. But that spring, Chinese contractors had completed a

* Bangladesh Poribesh Andolon.

five-mile-long, double-decked rail-and-highway bridge at one of the river's narrowest points. To withstand its strong currents required 400-foot pilings, the deepest in the world. After the World Bank and other multilateral lenders pulled out of the $4 billion project, alleging corruption, Bangladesh's prime minister vowed they'd finance it themselves. Despite the controversy over this extravagant step to develop the country's isolated southwest, Jamil and Kamal marveled that driving over the Padma cut their journey in half.

In Mongla, they boarded a riverboat to Banishanta, where men in lungis and women in bright saris whisked them into motorized rickshaws. Kamal's own sari was teal blue; Jamil wore a copper-colored panjabi. For kilometers they bounced atop a dike over a trail of hand-laid herringbone brick, along a leafy channel where bearded old men fished with cane poles, until they reached the throng awaiting them at the proposed dumping site.

On both sides of the dike, seas of green paddy rice spread to the horizon. "Does this look empty?" demanded a sinewy woman in an orange sari, brandishing a cell phone. "They want to snake a five-kilometer pipe here to bury this 'wasteland' with sediment. We won't permit it!"

An hour later, they mounted the speakers' platform. Hundreds gathered under a withering afternoon sun. "Pumping dredge spoils here contradicts government policy to save croplands. They say they'll pay compensation, but fields are priceless. Destroying fields means starvation," Kamal told them. "After 50 years of independence, people still aren't consulted. The authorities must reverse their decision!"

"They say they want to attract foreign investments to create employment," boomed Jamil. "But selling farmland to industries violates our laws. Bangladesh already loses 1 percent of its agricultural land every year." A galvanizing Bengali speaker, when he pointed north and yelled, "You know who this dredging is really for!" he had them roaring.

They could see it 10 kilometers away: the 900-foot smokestack of the just completed, not yet operational Rampal Power Station. To accommodate coal ships bringing fuel from Indonesia, Australia, and India, the river must be deepened.

"As long as we can hold off dredging, we delay implementation of Rampal," Jamil reminded them. "Delay is a tactic, because with a billion-dollar plant, every day matters."

He and Kamal had lost so many battles that any win for people and nature, however small or fleeting, was enough to keep them fighting. Originally slated, at over 6 gigawatts, to be the world's second-biggest coal-fired plant, international outrage over Rampal was so intense that several extension plants were canceled, cutting it back to 1.3 gigawatts. That was a victory of sorts, but Rampal's remaining capacity might still doom these villagers. The smokestack's colossal height, intended to keep fumes away, instead would make matters worse. With their groundwater now too saline, people had to harvest rooftop rainwater for drinking. With Rampal's chimney belching sulfur and heavy-metal fly ash directly to the clouds, what rains down would be poison.

If their fields became unusable for rice, there was still shrimp farming, already the fate of so much waterlogged Bangladeshi cropland. But with no potable water, how could they stay?

The Rampal Power Station symbolizes the dilemma in which Bangladesh is ensnared, which in turn reflects a quandary besetting all human civilization. The details are as complicated as powerful men want to make them—and they do, knowing that when print gets too fine or jargon too technical, people quit paying attention.

But the basics are simple enough that children can understand—and they do.

In Bangladesh, 700 rivers, including three of Earth's biggest and their distributaries, meet in the world's largest delta. The rivers are born in the world's highest mountain range, from icefields so huge that glaciologists call the Himalayas the Third Pole. As their ice melts, the rivers flow even harder.

Nearly four-fifths of Bangladesh lies in that delta, within a meter of sea level. Think Netherlands, only three and a half times bigger, with rivers many times the size of the Rhine. Bangladesh has its own Dutch-designed plan, Bangladesh Delta Plan 2100—BDP 2100—to hold back the seas and strike a workable truce with its rivers. But unlike the wealthy Netherlands, it can't afford its delta plan, ticketed at $39 billion by 2030.

Not yet. Bangladesh's grand scheme is that its delta plan will stimulate the economy enough to elevate its status from developing nation to upper-middle-income, so BDP 2100 in effect would pay for itself. If this sounds like perpetual motion or magical thinking, it's not far from the US approving oil pipelines to afford its transition to renewable energy, or the Dutch financing storm-surge gates with Rotterdam's oil and gas revenues.

It's actually the same: to speed development to stave off drowning, Bangladesh needs energy fast. Rather than massive investment in clean power, it goes the quickest, dirtiest route—although a child can grasp that burning fuels to afford protection from what results from burning doesn't add up. Especially since UNESCO's World Heritage Centre and the International Union for the Conservation of Nature had yet another reason to implore Bangladesh not to build Rampal—and why back in 2017 Al Gore cornered Bangladeshi prime minister Sheikh Hasina at Davos and begged her to stop it.

It's on the edge of the Sundarbans.

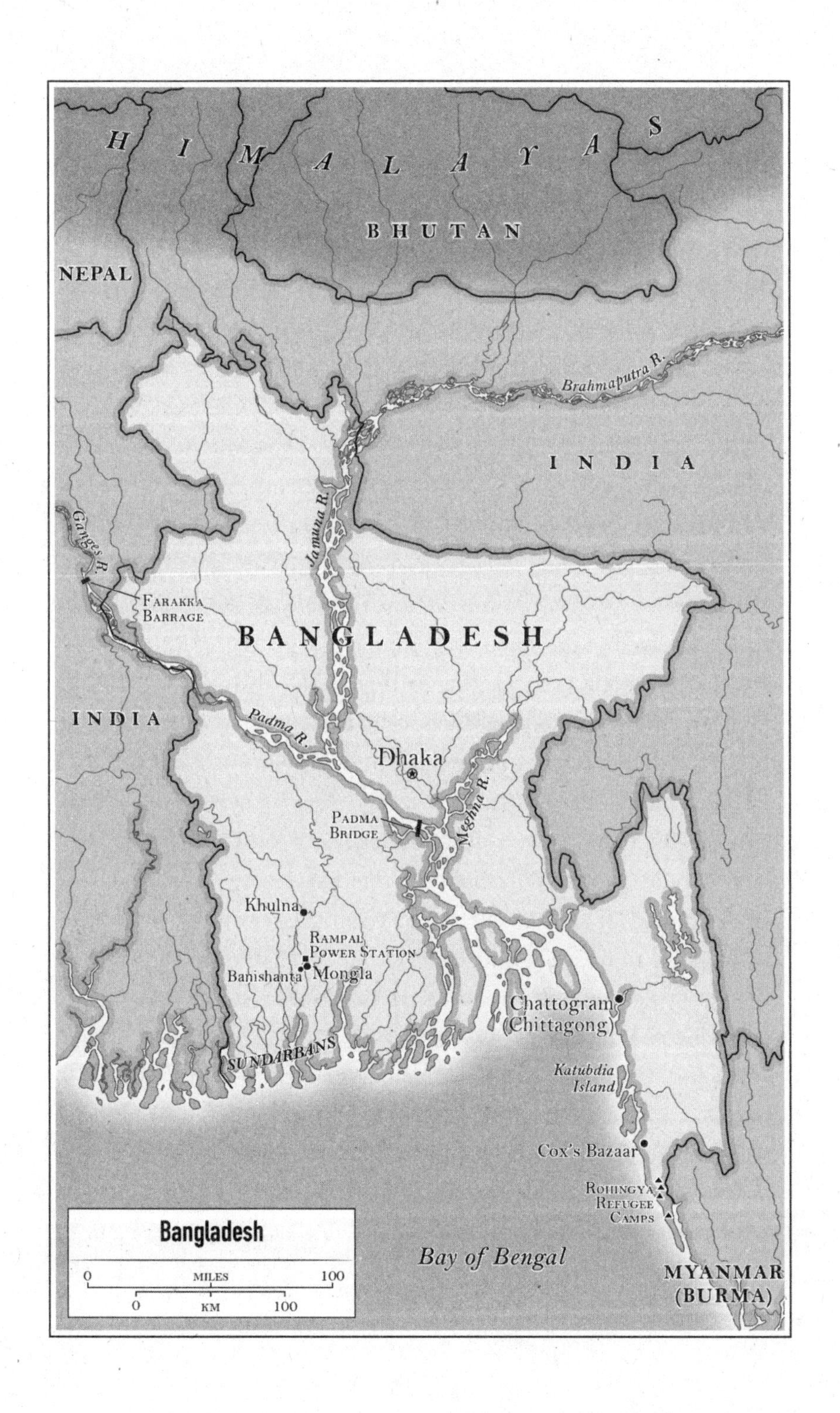
H I M A L A Y A S
BHUTAN
NEPAL
Brahmaputra R.
INDIA
Ganges R.
Jamuna R.
Farakka Barrage
BANGLADESH
INDIA
Padma R.
Dhaka
Meghna R.
Padma Bridge
Khulna
Rampal Power Station
Banishanta
Mongla
Chattogram (Chittagong)
Sundarbans
Katubdia Island
Cox's Bazaar
Rohingya Refugee Camps
Bangladesh
0 MILES 100
0 KM 100
Bay of Bengal
MYANMAR (BURMA)

ii. Mangroves

The Sundarbans, a sumptuously verdant tract three times the size of Rhode Island, is the world's biggest mangrove forest, divided by the India-Bangladesh border but otherwise unbroken, except by so many rivers that some go unnamed. Stretching from Banishanta to the sea, it's also the largest remaining continuous habitat for Bengal tigers.

The night before Mohammed Shah Nawaz Chowdhury leaves for a Sundarbans reconnaissance trip, he meets some of his University of Chittagong graduate students for dinner in a restaurant near his family's company offices. Ever since an ancestor was appointed local administrator by an invading Mogul sultan 600 years ago, Shah's family has lived in the same compound in the southeast Bangladeshi city of Chittagong.* His brother still runs the centuries-old family business, brokering foodstuffs, but an uncle who was minister of fisheries and livestock inspired Shah to study marine biology.

That uncle had once explained that the plentiful oysters in Bangladesh's fertile delta weren't exploited due to a local belief that they were haram, although millions of shellfish-loving Muslims elsewhere cite the Qur'an's assurance that anything from the sea is lawful food.† A nutritional and commercial opportunity awaited, but something else led Shah to do his doctorate on oyster culture at the Netherlands' Wageningen University.

Despite 175 million Bangladeshis being 10 times more numerous than 17.5 million Dutch (Dhaka alone, with 23 million, outnumbers them), in 2006 their two deltaic nations formalized a "twinning

* In 2018, the official spelling changed from Chittagong to Chattogram to more closely reflect Bengali, but Chittagong is still widely used, including by the university and Chittagong Port, Bangladesh's biggest.

† Surah 5:96.

agreement." A bilateral project to share Dutch water technology was proposed, and a Dutch shoreline defense program called Building with Nature caught Shah's attention. To blunt growing North Sea waves, the Netherlands was seeding submerged structures with oysters. Why not do that here, on the world's front line of climate change?

His test site was Kutubdia, an eroding coastal island south of Chittagong, a third of which vanished in a 1991 cyclone. Surrounding waters were too turbid for light-loving, reef-building corals, but clusters of oysters clung to Kutubdia's jetties. Enlisting several students, he sank inexpensive round concrete privy-hole liners 50 meters offshore in a staggered row 100 meters long, and waited. Within three months, free-floating oyster larvae had attached themselves. Within a year, layers of oysters encrusted them.

After six years, when he defended his Ph.D. dissertation, the shoreline directly opposite his thriving oyster reef had grown, adding nearly two meters of trapped sediments. Salt marsh vegetation had rooted, and Shah had planted mangroves. To either side, however, the unprotected shore had receded a meter: proof that wherever waves first hit porous oyster reefs, their erosive power diffused.

So far, the government hasn't shown interest. Rumors were that BDP 2100 would mainly build dams and more concrete river embankments, not natural protections—even though a similar effort urged by Dutch water ambassador Henk Ovink had become New York Harbor's acclaimed Billion Oyster Project, helping to stabilize the shorelines of five boroughs, filtering the water, and locking carbon into oyster shells.

Frustrated, Shah concentrated on teaching. Now in his 40s, slightly built with neat, short dark hair and round glasses, he doesn't look much older than the master's and Ph.D. candidates he meets for dinner. Their research includes how plankton adapt to acidification, how sea worms also build reefs, what increased jellyfish blooms portend, and how DNA barcoding can help assess biodiversity. Shah's hoping to get someone interested in the Sundarbans as a thesis project.

"The Sundarbans buffer storms like a massive reef," he tells them. "They're why millions survived Cyclone Aila in 2009. Now one of the world's biggest coal plants is just 10 kilometers north. Surely we should study its effects."

Everyone nods. "We're killing our chances to exist beautifully on this Earth," says a master's student with a red bindi on her forehead. "The planet always recovers, but it won't be suitable for us to live. We will suffer."

"It's not us," objects a woman in a pink hijab. "America and China have the most carbon emissions. It's hopeless because they won't do much."

"There's been no mitigation," agrees the doctoral candidate next to her, whose orange-trimmed sari ripples in the air-conditioned breeze. Her expertise, ecological modeling and GIS mapping, offers no consolation. "I'm 24. I've been scared since I was 16. Now 10-year-olds are scared."

A thin, bearded taxonomist seated opposite her has been assessing a Bay of Bengal biodiversity hot spot: St. Martin's Island, the southernmost point in Bangladesh. "Forty years ago, they counted 745 marine species. Now we find only 475."

"The IPCC now says 17 percent of our country will sink by 2050," adds a molecular biologist in a vine-embroidered shawl. "Twenty million of us will be climate refugees."

Two decades earlier, when Shah did his own master's here, he believed people could, and would, do something about this. "Do any of you still believe that?" he asks.

"I think people can do anything," the molecular biologist says. "But we have to become qualified, because government clearly isn't. We scientists must convince people who aren't aware."

Platters of fried hilsa shad, prawns, cucumbers, and rice appear. They stop talking.

Flying across his country to the Sundarbans affords Shah a God's-eye view of the GBM: the Ganges-Brahmaputra-Meghna delta. For almost the entire east-west length of the Himalayas, the Ganges and Brahmaputra parallel each other, one on either slope, around 250 miles apart. The Ganges is formed by two streams in northern India that converge 15,000 feet below their glacial headwaters. Dropping through the southern Himalayan foothills, Mother Ganga exits at Rishikesh, where her banks are lined with temples. Reaching the Indo-Gangetic Plain, she's joined by the turgid Yamuna, tainted with the detritus of Delhi, and dozens of tributaries that stream down from Nepal as the river curves east across the subcontinent.

Just before entering Bangladesh, where the Ganges's name changes to Padma, the Farakka Barrage diverts half its water back into India—but what remains is still staggering. In places, the Padma's far shore is beyond the horizon, and it resembles an inland sea.

The Brahmaputra begins on the northern Himalayan slope as Tibet's Tsangpo River, but must continue 500 miles farther east than the Ganges before finding a gap in the mountains. That pass, Tsangpo Gorge, is the world's deepest canyon, nearly 20,000 feet, and one of the longest. Descending in a series of torrents, the Tsangpo enters far eastern India, becomes the Brahmaputra, then backtracks west for nearly 400 miles, bearing all the sediment it scoured from Tsangpo Gorge. As it flattens, that load settles into berms and islets, and the river broadens into a maze of braided channels.

Where it crosses the Bangladeshi border, the Brahmaputra is

nearly 20 miles wide. Now called Jamuna,* it keeps dumping and dividing until it joins the Padma. Sixty kilometers later, they deposit their combined alluvium into the short, broad Meghna, its mouth at the Bay of Bengal lined with chars—silt islands—big enough for agriculture and villages.

All along these mega-rivers, silt-laden distributaries branch and fan into dozens of new channels. The GBM is constantly changing, land rising with each new monsoon's deposition—and then subsiding: the relentless loads of sediment are so heavy that Bangladesh's vast delta is actually sinking.

In Khulna, a port on the Rupsa River, a Padma distributary, Shah joins his traveling companion, Bablu Hajra, one of Bangladesh's first aquaculturists; during his university years, Shah learned to raise tiger shrimp and giant prawns from him. His whitened sparse hair and curly beard now hennaed orange like the Prophet Muhammad's, Hajra is retired from the crustacean business but still owns a trawler that gill-nets hilsa and croakers.

The boat they've chartered, the *M.B. Dolphin*, has a canvas-shaded upper deck and four berths below. Provisioned with fish, fruit, rice, and vegetables, it heads south into heavy barge traffic up from India bearing loads of fly ash for cement factories held under bright blue tarps. Small whitecaps break the choppy water, gray with silt.

"A tropical depression is forming in the Bay," Kader, the bare-chested skipper, tells them. "We're watching for storm surges."

Presently the ghostly spike of Rampal appears, looming over mangroves fringing the Rupsa. Although Rampal's mammoth power plant hasn't fired up yet, coal boats from Indonesia are already lining up beneath two huge shovels that dip their jaws like mechanical theropods, then cough black clouds of soot into the breeze as they unload over chutes that lead from the jetty to four long Quonset coal sheds.

---

* Not to be confused with India's Yamuna, the drain of Delhi.

As the *Dolphin* passes, they can hear from the workers' accents that they are Indian: Rampal is a project of the Bangladesh India Friendship Power Company. The night before, Shah had visited the cramped, bookshelf-stuffed Khulna office of CLEAN, the Coastal Livelihood and Environmental Action Network.

"Bangladesh didn't need any coal-fired plant," its skull-shaved director, Hasan Mehedi, told him. "Our government built Rampal to keep its big friend happy."

India is planning a rail line from Kolkata to Mongla, he explained, to develop a special economic zone like the Japanese have in Chittagong and the Chinese have near the Padma Bridge. "They want an uninterrupted power supply."

Bangladesh was already locked into buying energy from one of Indian prime minister Narendra Modi's closest friends, Gautam Adani, India's richest man and the world's biggest coal baron, who touted his 60-year plan to electrify 100 million subcontinent families with coal from his giant Carmichael mine in Australia. Rampal, which will burn 12,000 tons of coal a day, was sited next to the Sundarbans instead of 60 kilometers north, as environmental groups begged, simply because it's that much closer to India, saving shipping costs.

"They're planning a harbor in the Sundarbans," Mehedi had said. "Imagine hundreds of coal ships and barges navigating the forest's rivers." Just in the past decade, eight ships carrying coal, furnace oil, and fertilizer have sunk in Sundarbans channels.

"This will soon get very dirty," remarks Shah, craning his neck to see the top of the chimney.

"But people want energy," says Bablu. "How could solar run all these industries?" he says, waving at the factories lining the shore.

Although the Bay of Bengal is 50 miles away, they've timed their departure to run with its outgoing tide, which is far stronger than the current. At Mongla, the Rupsa's name changes to Pashur, its banks piled with sandbags to hold back the rising river. They pass fertilizer plants, a blackened coal-fired generator, and new spherical tanks of LNG: liquefied natural gas, the latest fossil fuel driving the Bangladeshi economy's race against the thermometer. Just beyond are the sagging stilt huts of the hundred or so Banishanta prostitutes left after the pandemic, most Hindu-minority girls. The huts' foundations are dissolving; two docks are halfway collapsed.

Where the brothel ends, a broad stream enters from the west. Across it begins the impenetrable Sundarbans—although civilization still claws the east bank for six more miles, ending at a tall grain elevator where Indian ships buy Bangladeshi rice.

Past it, they turn into a side channel and immediately are walled on each side by towering vegetation. Most mangrove forests are low assemblages of two or three saline-resilient species growing along shorelines. Here, the tallest of the Sundarbans' 27 mangrove species, its namesake *sundari*, stands 80 feet high.

The last time Shah came here, 2009, was just after Aila flattened the Sundarbans, but mangroves evolved with cyclones and little suggests that a devastating storm passed through in this century. "See the roots holding the sediments intact?" he says to Bablu, pointing at the thick living matrix. "Mangroves are even better engineers than oyster reefs."

Along the riverbanks they see small spotted chital deer; 6-foot monitor lizards; tall white herons with yellow gorgets, fishing for scarlet fiddler crabs; red-rumped rhesus macaques; otters; boars; and all sizes of *Crocodilia*: alligators, crocodiles, caimans, and needle-nosed gharials. Kingfishers with resplendent red beaks zoom across their bow; white-headed brahminy kites wheel overhead. They wonder if native Burmese pythons and royal Bengal tigers, both strong swimmers, ever catch the Irrawaddy dolphins accompanying their boat.

For three days, they'll see no humans except for each other, the crew, and occasional fishermen in sampans who sell them mullet, catfish, carp, and gobis. The thick heat buzzes; honey collecting is a valuable enterprise here (for those who dare, because tigers have learned that beehives attract humans, and sometimes lurk nearby).

In a shallow-bottomed punt they float into placid streams where birdsong bubbles up and down the scale. Black mangrove pods bob on the surface; pink and white bellflowers hang above the waterline, bombarded by scarlet-backed flowerpeckers and yellow swarms from beehives the size of bushel baskets. The air smells tangy and deliciously clean. Only an occasional plastic bag snagged by a root reminds them that the world they've left behind is not this perfect.

At twilight, the wind calms. Mangroves being so well anchored, no floating logs prevent navigation by night. Their plan is to awaken at the mouth to the sea. Over dinner platters of dal, mullet fried with cauliflower, and Malabar spinach berries cooked with tiny shrimp, Bablu Hajra keeps shaking his orange head.

"I came to remember how important this is," says Bablu, who's a district assemblyman with the opposition party. "We tried to tell the government not to put the power plant so close. If they can risk destroying nature, they don't love our country enough. But to stay in power these days, you must indulge your influential neighbors."

Or influential Europeans. "The government held Delta Plan workshops with the Dutch engineers in Chittagong," Shah tells him, "but invited no one from the university. Our institute has more than 50 years' experience, but none of us is involved with the Delta Plan."

He scoops out some sweet white pulp from a dragon fruit. "I studied in the Netherlands; I know good Dutch engineers who worked on BDP 2100. But this isn't just an engineering problem. As a scientist, I have to listen to locals. To create a wise plan, we need to feel what people on the bottom want."

The next morning they land on a beach three kilometers above the Bay of Bengal and continue on foot. Asad, the forest guard they were required to hire, accompanies them, carrying a Chinese bolt-action .22-caliber rifle that he has yet to fire at a tiger. Glasswing butterflies, transparent except for orange-trimmed hind wings, flutter at eye level. They flush a dozen boar large as tapirs, a skittering band of rhesus monkeys, and a flock of orange-billed lapwings. They can hear waves crashing. In the dark understory of leathery ferns amid the mangrove tangle, a flash of red turns out to be a thong sandal. Abruptly the clay forest floor is strewn with broken Styrofoam, gill-net floats, plastic bottles, and a rubbery rainbow of flip-flops, heaved up by spring tides and storm surges. For the next half hour, they walk through this testament of plastic's permanence and the sea's growing reach.

At the Bay of Bengal's shore, they find a violent tumble of thick bleached downed mangrove trunks. Beyond the crashing breakers, a few dead trees still stand under a roiling sky. As if being the Himalayas' drain weren't enough, Bangladesh's entire coast is a funnel for cyclones. But Asad, shouting above the surf, says that most of these trees came down in the last monsoon. "There were unusual winds. And each year the sea comes closer and takes more trees."

The moist clay shore is patterned with radiating spirals, perfectly machined by sand bubbler crabs. They wade in surf warm as a baby's bath. In the mist, no fishing boats are visible, but these waters are rich with penaeid shrimp that breed in open ocean, then move up Sundarbans rivers in their juvenile stage. Commercial fish like sea bass do the opposite: spawn in the Sundarbans, then return to sea.

Shah jots in a notepad, "We need a doctoral study to analyze whether changes in fish harvests are due to climate or to other disruptions to the Sundarbans."

For lunch they pass a bag of jujube fruit. Bablu elects to return to the boat, while Shah, three crew members, and their armed guard continue inland through tall keora mangroves toward a place called Tiger Spot. Just past spring tide and monsoon season, the waterlogged forest floor beneath the canopy is nearly quicksand. Several sandals have to be rescued, and slogging barefoot doesn't decrease the gray muck's suction. They hope Asad is a good shot, because running from a tiger would be ludicrous. Except for an occasional piercing drongo's whistle, it's ominously quiet.

It's also cooler and darker. Dimly they see hundreds of dainty spotted deer ahead, bounding effortlessly and somehow soundlessly: four-point bucks, juveniles with velvet antlers, does, and fawns, all frequently pausing to browse on dark keora fruit called mangrove apples. Their feces stain the ground purple.

The pace is so slow, it's already 4:30 p.m. when the skipper announces, "We're here."

Nothing looks very different, except they're in a small clearing, maybe 10 meters across. In the middle is a scattering of orange ceramic pottery fragments. Captain Kader says that the land is almost imperceptibly higher here, maybe a half meter, an artifact from when poachers would heap earth to camp above sudden floods. When they weren't hunting tigers—or being hunted by tigers; until recently, at least 20 humans annually here became cat food—they evaporated seawater in ceramic pots to make salt, gathered honey, or harvested golpata palmettos for thatching. But tiger pelts paid the most, and in 2013 the Bangladesh Forest Department finally got serious, forcing out the poachers and licensing golpata and honey foragers. The Sundarbans tiger population, which had dropped to 96, stopped declining. The last three-year camera-trap survey, Kader heard, found 114, but the numbers haven't risen as much as hoped.

Shah notices something else: the trees. "All the keoras are the same size. We aren't seeing new generations."

There are two other Tiger Spots ahead—so-called because tigers seek higher ground when giving birth—but given the sodden conditions and the hour, they trudge back to the boat, managing not to get eaten. Over a dinner of cabbage pakora, sea bass, and papaya yogurt, they begin motoring north along a sinuous moonlit route, via nameless channels wider than most European rivers.

Another day spent floating through hushed beauty. On their final morning, they stop at the Karamjal Wildlife Breeding Center at the northern end of the Sundarbans. There are mangrove exhibits and a pipe corral where children feed greens to red spotted chital deer. Charmed, Bablu Hajra joins them, while Shah talks to the Forest Department's officer-in-charge, Howlader Azad Kabir. A soft-spoken man in his 50s, he says the ecosystem they've just seen is deeply stressed.

"People aren't abusing the Sundarbans anymore, but salinity intrusion has advanced. We're losing plant diversity because there is less fresh water coming. India takes so much with their Farakka Barrage. Mangroves need a mix of fresh and brackish water, so they're stunted and not reproducing. All our scavengers, from porcupine, boar, and deer down to nematodes, have less to graze."

"And tigers?"

Kabir stoops and runs a finger through the mud. "Unflooded areas are fewer, meaning fewer breeding opportunities. Crocodiles are even more vulnerable. They must lay eggs on dry land. Their population is way down."

Because crocodiles don't migrate and are territorial, his staff can count them at winter drinking sources. "We think there are only 250. There used to be more than 1,000. Pythons, same problem. And monitor lizards. They all need higher ground. That's why we've started a breeding program."

He shows them a cage of new baby alligators. There are pools divided by chain-link fencing for different ages of crocodiles, to stagger their releases into the wild, mimicking reproduction seasons. "Their last month here, we acclimate them to a more natural setting, but it's unclear how they do when released in the wild."

Also worrisome is heat. "Crocodile gender is temperature-sensitive. Ours evolved to reproduce at around 32°C.* Below 30°C,† you get only females. Above 34°C,‡ more males."

Even tiger gender seems to be affected. Normally, male tigers have three mates, but the ratio is now 1-to-5. Every morning, Kabir finds tiger prints around the water hole on the wildlife center's nature walk, meaning they're seeking sustenance closer to where humans live and work. He keeps asking the government for money to build more tiger spots but gets no response.

Leaning on a railing, he regards a river terrapin sunning by a pond. "I've sacrificed living with my family to protect this forest. I'm willing to do it. But we need help. Sundarbans isn't just Bangladesh's. It belongs to the world."

As they leave the mangroves behind, white smoke is billowing upriver, although Rampal shouldn't be operating yet. The smell resembles cooking. At the far end of Banishanta, beyond the busy Chinese dredger, they pass a Hindu funeral, the flaming corpse laid on a pyre of mangrove trunks.

iii. The Plan

Bangladesh's humans also need higher ground. The Bangladesh Delta Plan 2100 can't manufacture mountains like plate tectonics

* 32°C = 89.6°F.
† 30°C = 86°F.
‡ 34°C = 93.2°F.

keep doing in the Himalayas, but the bilateral Dutch-Bangladeshi agreement intends to do the next best thing: make people safe where they live. For two weeks in October 2022, BDP director Jaap de Heer has met with the government in Dhaka, working through the evenings over sandwiches and beer in the expat Dutch Club's garden café, among the few places to drink legally in this Muslim-majority country. Jaap, a civil engineer with TwynstraGudde, the Dutch consultancy guiding BDP 2100, has been coming to Bangladesh since 2001. This is his second time back since COVID, but he hasn't gotten past the capital.

Finally, one Saturday he escapes with a young colleague, Joost Meijer, who will assess existing infrastructure to see what progress there has been made thus far, which isn't much. Jaap has their driver head to the new Padma River bridge. "I want to see how the Chinese protected the banks."

It's only 25 miles, but the drive through strangled central Dhaka takes two hours. It's Joost's first time here. He's awed, first by the sheer mass of humans, each hawking something, and then by a river five miles wide. It's 11:00 a.m. when they park near the bridge's north bank. The few cloud wisps do nothing to soften the sun's punch. Jaap, in jeans, shades his pale cheeks with a canvas hat, a Nikon dangling on his belly. Joost is in a blue ball cap, sunglasses, and a moisture-wicking T-shirt. For miles in each direction, the riverbank is now hardened with rectangular tan paving stones that form a flat bench 50 feet across, then slope to the water, topped with staggered rows of raised blocks—"to break the wave energy," says Jaap approvingly.

Vast amounts of CO_2-intensive cement went into paving this apron on either side of the bridge, on both banks. "That's the negative part," says Jaap. "The positive is protection against flooding, and guiding the river in the right direction."

They are 30 miles downstream from where the Brahmaputra-Jamuna and Ganges-Padma join. "The tectonic plate below the

Brahmaputra tilts a bit, so it drifts westward. Rivers here wander away from bridges. That's why they had to stabilize it."

The paving blocks, numbering in the millions, were all laid by hand, possibly by the same people they see drying dung for fuel on the hot new pavement. "They must have worked right through the pandemic," Joost remarks.

China both built and financed this double-decked, road-and-rail bridge as part of its Belt and Road Initiative to link East Asia and Europe—the modern echo of its ancient Silk Road. Another leg runs from Kunming, an industrial city in southwestern China, through Myanmar to Bangladesh's Chittagong Port: thousands of miles shorter than shipping goods to Europe through Shanghai or Hong Kong. Upon learning that Chittagong's factory-choked port had no more space, China simply offered to build and manage another one—for free.

For Jaap, it's frustrating to watch China spend things into reality blindingly fast, while BDP 2100 still awaits the $38 billion loan the World Bank promised in 2018.

"Four years after the plan was approved, the Bank's still discussing these things. They could easily allocate billions. But they're not doing it." Even a project the World Bank advanced—dredging two main channels into the braided Brahmaputra for an economic zone on its north shore—remains on hold.

It's more frustrating for Bangladesh. Along with safety from flood and climate disasters and instituting water governance like the Dutch, Bangladesh has its three "Higher Goals":

> Eliminate extreme poverty,
> Become upper-middle-income by 2030,
> And by 2041, become prosperous.

Maybe that seemed achievable in 2018, but now 2030 is nearing. The stalling, Jaap suspects, is because BDP 2100 is a climate adaptation strategy, and the World Bank's president won't admit that fossil fuels alter the climate.*

"The Delta Plan is Bangladesh's only hope," says Jaap, returning to the car. "If nothing is done, there will be huge numbers of climate refugees. The prime minister thinks 35 million people could be impacted. That figure seems low."

But the World Bank hadn't even funded feasibility studies for BDP 2100's 65 separate engineering projects—less than $100 million total, pocket change to the Bank. To demonstrate its earnestness, Bangladesh volunteered to finance two Plan projects with its own money, at $60 million apiece, including a cross dam to link two char islands with the mainland, to trap more river silt and speed sedimentation.

"The cross dam uses estuary dynamics: after a storm, suddenly there's a new sandbank. Bangladeshis call themselves a land-hungry nation. You coproduce land with nature," says Jaap. A BDP 2100 map of similar land reclamation scenarios shows the future coastline growing dramatically toward the south. But to avoid the UN's warning that 17 percent of the country could drown by midcentury, he cautions, "Sediment must not be sold to Indonesia"—a growing illegal industry.

The two-volume *Bangladesh Delta Plan 2100*—strategy and financing, 800 and 900 pages, respectively, plus a 3,000-page appendix of 26 baseline studies—is the result of Jaap traveling for years from region to region, sometimes in a convoy of police cars, to meet with 2,500 businessmen, farmers, academics, and politicians in workshops the government organized, listening for ideas.

"Also the environmentalists. They were quite critical."

Those critiques were shared by the UN's chief of development research, Bangladeshi economist Nazrul Islam, who objected to chan-

* Widely accused of climate denial, World Bank president David Malpass resigned in June 2023, a year early.

neling rivers rather than letting them replenish soil through seasonal flooding, as nature intended. People farmed like that for millennia until American and Dutch advisors appeared in the 1960s, when Bangladesh was still East Pakistan, with schemes for concrete embankments and floodwalls, dikes and polders. Dikes did bring flood protection—and, as they connected, the novelty of roads. But instead of nourishing fields in the floodplain, silt stayed in the rivers, where it now has to be dredged.

"Bangladeshi farmers like living with some flooding," acknowledges Jaap, as they board the scorching SUV. "The Nile effect brings nutrients. You don't need to buy much fertilizer."

He knows environmentalists hate polders. BDP 2100, he says, is designed to be adjusted every five years. "We could open polders to flooding for a couple of years, let them silt up so they rise with the sea level, then close them again. But they talk about rivers like they're human beings: 'The river doesn't allow it, the river doesn't like it.' They want to live with the river, which is understandable. But they also want dry feet. And the government wants dry land. More and more dry land."

They drive north to a BDP 2100 risk management project funded by a $15 million grant from the Netherlands and a $250 million Asian Development Bank loan. Mortared pavers, made locally from crushed brick, slope to the river from an adjacent brick-paved earthen dike along the Padma. Each is initialed and dated. "These people also worked during the pandemic," Joost realizes.

For an hour they walk upstream, admiring the labor. The last time Jaap was here, portions of this bank were 20 meters farther out. "We call this curve an attack zone." He indicates where the combined force of the Padma and Brahmaputra has suddenly carved away chunks of riverbank without warning. Stooping to inspect the masonry, he sees a few pavers are spaced too widely, inviting plant roots to undermine them, but he's satisfied that the river has been tamed here—as, apparently, are people whose tin houses crowd the dike.

They come upon two wooden barges. A dozen men balancing bushel baskets of sand on their heads descend gangplanks to dump their loads where more bricks will be made, then return for more. Just beyond them, the pavement ends near a slump-block shed. Vines with heart-shaped leaves cover the bank, their exposed roots dangling. "It will continue washing away," says Jaap. "That building will collapse in a couple of years."

So they'll keep paving, even though not letting the river attack here means that its energy simply gets pushed elsewhere—another reason why this will be a perpetual Plan, because fixes in one place cause changes in another—and as glaciers melt and storms intensify, the rivers will only become wilder.

They meet their driver at a tin-covered market, where wicker baskets display groundnuts, cucumbers, plantains, ginger, shallots, papayas, chilies, red lentils, squash, and several potato varieties.

"When I started coming here," says Jaap, "it was mainly rice."

There are also racks of packaged snack foods, a sign that people have money to spend. Poor as they may be, Jaap notes, they're in clothes, not rags, and no one looks malnourished. "I was expecting much worse poverty," says Joost. "There's quite a lot of entrepreneurial energy."

Jaap nods. "Becoming a prosperous country by 2041 made me question the Plan's reality, because none of those figures seemed believable. But there is spirit here. Bangladesh is now self-supporting in food and exports high-value crops like flowers. Their garment sector is bigger than China's and better quality—Nike and other world brands, made in Bangladesh. There's shipbuilding, not just ship dismantling."

Energy, but such uncertainty. The Plan has multiple future scenarios: low, moderate, or high versions of three unknowables—climate change, economic growth, and, increasingly unnerving, hijacked water.

India now hints at deviating upstream Brahmaputra water into the Ganges. "Meaning they'll keep it. If that happens, the Delta Plan must recalculate what Bangladesh gets in the dry season." They'd already been assuming drier winters, as China keeps taking more Tsangpo-Brahmaputra water on the north Himalayan side for hydropower.

"What can you do? Bangladesh is a downstream country. These are mighty powers."

All of that impacts one Delta Plan project that Jaap considers among the most critical, but also one of the most controversial: a proposed second barrage on the Ganges. Farmers are wary, knowing how the first one, Farakka, hurt pre-monsoon harvests. But this one, which would divert Ganges-Padma water into a river that after several name changes becomes the Pashur, would affect much more than agriculture.

Joost hasn't seen the Sundarbans yet, but he knows that it's more than tigers and tourists. Without the biggest mangrove forest in the world to absorb the fury of ever-more-dangerous cyclones, much of coastal Bangladesh would be as defenseless as Florida.

With less fresh water coming down the Pashur in this low monsoon year, the Sundarbans are dangerously stressed, explains Jaap as they sit in deranged Dhaka traffic, amid surging rivers of pedestrians and pedal rickshaws.

"If this barrage isn't constructed in time, the Sundarbans will be gone. With land each year subsiding five millimeters and sea level rising maybe a centimeter per year, the mangroves are sinking. With another Ganges barrage, you can increase the fresh water. We'll need to adjust the embankments and keep the Pashur dredged. Because if we don't get extra fresh water to the mangroves, they are going to die. So we must."

Bangladesh's Delta Plan has become Jaap de Heer's life's mission. "But time is short. If nothing is done, large portions of the land will be underwater and climate refugees will . . ." He doesn't want to think of it: tens of millions of refugees, spilling out of Bangladesh—where? Who will be able to take them, let alone want them?

iv. Sunlight

The world's first climate refugee village was built across the channel from Shah Chowdhury's Kutubdia Island oyster reef, the year after nearly 30,000 people there died in a 1991 cyclone, among Bangladesh's deadliest ever. Thousands of survivors fled to the mainland, where they camped on the shore for a year until the government agreed to erect five-story concrete dwellings on an abandoned shrimp farm.

The blocky, echoey apartment complex, reminiscent of Soviet-era architecture, couldn't be more dissimilar from the thatched-bamboo homes the residents left behind. Yet consoled by reinforced concrete and free rent, they've grown to accept them—except when electricity fails, fans stop turning, and they suffocate. No one contemplates returning to Kutubdia, where 30-meter sections of coastline sometimes disappear overnight, and where they precariously fished, made salt, and raised shrimp. Here there's plenty of construction work on the 119 additional buildings surrounding the original 20 that Bangladesh's government is readying for the next disaster.

Twenty-five miles to the south, the latest calamity isn't climate-driven, but it's emblematic of other pressures squeezing 21st-century human civilization—sectarian cleansing: people of different persuasions murdering each other over limited space and resources. In August 2017, when their Buddhist nationalist persecutors turned genocidal, most of Myanmar's Muslim Rohingyas fled across the Naf River into Bangladesh. For a year, the refugees camped in a miserable, exposed enclave on the world's longest beach, Cox's Bazar, until 5,000 acres were requisitioned for them in Bangladesh's farthest southeastern corner, against the Myanmar border.

The land, a hilly forest preserve, was home to elephants, deer, boars, and monkeys. Bamboo towers throughout the refugee compounds, population now 1.2 million humans, are manned by the Rohingya themselves to watch for elephants intent on reclaiming their territory. They live in one-room bamboo shacks wrapped with thatch and black plastic, on concrete slabs poured by UNHCR, the United Nations High Commissioner for Refugees. There's no plumbing—twice a day, they wait in long queues at spigots. No electricity, either—until recently, in a sector called Camp 26.

Like the others, Camp 26 is surrounded by prison fencing: barbed wire above, razor wire below. Each camp contains nine blocks of 1,200 huts crowded along dirt paths, averaging eight occupants. Unlike other camps, however, Camp 26 has a small cell-phone tower behind a blue concrete minaret. In one cluster of 50 households, from every fifth shack emerges a bamboo pole with a small wireless relay and an antenna that doubles as a lightning rod (a year earlier, a thunderbolt ignited an entire neighboring camp, leaving 100,000 homeless).

Also, atop every house is a compact 150-watt solar panel.

"How's everything working?" asks Abu Hasan, ducking to enter the hut where Abdul Haque lives with his wife, Noor, his son and daughter-in-law, and four grandchildren, two of them born here.

Haque, 75, sits in blue pajama pants on a red cushion, the only furniture, his cloud of chin-beard obscuring the logo on his stained

UNHCR T-shirt. He points to the rotating 16-inch desktop fan on the floor next to the two youngest kids. "This helps so much with that," he says, indicating the black plastic roofing stretched over bamboo lattice, which keeps them dry during monsoons but absorbs stifling heat.

Hasan, in his early 30s, his own beard just a hint on his nut-brown face, has worked in the solar industry for a decade, but now he's a field engineer with SOLshare, a Dhaka company with an entirely different approach. Via a peer-to-peer WiFi microgrid, it connects surrounding households, each provided with a solar panel, two 14-inch fluorescent lights, a fan, a battery, and the key component: a lunch-box-sized SOLbox. Instead of bewildering users with kilowatt-hours, its digital display shows, in Bangladeshi takas, how much the electricity remaining in the battery is worth.

When the sun shines, everyone's batteries are full, but as they discharge at night, anyone needing more can press a buy button to purchase extra juice from the neighbors. Or, if they've already charged their phones, the kids have finished their tablet homework, and there's still electricity remaining, there's a sell button to offer it. There's also an auto mode, in which a battery automatically starts selling once it's full, so the excess solar energy isn't lost; when its charge goes below a certain threshold, it stops selling. Households here can earn extra money this way, to augment the monthly stipend of 100 takas (about 85 cents US) that UNICEF provides.

Each microgrid of linked households is then linked to the next microgrid—in a month, SOLshare has connected 50 households into a wireless, internet-enabled local network with a shared common pool of solar energy.

"Are you using your SOLbox?" asks Hasan, whose orange SOLshare T-shirt is printed with a white solar grid.

Haque points to the cord switches for the lights and the fan. "I use those. My grandchildren understand the box."

Hasan laughs, but Haque is expressionless. He came with a hundred others, piled into two boats to cross the Naf into Bangladesh. One sank, and 23 died. When Myanmar was still Burma, his family had farmed rice, peanuts, and peppers on the same land for 400 years.

"I thought we were coming for a few weeks. That was five years ago. I would go home in a minute if they wouldn't kill us," he tells Hasan. "I want to die there."

Now they only have this. But on the wall opposite him, attached to a bamboo strut, glows a fluorescent tube. "Now we don't have to buy oil and breathe lamp smoke. Thank you," he murmurs.

Hasan moves to the next house, where Karimula and Moustuv Hatun live alone; they don't know if any of their two sons and five daughters are still alive. Their neighbor Sabuda is visiting. Everyone knows how to turn on the fans and lights—they don't bother with the box. "But without the fan," says Karimula, "it was horrible in here before." It's still warm—he is naked to the waist, although his wife is swaddled in a green-and-black burqa.

"Allah should help us," spits Sabuda, her gaunt face glowering beneath a red-orange shawl. She is deranged, Karimula whispers, ever since soldiers murdered her six children. She came alone.

"Everything works," says Karimula. "Thank you."

There are utility poles in the Rohingya camps, but only to power UNHCR posts and the security lights that fail to deter drug smugglers from across the river. The current regularly drops. "But our solar system microgrid is always on, even when national grid is load-shedding," says Hasan.

In Dhaka, where SOLshare is headquartered, despite an ordinance requiring all new structures to have solar panels, load-shedding—planned interruptions—is nearly continuous, to the supreme frustration of SOLshare CEO Sebastian Groh.

The SOLshare office is airy, with tables and group work benches instead of cubicles, open-frame dividers instead of walls, vines curling around white-painted studs, and throw-pillow meeting nooks. Outside, however, the windowpane is coated with the murk of Dhaka, which regularly alternates with New Delhi as the most polluted city on Earth.

As blackouts roll across town, the city drones with auxiliary diesel generators that foul the air even more. It drives Groh, a boyish German in his mid-30s with a dark scraggle of beard, crazy, because all the technology is in place to fix it.

"Go on the roof of any apartment building," he says, padding around a green carpeted lounge area in an untucked SOLshare T-shirt and stocking feet, holding a mug of tea. "There's a solar panel, because that's the law for the last 10 years. But none were installed to actually function. Some are even upside down. They're only there because the building code says if you don't have a certain percentage of renewable energy, you don't get a gas connection. And without gas, you can't cook. But they aren't connected to the building's main circuit. Maybe they power a security light. That's it."

It's a maddening glitch in a country that has a legitimate claim to being the world's pioneer in mass-scale, off-grid solar electricity installation, which Groh discovered as a student in 2007 at Mannheim University while bent on a career in investment banking. He'd attend events that firms like Goldman Sachs held there, fishing for young talent, but one day the Club of Rome offered something rather different: a seminar called "Doing Business with the Poor."

One presentation was by a research group from Berlin, MicroEnergy Systems, whose director had visited Bangladesh in 2001 to see the work of Grameen Shakti, a social enterprises spin-off of the nonprofit Grameen Bank, founded by University of Chittagong professor Muhammad Yunus, whose creation of microlending won him the Nobel Peace Prize.

In 1996, when Grameen Shakti began with initial funding from the Rockefeller Foundation, two-thirds of Bangladeshis lacked electricity. Over the next 17 years, it installed so many home solar systems in rural areas with no grid, offering microloans payable by the taka equivalent of $5 per month, that in 2013, of 4 million household solar systems worldwide, 3 million were in Bangladesh.

Sebastian Groh interned with MicroEnergy Systems to learn about combining microfinance with decentralized energy technologies. The Bangladesh model became the basis of his Ph.D. thesis—in a single month in 2013, 85,000 homes were solarized. That year, Groh won an Ignite fellowship to Stanford's Graduate School of Business, a program to help budding entrepreneurs launch innovative concepts. By now, he had no interest in being an entrepreneur, but he had an original idea. This was the early days of the Uber sharing economy, and it occurred to him how to share solar electricity.

His project—to connect the households that Grameen Shakti had solarized via WiFi into microgrids so that they could pool and trade their energy—was chosen for developing into a business plan. In late August 2013, he and five Stanford classmates presented it to a Silicon Valley mock jury, which called it fantastic. Thus SOLshare was born. "What's Bangladesh like?" a juror asked him.

"I've never been there," he admitted, which earned him a zero on authenticity.

"Go, Sebastian," the head juror said. "Then come back and talk to us again."

He arrived just after the national elections of January 5, 2014. His timing couldn't have been worse. The incumbent prime minister, Sheikh Hasina, won in a landslide after jailing her opponent and several more opposition candidates. Every party except her own boycotted the voting.* The following weeks brought crackdowns, police killings, and strikes. At one point, Dhaka was sealed off. Some who risked traveling had Molotov cocktails rolled under their cars.

The national paralysis halted photovoltaic installations and precipitated a crash in the solar market that was already looming. For more than a decade, Grameen Shakti's program had been a gigantic success, largely because all development donations in Bangladesh went through a government agency whose director insisted on high-quality materials. Unlike programs for the poor that typically distribute cheap goods that don't last, a battery with a solid five-year warranty guaranteed that borrowers with affordable three-year loans would likely repay them.

But after the elections, the agency was corrupted by politicians who leveraged its monopoly on dispensing donations for their personal gain, canceling debts in exchange for votes or offering loans to people unlikely to pay them back. When the program collapsed, Grameen Shakti's large contingent of installers had no work—which was when Groh went to them with his idea.

When he could finally travel to villages with solar systems, he'd noticed that by afternoon, every home's batteries were fully charged. Up to 30 percent of that power would be wasted, he calculated. Suppose those batteries instead could offer the excess for sale to anyone who needed it? Normally, people might be reluctant to sell energy—unless they were reassured they'd never run out, because if an entire

* A decade later, Hasina, daughter of Bangladesh's founding president, was still in power. Her ongoing vendettas against rivals had included charges of tax evasion and labor violations against Muhammad Yunus, whom she'd repeatedly attacked ever since he won the Nobel Prize. But in mid-2024, after hundreds died when she ordered lethal force to quell peaceful student protests over government job quotas, a popular uprising finally overthrew her, and Muhammad Yunus became interim head of government.

neighborhood pooled its energy, chances were excellent that someone's extra would always be there to power someone's need.

In 2017, the United Nations awarded Grameen Shakti, in partnership with SOLshare, $1 million to implement a two-year rural electrification project by creating the world's first solar network linked wirelessly by smart peer-to-peer sharing. Financed also through microloans to poor families, it brought electricity to thousands of households, connecting them from the bottom up, no transmission lines required.

Some European investors doubted that the SOLbox was a viable product, because they didn't think anyone would ever press a button.

"They press the button all the time," a SOLshare engineer replied. Even users not ready to pay $5 per month for solar panels could be served; with an inexpensive SOLbox, they could purchase surplus energy from their neighbors.

By 2019, they had built 116 microgrids. Twelve percent of Bangladesh now had photovoltaic energy. SOLshare could have kept connecting existing stand-alone solarized households into peer-to-peer microgrids. It also could have kept solarizing more remote rural villages.

Except everything changed.

"I hadn't counted on Bangladesh being flooded with money from China, India, and the US," says Groh. "But this is a very strategic location." China had its Belt and Road Initiative, India was flexing its regional muscle, and the US was worried about its world dominance slipping away.

"Eat," says SOLshare communications director Salma Islam, passing him take-out chicken biryani.

SOLshare's conference room doubles as a staff lunchroom. "They took that money and extended the national grid, against any economic acumen. Little islands on rivers got grid-connected. The cost per household was insane, but they did it, and at a subsidized rate—

the government is losing nearly $1.5 billion per year. That destroyed our business model where people trade electricity among each other, because the price can't compete with the subsidized grid."

At Stanford, he'd learned that entrepreneurs must be nimble when winds shift. One place Bangladesh had no interest in extending its grid was the camps housing a million Rohingya refugees they wished would go home. When UNHCR put out a tender for solar panels, SOLshare's peer-to-peer network was the winner.

It also seized the opportunity when the government finally passed net metering, meaning an industrial building's rooftop solar panels can connect to the national grid via a meter similar to a SOLbox on auto mode. Suddenly, consumers with photovoltaic systems became *prosumers,* selling their excess sunshine.

"So when the garment factories where we've solarized have more energy than they need," says Groh, scooping more rice, "their meters run backwards. Their $100,000 monthly utility bill dropped to $30,000. Unfortunately, so far only commercial is eligible. No residential."

The government capped grid-tied prosumption at 70 percent—supposedly to phase in net-metering gradually. The real reason, everyone knew, was pressure from centralized utilities that feared being replaced by efficient systems that mine free energy from the sun. Those utilities were already in a panic. In its rush to afford the Delta Plan, Bangladesh had authorized 20 new coal-fired plants, but had to cancel half because fewer lenders were willing to finance coal. With war in Ukraine, fuel costs for its new liquefied natural gas plants had soared. LNG container ships from Qatar were turning around in the Bay of Bengal, eating the penalty for defaulting on Bangladesh contracts because they could make much more in Europe.

Rolling blackouts became constant. "My electricity went out five times yesterday," says Salma, re-pinning her lavender hijab. "Each time it was gone for an hour." The previous weekend, the entire nation plunged into a three-day blackout. "I have a generator, so I can get through that. But others . . ." Each new outage brought more heatstrokes.

"Every house running diesel generators—more pollution and so much noise!" wails SOLshare projects manager Mashiat Fariha Alam, hennaed hands to her ears.

In 2019, SOLshare had been allowed to hook one village's residences to the grid as a test case. It worked flawlessly through COVID. Connecting all the useless solar panels atop apartment buildings to the national grid would revolutionize Bangladesh's electricity sector.

Instead, war in Ukraine had siphoned away more than half of UNHCR's Rohingya budget, so solarizing the rest of the camp's million refugees was on hold. Salma spent much of her time seeking money to keep field engineer Abu Hasan maintaining the two existing microgrids, and exploring with UNHCR ways to fundraise jointly.

Donors might be reluctant to fund infrastructure in temporary refugee camps, but in a world with more adrift people each year, the average age of refugee camps is now 17, and some, like in Palestine, have effectively become permanent.

Central African refugee camps for South Sudanese, as big as the Rohingyas', have emulated SOLshare's peer-to-peer system, creating giant interlinked networks. If refugees there ever get home, they'll leave something valuable behind for the host countries: a giant solar battery for locals to plug into.

v. Wheels

In Dhaka, often called the rickshaw capital of the world, pedaled three-wheelers line the edges of roads filled with at least 3 million more battery-powered versions. Made in local shops, some hold up to six passengers. One day while stuck in stupefying traffic, it occurred to Sebastian Groh that the EV revolution will not be driven by cars but by scooters. Dhaka's electric tuk-tuks alone far outnumber the world's Teslas.

That got him thinking: When tuk-tuk owners drive to the garage each night to plug in, there's still juice in their batteries. "That's around 8:00 p.m.," he told his SOLshare colleagues, "when the grid is totally stressed because everyone's home switching on appliances. If we could push whatever's left in those 3 million batteries into the grid, and then recharge at midnight when the grid has surplus . . ."

To do that, the country's 25,000 rickshaw-charging garages would need two-way net meters to feed into the grid, and tuk-tuks would need smart batteries. Nearly all currently had lead-acid batteries designed for stationary use; they leaked, corroding vehicle floors and spattering drivers' skin. They also lasted barely a year: just as drivers paid off the loan, the battery would die. Lithium-ion batteries would cost more, but would make up the difference fast in extended range, with three times the life cycle. They also recharge in half the time: on a two-hour lunch break, a driver could kick his smart battery up by another 50 percent. Shorter charging times could increase both drivers' and garage owners' earnings by at least 30 percent.

But it was one thing to permit the garment industry to sell surplus solar energy to the grid—with nearly half the world's new clothes now made in Bangladesh, textile tycoons can drive policy. Charging garages were small businesses, often with bamboo walls and tin roofs.

Still, Groh calculated, the market for EV scooters and chargers was growing by 30 percent annually.

"In five years, maybe 100,000 garages," he argued. "Say a garage can charge 50 tuk-tuks. Its roof can fit 20 kilowatts of solar panels. Multiply by 100,000, you suddenly have 2.5 gigawatts. That's 17 percent of the national grid's peak load right there. No gigantic power plant, and no land. Just those garages."

June 13, 2023, was sweltering, a pre-monsoon day in northwest Dhaka, made even worse because the garage was enclosed on three sides. SOLshare's Salma Islam had arranged for fans, but she still was stifling in her black abaya, and worried about the stakeholders attending: Prime Minister Hasina's energy advisor; the new British high commissioner (whose residence SOLshare had solarized); European Commission representatives; government energy and finance institutes; and an executive from bKash, the mobile financial platform that Bangladeshis rich and poor use for transactions large or small.

As the *Dhaka Tribune* later confirmed, Sheikh Hasina's energy advisor, Tawfiq-e-Elahi Chowdhury, positively gushed. "It brings me immense joy to see how you have embraced Bangladesh's EVs—'Beslas'—and empowered them with an innovative Internet-of-Things application. May your impactful work continue to inspire and spread across the nation. I am very proud that this is a homegrown effort. I want you"—SOLshare—"to go forward, leapfrogging the experiences of the West."

Pointing to the assembled government agencies, he added, "They will help you finance this."

SOLshare's technicians had invented a SOLdongle—the mobile equivalent of the SOLbox—to connect each rickshaw's new lithium-ion battery to the internet. That allowed battery data to display on their web application: how much charge remains; whose vehicle, and how long until its battery lease expires. Among the greatest advantages was that drivers wouldn't have to buy batteries. It was a pay-as-you-go leasing model: via bKash, they paid for charging—but only for the actual time they use the battery.

Garages and drivers from Dhaka to Cox's Bazar had begun adopting the new "SOLmobility" technology. If the whole three-wheel fleet followed suit, it would be an end to load-shedding. But for SOLshare, the real energy future for Bangladesh and all South Asian countries is a garage they've outfitted in Rajshahi, a city on the Padma River near the Indian border: Bangladesh's first all-solar EV charging station. It's the pilot that they hope will drive the need for fossil fuels out of the country.

"Nuclear and coal power plants take hours to ramp up," says Groh. "A rickshaw solar power plant can reach full power in seconds."

In 2023, he and the director of Grameen Shakti spoke on a panel at the United Nations, followed by another in which Christiana Figueres, former head of the UN Framework Convention on Climate Change, interviewed Bill Gates, who talked about fusion and nuclear energy projects his Breakthrough Energy Ventures was funding.

All the time Groh was thinking: Any of those big things need 10 to 20 years to become commercially available, but the world no longer has that kind of time to wait. The UN was now warning that we need to radically cut carbon-based fuel usage by *2027* for a chance to rein in runaway climate changes.

"We should invest in things that work today—and which come from countries where they know about the challenges lying ahead."

Groh's wife is Bangladeshi; his operational and technical staff are nearly all Bangladeshi, and this is now home.

"From peer-to-peer groups in Africa to the UN, our work is spreading around the world. If the country that stands to most suffer from climate consequences can show a pathway out, the world should listen. If we can do it here, you can do it anywhere."

CHAPTER TEN

PIVOTING

If making food from thin air seems unnatural, it's because nature doesn't do it that way, and neither do we. Regardless of what DARPA adds to our diet, we'll likely always grow food from soil. The challenge is how, lest the way we now do it leads to our undoing.

i. Corn Rows

The killing frost in mid-September 2020 was the earliest upstate New York's Black River Valley had seen since 1943. Overnight it yellowed the birch hedgerows and reddened the maples that line two-lane roads north of Utica, stitching even more color into the rolling patchwork farmland, already aglow with goldenrod and purple asters.

Any corn, however, was dead. More than 200 dairies here grow hay and silage corn to feed their cows, which live yoked to troughs and milking machines in long, open-sided, metal-roofed barns. The biggest dairy has 4,000 cows, and cornfields that stretch 70 miles.

The Conway Dairy Farm, just outside the hamlet of Turin, is midsized: 700 acres planted in a half dozen corn hybrids and another 700 in hay and alfalfa, to feed 550 Holsteins. Its daily milk output, 5,000 gallons, is collected every evening by its sole client, the Philadelphia Cream Cheese plant 12 miles up the road.

A few days after the frost, Jake Conway, a hefty 32-year-old with a sandy mustache and sparse beard, climbed into the cab of his family's red Magnum 310 tractor to pull a 26-foot silage trailer to their last unharvested field. The corn was now dry enough to chop before more cold snaps drove its nutritional sugars back down into the roots.

It had warmed again; Jake was working in a blue T-shirt printed with the farm's cow logo and a green mesh ball cap. His great-grandfather started this operation around 1948. His dad, Randal, has run it since his own father retired. When Jake and his older brother, Derek, went to study at the State University of New York at Morrisville, Randal advised them to become lawyers or doctors, but both loved pulling food out of this land and now run it with him.

"It's satisfying. Not many people get to work with their father and brother every day." Should Jake's newborn son, Oakley, also be seduced by the smell of manure, he'll be a fifth-generation farmer.

After bumping across a bridge over the creek that runs through their land, he met up with the John Deere chopper driven by one of their hands. Soon it was ingesting eight rows per pass, chewing and spitting the entire plant—ears plus the chlorophyll-rich leaves and stalk—through a 15-foot spout into the silage trailer as Jake kept pace alongside. The load would be packed into 500-foot-long white plastic silage bags, 12 feet in diameter. Between corn silage and haylage, they fill a couple dozen a year, which resemble colossal worms crossing their land behind the cow barn.

The 40-acre field they were chopping had been fertilized with their cow manure after planting. Mid-growing season, they'd sprayed a sidedressing of UAN—a mixture of urea, ammonia, and nitric acid: liquid synthetic nitrogen fertilizer. In a nearby field, however, they'd been persuaded that year by Andy Mower to try something different. Jake has known Andy all his life; the Conway family has bought seed from him for decades. With farming, every year is a gamble and a prayer that your harvest is worth more than the cost of seed and fertilizer. On a 1,500-acre American farm, fertilizer alone can run more than $100,000 annually, so when Andy came by, excited to be the first sales representative in New York state for a new company in California that had found a way to slash nitrogen applications, they trusted him enough to listen.

The company, Pivot Bio, had been testing its product, Proven, for three years in the Midwest corn belt, with promising results. Farmers from Oklahoma to Minnesota were not only lowering nitrogen usage by 25 pounds per acre—between a sixth and half of what's usually needed, depending on soil and rotation—they were also increasing yields by up to 13 percent. Pivot Bio was now trying it in the tougher soils of the northeastern US and Canada, and they'd invited Andy to be a distributor. The Conways agreed to apply it to half a field and compare the results.

Proven was a fermented microbial brew, injected into furrows during planting. Its benign, soil-derived microbes supposedly captured atmospheric nitrogen and converted it to ammonia plant food. As seeds germinated, the microbes colonized and spread with their roots, fueling themselves on the plants' sugars and in turn continuously feeding them. In effect, Proven was mimicking the symbiotic relationship of legumes like beans or clover, whose roots host nitrogen-fixing bacteria—turning grains into nitrogen fixers.

In mid-July, Jake pulled samples. The root masses from the side treated with Proven were notably bigger and longer. Aboveground, "it's greener," Jake reported to Andy. "And a bit taller. It seems like it's handling the stress better."

The stress was a midsummer drought that was visibly firing—browning—the untreated half, indicating nitrogen deficiency. But the corn with Proven stayed bright green. "Some places, it looks a week ahead. The treated plants already have tassels. The others don't."

"It's handling the drought better because it's getting constantly fed," said Andy.

"Thanks for feeding America!" Andy Mower greeted them as they brought in the last chopped corn. Now in his late 50s, under his blue Proven cap Andy's chunky face had a friendly salesman's smile. "I've got some figures for you."

Two weeks earlier, he'd come by to core silage samples to test for nutritional content and neutral detergent fiber, which determines digestibility. In the lab, the cores are heated to force fermentation, then tested every 12 hours for three days. The lower the number, the easier to digest. "Take a look—15.1 for forage treated with Proven; 19.8 without."

Jake's father, Randal, came over as Andy added, "And look at the starch."

"That's energy for the cow," approved Randal, who wore khaki overalls and ear protectors over a sawdust-coated ball cap.

"Between yield and nutritional value, we're looking at a 5 percent improvement, minimum."

Jake thought it was more. He'd already roughed his own estimates. "My initial check was near a ton more per acre. That's huge. We're running around 700 acres, so potentially at least 600 more tons."

"Producing more energy on less, that's our goal," said Andy. "Because this stuff attaches to the roots, plants utilize all of it. When you run your sprayer, you lose liquid nitrogen that volatizes in the air. Or rain washes it away."

A sudden gust ripped the results from Andy's hand. "We got some hurricane winds coming," he said, grabbing at the papers. Dust was blowing; stray corn kernels whipped around them. As if farming weren't hard enough, and as if COVID hadn't slammed milk prices, weather had become the wildest card in a deck already full of jokers. This year they'd planted on May 6. On May 10, they got four inches of snow.

"And then we got a dry summer, so liquid nitrogen leaches away. That's not good for the environment, or us."

During Andy's drive back through a brilliant afternoon to Performance Premixes, his seed business, the whipping winds finally eased. Climate changes were becoming hard to deny, even though all the Trump signs he passed suggested denial was prevalent in this dairy country—mid-pandemic, farmers here prided themselves on living in a "mask-free zone." Andy had 14 customers trying Proven, and like Randal Conway, their main reason wasn't environmental but economic: nitrogen fertilizer cost around $22 an acre; Proven, $13.

Yet increasingly, including from Jake, he was hearing talk similar to what had inspired the Pivot Bio people to find a substitute for nitrogen derived with fossil fuel: agriculture is a major source of greenhouse gas, and nitrogen fertilizer is the biggest reason why.

Andy's son Dan, who works with him, was waiting in the garage; they had an event that evening at their Baptist church. Dan's wife is a Russian immigrant; her whole family is here. They both hoped for a life as good as his parents have had, but Dan wasn't blind to what was going on. He knew that this new product, just in its first iteration, was a long way from totally replacing synthetic fertilizer, although Pivot Bio was telling them that the next generation looked to be twice as effective. Would eliminating fertilizer, though, be enough?

He shrugged his lean shoulders—it was out of humanity's hands. "It's all spelled out in Scriptures. Revelation tells how the world will end. Thessalonians says the Lord will come like a thief in the night. No one can know the exact time, but as we get closer, we'll see the sequence of events adding up."

ii. Step Change

In every sense, Pivot Bio was at the country's opposite pole: Berkeley, California. All that its team had in common with these upstate New York farmers was a passion for growing things. In a state replete with meteoric startups, Pivot Bio was one of California's hottest, even though its product had to do with the heartland, not silicon.

One of its cofounders, Alvin Tamsir, grew up in Indonesia, surrounded by rice in all directions. Top elementary school grades made him eligible for a high school entrance exam. Scoring in the top 20 among thousands of applicants, he was just 14 when he headed to Singapore, which was luring young minds to come study and maybe stay.

High school taught him that he wanted to be a scientist, but for that he wanted to be elsewhere. For a foreigner to get into a prestigious US university, he'd heard, the path starts in junior college. After excelling at Pasadena City College, he transferred to UC Berkeley, where he spent hours in the botany lab, cross-pollinating flowering species. Working with DNA-bearing pollen grains drew him down to microscopic levels, into a doctoral program in molecular biology at UC San Francisco.

His lab partner there, Karsten Temme, was from Wyoming but did his bachelor's at the University of Iowa because of its biomedical engineering school and its marching band. After four years of playing the trumpet at Hawkeye games, studying neurology and thermodynamics, and setting up WiFi internet for farmers who lived far from cable service, he earned a National Science Foundation fellowship. Within a year, he'd finished his master's thesis on using AI to model how to preserve donated organs cryogenically. For his Ph.D., the emerging field of synthetic biology intrigued him: the premise that editing DNA code could allow microbes to be programmed like computers.

He and Tamsir discovered they shared the same goal: tackle a big challenge that impacts the globe. One November day in 2011, a few months before receiving their doctorates, they crossed the street from their lab to Starbucks. It was chilly but sunny, so they took their coffee back to UCSF's Koret Quad and sat on the grass.

"So what do *you* want us to do with all the microbial engineering we've helped develop?" asked Tamsir.

In that moment, they realized they were more than lab partners: good-enough friends to contemplate going into business together. Temme patted the fescue grass carpet. "Right under us, photosynthesis is delivering carbon to these grasses. Their roots exude sugars to feed microbes, which make enzymes to fix nitrogen into the ammonia it feeds to the plant. It was a balanced cycle—"

"Until people started breeding crops to be bigger," Tamsir finished.

At the time, of course, bigger seemed better, indisputably. When *Homo sapiens* began selecting crops to be more productive, the results often proved miraculous. About 9,000 years back, ancestors of today's Mexicans chose kernels from the most robust stalks of teosinte, a wild grass growing in the central plateau, its ears no larger than a sprig of wheat's. At some point, geneticists now know, a single mutation in one letter of teosinte's DNA code flipped teosinte's hard seed casing, resulting in an expanded inner cob studded with fleshy grains.

Over succeeding millennia, selection of that one mutated strain led to our now familiar big ears of corn. Today, their kernels are second only to rice in providing our species with calories.

Wild teosinte still thrives in Mexico's stony highlands, in balance with naturally occurring nitrogen produced by local soil microbes. But mass cultivation of improved strains of corn, wheat, and rice worldwide have created a nutrient need far greater than the soil microbiome can provide.

"We outpaced evolution," mused Temme. The way humans had strong-armed nature with the Haber-Bosch process had allowed 8 billion of us to be alive at the same time, but had also chemically undermined air, water, and soil—along with the weather. In addition to the CO_2 that Haber-Bosch spews, because farmers overapply synthetic fertilizer to cover losses from runoff or evaporation, the excess nitrates feed soil microbes that belch potent nitrous oxide. Agricultural N_2O emissions, the most potent greenhouse culprit after CO_2 and methane, were also now the biggest contributors to stratospheric ozone depletion.

Revolutionizing agriculture, they agreed, was the problem they wanted to solve: specifically, finding a way for grains to fix airborne nitrogen themselves. There on Koret Quad, Pivot Bio was born.

Tamsir's doctoral dissertation was on engineering microbes. Temme's was on making transgenic cereal crops that could self-fertilize. At first they thought that tinkering with DNA was where to begin. Together they mapped genomes of nitrogen-fixing microbes. Their first grant, from the Bill and Melinda Gates Foundation, was to insert those gene clusters directly into grain plants, enabling them to produce their own ammonia, like soybeans.

They quickly discovered how daunting genetic engineering is. "We're still barely able to insert a single gene bearing desired traits into biotech plants," Temme later told investors, "let alone 30 complex gene clusters that must be exquisitely balanced in order for nitrogen fixation to work."

Besides the public distrust of transgenic crops, if they changed a plant's DNA, they'd have to wait for it to grow to see if that worked; then, in order to fix something, go back to change its DNA again—potentially needing decades. At best, they figured, perfecting such a

GMO was probably 20 years off, and by then there'd be nearly 2 billion more to feed. That confirmed it: they should get soil microbes, which grow in hours, to do the work.

Our ancestors selected for appearance aboveground, not traits below the surface. They neither knew how roots and microbes interacted nor that microbes even existed. Identifying microbes and tapping their potential became the *pivot* in Alvin Tamsir and Karsten Temme's fledgling company's name.

Karsten's years of driving through Iowa cornfields partly influenced their decision to begin with corn—although between humans, animals, and ethanol, it was also the biggest crop in the US. They raised $750,000 in seed money—literally—by convincing Bayer-Monsanto it was worth investigating replacing agricultural chemistry with microbial biology. Renting a warehouse near the waterfront in Berkeley, they mined Bay Area universities for a team of researchers, who gathered buckets of soil from their friends' and families' gardens. In each, they planted unfertilized seedlings to see which microbes were attracted to the sugars their roots leaked, then isolated the most interesting ones in a nitrogen-free goop made of agar. Any that survived must be pulling nitrogen from the air.

One was unearthed by the father of molecular biochemist Sarah Bloch from a plot he farmed in Missouri, near the Mississippi River. That microbe, *Klebsiella variicola,* proved the most robust colonizer of the corn roots, feeding them ammonia and spreading with them as they grew. It became their inaugural product, Proven.

Identifying nitrogen-fixers was one step. The next—their big breakthrough—was realizing that if soil microbes like *K. variicola* sensed there was already activated nitrogen present, they'd stop producing nitrogenase, the key enzyme. The reason was simply to save the energy it takes to tear apart the two atoms in N_2 and then combine them with hydrogen to form plant-nourishing ammonia, NH_3.

This meant that whenever farmers added synthetic nitrogen fertilizer, natural nitrogen-grabbing microbes went into hibernation.

"It explains why soil health has been degrading over the last century," said Karsten. "Using fertilizer becomes self-reinforcing. The natural microbes stop contributing nitrogen, requiring more and more fertilizer, creating more and more pollution."

Since instantly replacing all synthetic fertilizer was unrealistic, how could they coax natural microbes to keep fixing in the presence of nitrogen?

Another obstacle: they'd be taking on a global business worth nearly $200 billion annually, which wouldn't relinquish its grip on the market willingly. They would have to gradually wean farmers, and then the whole industrial agricultural system, away from what Haber and Bosch had wrought.

Reverse engineering is what enemies do when they steal each other's weapons: take them apart to deduce precisely how they work, either to copy or improve upon them. That was how they approached the *K. variicola* genome: dissecting its DNA to first learn which gene triggered its impulse to snatch nitrogen out of the air, then figuring out how to trick that gene to keep firing even when nitrogen was already present.

Disrupting the link between how it detects fertilizer and when it decides to make nitrogen was now possible due to techniques to edit a genome's DNA. But in this case, they intended instead to program the microbe to sense how much nitrogen its host plant needed, and deliver accordingly.

Because there was no genetic modification either to the plant or to the 2,000ths-of-a-millimeter-long microbial bacterium with nitrogen-fixing genes, just to its timing mechanism, Pivot Bio corn was not considered GMO in the US (although the timing tweak might not fly in Europe). They had other ideas of how to satisfy the European Union, but the domestic market was where they would first be testing. To produce enough for that market involved the same technique used by bakeries and the wine industry: fermentation. As with yeast, all they needed was a small drop of microbes and big containers of sugar water for them to multiply into trillions.

Discovering how to do all that, then doing the necessary field testing of their first product, took seven years. It also took money for laboratories and field tests, and for luring highly skilled talent; like microbes in a vat doubling daily, Pivot Bio's staff soon was doubling annually. Fortunately—if a global environmental crisis can be fortunate—the climate fallout of the Haber-Bosch process was increasingly on the radar of venture capital investors, whose job was gambling billions on the future. In the 21st century, that was increasingly a high-wire act without a net. With agriculture producing a third of the greenhouse effect that was threatening their other investments, Pivot Bio became one of the century's rising new biotech firms.

By 2018, before they even had a product to sell, they'd raised nearly $90 million, much of that from Breakthrough Energy Ventures—a consortium including Bill Gates, Jeff Bezos, Mark Zuckerberg, Richard Branson, George Soros, Michael Bloomberg, and Marc Benioff. The following year, Pivot Bio had Proven available in select states for spring planting—Randal and Jake Conway were among their first customers.

Proven's launch was auspicious enough to merit a further $100 million blessing from Breakthrough in 2020. The following year, their final Series D round gave Pivot Bio a valuation of nearly $2 billion—serious attention for fertilizer replacement.

In 2023, Andy Mower couldn't have been more pleased. In his mask-free upstate New York bubble, none of his family or customers were felled by COVID. Instead, Pivot Bio was spreading like a good virus. "From 1,800 acres in 2020 to well over 200,000 acres—that's in just New York alone!" His son, Dan, was now a Pivot Bio territory sales manager, and his daughter, Jennifer, their marketing event coordinator in Pennsylvania.

Nationwide, more than 3 million acres were being treated with their products. Their second-generation Proven 40 added a second potent microbe, *Kosakonia sacchari*, discovered in soil from California's San Joaquin County. It was advertised to eliminate 40 pounds of fertilizer per acre, up from 25. A gallon of Proven 40, a company researcher wrote, could replace a railroad car full of anhydrous ammonia fertilizer.

At a dairymen's conference in Las Vegas, an Oklahoma farmer growing 3,000 acres of corn told Andy Mower their input costs had dropped by $204,000. One of his own customers, after switching to Proven 40 on 100 percent of his crops, saved enough to buy a second farm. The Conways were now treating all their corn with Proven 40 and had completely eliminated the midsummer sidedress pass, when corn usually needs nitrogen the most. That meant saving nearly $56,000 on fertilizer alone, Jake said, plus a week's labor—"an 80-hour week, plus saving diesel going up and down all those passes, saving machinery wear and tear, and reducing soil compaction." The Conways were the first in their county to try Pivot Bio; now, 10 more surrounding farmers used Proven 40.

Pivot Bio had also introduced Return, for smaller grains like barley, millet, oats, sunflower, and spring wheat. The first trials, with sorghum, actually produced more bushels per acre of test plots than the fully fertilized control plots.

But their big innovation was the option to offer Proven 40 already coated on the seed, rather than as an in-furrow liquid inoculant. Twenty-five hundred pounds of seed corn could be treated in four minutes right on-site—saving farmers the expense of in-furrow application equipment, and losses to evaporation or leaching. All they needed was to load their planter and go. The nitrogen was produced right on the plant's roots, not in a factory.

"A box of treated seed for 125 acres equals five tons of urea you're saving, or a whole big tank of anhydrous ammonia that you won't have to apply to the ground," Andy Mower told his customers. "That's huge."

Especially since in a single year, between drought and war in Ukraine, the cost per ton of anhydrous ammonia, *The Des Moines Register* reported, went from $600 to $1,600—a 167 percent increase. To reduce reliance on Russian fertilizer* and to meet demand—Pivot Bio was selling out weeks before planting season—they were expanding, with a completely rebuilt headquarters in California, a new production facility in St. Louis, and a second distribution center at Iowa State University's Research Park.

New climate data revealed that two-thirds of synthetic fertilizer's emissions drift skyward after it is applied, accounting for 38 percent more greenhouse gases than commercial aviation. Downstream nitrate pollution was now blamed not just for dead zones at river mouths, but for the sargassum fouling Florida's beaches and the algae killing California sea lions. With its nitrogen produced underground and staying there, Pivot Bio seemed exquisitely positioned. Unlike other green products, its fertilizers appealed to customers without environmental inclinations but with an understanding of their bottom line. And unlike solar, wind, induction stoves, or heat

* Russia is the world's biggest fertilizer exporter.

pumps, Proven 40 and Return involved no initial outlays for new equipment, just immediate savings for the consumer.

By 2023 Pivot Bio's annual earnings topped $50 million. It now employed hundreds of researchers and sales representatives, was expanding into Canada and Brazil, and was in discussions with the EU and the United Arab Emirates. Besides Facebook and Twitter, its ads were a fixture on the 24-hour farming and agribusiness channel RFD-TV, including Pivot Bio "Originals," its series of agro-comedy and game shows, agro-science documentaries, farm truck giveaways, and a twangy music video:

> For greener plants in a healthy land
> use our predictable nitrogen.
> Wherever farmers grow—
> that's where we go.

But could it really make a difference?

Pivot Bio was earning good profits by substituting around a fourth of a field's nitrogen needs—significant for its customers, but barely cracking the global climate crisis. There were also questions over how effective its approach really was. Research that Pivot Bio sponsored at Iowa State, Nebraska, Purdue, Illinois, Georgia, and North Carolina State showed increases in nitrogen, potassium, and phosphorus uptake, and corresponding plant health: greater biomass (an indicator of ear size), and several instances of higher yields. But other studies, including one by emeritus University of Illinois extension agronomist Emerson Nafziger, showed mixed results from applying Proven 40.

"This is a company that got a huge amount of venture capital funding, over a half a billion dollars. They've been investing a huge amount in advertising and public relations. You can't compete with a company that's spending millions of dollars to generate enough results so they can sort out the ones favorable to them and tell you this is the truth. I've seen some of their study summaries," said Nafziger, whose own studies were funded by the Illinois Fertilizer and Chemical Association. "Some are not about yield at all. They're about nitrogen content, or size of plants during the growing season. Nobody pays you for that."

That misses the point, Karsten Temme would patiently respond. Pivot Bio wasn't intended to increase yields, but to maintain them with fewer damaging, expensive soil inputs.

And something else: "In the past," he'd say, "we'd evaluate fertilizer by how much went on the soil, and how much yield came out the other end. Increasingly, that's too unpredictable because of weather. What captivates our growers is having a more dependable path to their target. Our microbes don't get washed away by rain. Nitrogen is stored directly on the roots, available for different growth spurts during the plant's life cycle. It means being less likely of running short of fertilizer throughout the season. And it eliminates excess N_2O emissions and excess runoff, and saves time and expense. We have 90 percent returning customers. We've got to be doing something right by them."

iii. Pop

Giltner, Nebraska, population 800, resembles dozens of towns dotting the Platte River valley of southeastern Nebraska: two churches, two bars, a body repair shop, a water tower, and rising above them, the grain elevator. In every direction, it's corn and soybean fields, then more corn and soybeans.

Until the 1980s, the fields were surrounded by windbreaks planted during the Dust Bowl years: elms, ash, and eastern red cedar. The first two succumbed to disease, and the third has lately become an opportunistic prairie invader, its propagation discouraged. Windbreaks also conflict with the reach of center pivot irrigation, which have turned America's flyover heartland from patchwork quilt to something resembling a viral circular rash. This is pump-irrigation country; the center pivots are easier on the diminishing Ogallala Aquifer than flood irrigation.

But lately the wind is blowing harder. "I'm not saying it's the 1930s again," says Giltner farmer Brandon Hunnicutt, "but combined with how dry the last couple of years have been, we're glimpsing what it was like."

Hunnicutt, at 50 strapping and thick-chested, is a fifth-generation farmer on both sides—and with seven kids, there are decent odds for a sixth generation. He's served on boards of the National Corn Growers Association and Nebraska's chapter of the US Grains Council, and has gone on trade missions to Africa and Southeast Asia.

The family farm that he, his dad, and his brother run is 120 years old; along with soybeans and seed corn, they raise several popcorn varieties: classic platinum for the movie theater market, roundish mushroom for caramel corn, fluffy yellow butterfly for gourmet, and sometimes organic. A few years ago, he was at an annual summit for Field to Market, a sustainable agriculture alliance whose board he's also chaired. Wandering the trade show area, he saw Pivot Bio's booth, stopped to talk, and left thinking, Well, that's really interesting.

A year later, at the 2020 conference, four people from the company approached him. "We've got this idea about trying to raise popcorn with reduced nitrogen, using Proven."

They were thinking, they told him, of approaching the giant brand Orville Redenbacher's.

"Over my dead body," he replied. His dad was a part owner of Preferred Popcorn, headquartered 17 miles north of his farm: a 100-farm cooperative distributor that ships to 70 countries. "There's no way I'm going to let you do that"—quickly leading to the deal they were hoping for.

They'd calculated needing about 80 acres. "That's a lot of popcorn," he told them. "It'll be coming out of your ears."

They needed a lot of data, they explained. They wanted to plant half the field with their product and half without, to compare yields—which he'd planned to do anyway.

For any popcorn farmer who'd plant a side-by-side comparison, they'd pay a site visit during the growing season to sample leaves attached to ears; dry, chop, and mash them; then use a spectrometer to measure the percentage of nitrogen. With drones and sometimes even satellites, they'd take spectral measurements of plant color—the greener, the better.

Pivot Bio was approaching a million commercial acres, and nearly all of it, they claimed, averaged 14 percent more nitrogen content over synthetically fertilized plots and 12 percent more plant biomass. The tests on Hunnicutt's 80 acres, however, had to be scratched, because right after planting with the new Proven 40, a huge wind storm came through. He couldn't keep fertilizing the control patch, because the corn was leaning too far to bring in a tractor. But in the end, things worked out. "Even with wind damage, we still got over 50 bushels, about 3,500 pounds, per acre."

Fortuitously for Pivot Bio, that was enough to let them market Connect Gourmet Butterfly Yellow Popcorn, grown with *no* synthetic nitrogen. At the next National Corn Growers Association meeting, Pivot Bio gave out bags of Connect to the attendees, both a stovetop version and non-plastic, recyclable microwave packets. It also sold the popcorn online. People kept telling Hunnicutt it was the best they'd ever had and that every kernel popped.

So he was dismayed when Pivot Bio didn't renew the program the following year, saying that they'd realized that they weren't set up to be a food retailer—yet. "Stay tuned," said Karsten Temme.

Disappointed, but encouraged by the synthetic chemistry-free results, Hunnicutt stuck with Pivot Bio, though he also tested some of the growing competition in biological nitrogen replacement, including a product called Source, which purports to coax preexisting soil microbes to keep producing ammonia even when human-made nitrogen has been added.

And then there were the various companies trying to replicate the phenomenon well known to farmers when thunder and lightning rain down free fertilizer. In Australia, Sydney-based PlasmaLeap Technologies was testing an ionizing plasma-arc device that zaps air bubbling in a large beaker of water, filling it with nitrates to irrigate grapevines. Hunnicutt bought a machine to try it, but first checked with the company to make sure he wasn't just lacing more water with nitrates, the last thing Nebraska needed.

No one knew that better than the father of seven children. Across the state, lakes were closed to swimming and wells shut for drinking because of nitrate contamination. The cause was undisputed: one-third of all fertilizer applied to cornfields leaches into the water table.

It is also undisputed why Nebraska has one of the highest pediatric cancer rates in the US. The highest per capita rate is in Hunnicutt's county. He's working with University of Nebraska cancer doctors.

"They don't understand farming, and I don't understand cancer. But we're trying to reduce nitrates in the groundwater and deal with climate change at the same time."

He knows what needs to be done; it's just hard in a state whose flagship football team is the Cornhuskers—and whose livelihood and growth of Omaha and Lincoln owe so much to chemical fertilizer. Munitions factories that pulled the US out of the Depression during World War II kept the prosperity going postwar by converting to fertilizer production. But in large part due to the Haber-Bosch process that begat it, Nebraska now has a double water problem. Its shallow, braided Platte River used to run dry in midsummer: now it's March or April. And its groundwater is suspected of giving children brain cancer.

"Applying the feed to the seed may be a bit of the Holy Grail," Hunnicutt says, "because then there's nothing to leach. I want to just plant it and know I'm reducing nitrogen. I don't want to think about it anymore."

iv. The Grail

Imagine a world with total replacement for synthetic nitrogen—not just a quarter. We might still have industrial monocultures, vulnerable to entire harvests being lost to disease and devoid of insects due to pesticides that also kill billions of birds. But some of the worst pollution on Earth—carcinogenic to humans; murderous to the planet's most fertile places, river deltas; with greenhouse gases coming and going from its energy-guzzling production to its gush of nitrous oxide—all that would vanish. Children and marine life, still beset by many remaining challenges, would have one less—a huge one.

Dr. Trenton Roberts, an early Pivot Bio tester, holds an endowed chair in soil fertility at the University of Arkansas's Agricultural Experiment Station. "It has the potential to be a genuine game changer," he believes. But Roberts, who's in his 40s, with close-cropped auburn hair and a goatee shot with gray, adds that they'll need to overcome a key limitation. "Energetics."

The Haber-Bosch process, he explains, is so energy-intensive because N_2's stable bond—two nitrogen atoms sharing each other's electrons—is hard to sever. Soy and other legumes, in their symbiotic relationship with microbes capable of doing that, evolved tiny nodules on their roots to house them, assuring that no rival microbe could steal the sugars they feed them.

Grains, however, lack those protective nodules, so the fermented microbes that Pivot Bio provides must compete with everything else in the soil for the sugars oozing from, for example, corn's roots. Since there's no co-evolved symbiotic cooperation, corn can't sense how much food Pivot Bio's microbes need. "So billions of other microbes are also living off those root exudates."

The next step, he thinks, "will require CRISPR-type genetic tinkering, both on the microbial side as well as the corn plant side, to get closer to the system soybeans use. Unless we alter the anatomy and the metabolism of the corn plant, we won't get to full replacement."

Pivot Bio's scientists can now trick a microbe's own DNA to keep fixing nitrogen even when fertilizer is present, but thus far they've avoided inserting another microbe's genes, because there is public resistance to GMOs—even though 90 percent of corn and soy is already genetically modified to resist pests and tolerate herbicides. Also, there's that problem of needing two decades to engineer a new viable plant. "You'd be taking whole anatomical structures and enzymatic pathways from crops like soybeans and inserting them into the corn genome," Roberts acknowledges. "Those are major changes."

How to feed billions more born by midcentury is one of the most vexing problems on Earth—but not for Trenton Roberts. "We already produce more than enough calories to feed projected populations well into the future. Shifting *what* we produce solves the problem. In perpetuity."

In a red Nike cap with "ARKANSAS" embroidered above the visor, Roberts gestures at the former piney woodlands surrounding his ag station where ivory-billed woodpeckers once nested—now soybeans and grain to the horizon.

"If we quit feeding corn to animals, that solves things forever, because the efficiency and energy lost through meat production is huge"—including the one-third of all human-related methane that cows expel. "Food production doesn't concern me. We just may have to change what we eat."

With his U of A Soil Extension shirt stretching over his belt, Roberts doesn't look like a vegan, but he is a scientist, citing irrefutable math. Yet although growing numbers agree with him, *Homo sapiens* evolved as an omnivore. As demographers and marketers know, the more people can afford meat, the more of it they tend to eat.

He acknowledges that. It's hard to blame people for just being human. So he focuses on humanity's and livestock's chief accomplice in the ecological hijacking of Earth.

"We grow so much corn because we're really good at it. People make a lot of money doing it. If they grew something else, most of our nitrous oxide problem would suddenly be solved, too. But until there's a reason to change, they're going to. What Pivot Bio's done so far is great. But we must keep improving."

CHAPTER ELEVEN

XNOIS

i. Comcáac

Since the time of the dinosaurs, tectonic plates have been in a tug-of-war along the American edge of the Pacific Rim. Over eons, the ancient Farallon oceanic plate, shoved from the west by the newer Pacific plate, gradually slid beneath what is now North America, pushing up the Rocky Mountains and the Sierra Madres. As it vanished into Earth's maw, islands floating atop the Farallon and accretions from smaller subducted plates stuck to the North American plate's coastal crust. Today they amount to almost everything west of Utah.

By 7 million years ago—around when two descendent lineages of an African primate, *Pan prior*, branched in separate ways, one staying in the trees while the other descended into the savannas—the Pacific plate was grinding right against North America. As it moved northwest, it tugged away an accreted section of coast now called the Baja California peninsula, opening a rift along the Pacific coast of today's Mexico.

As the breach widened,* the Pacific flowed in, forming a narrow gulf with extreme tides. To the north, clouds of Pacific moisture colliding with the cordilleras that had bulged upward dropped

* It still does: each year the Pacific plate pulls Baja California nearly two inches farther away.

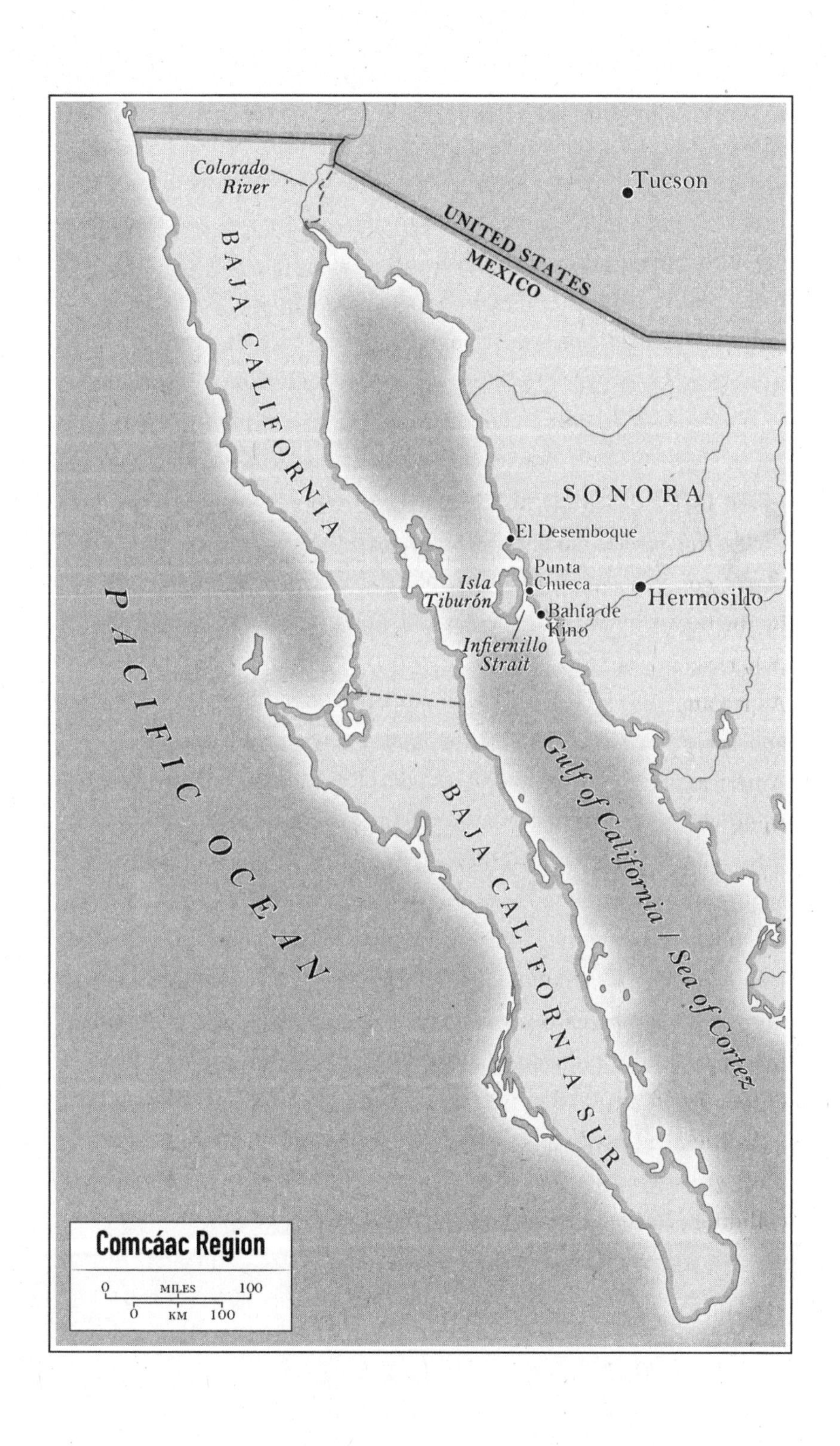
Colorado River
Tucson
UNITED STATES
MEXICO
BAJA CALIFORNIA
SONORA
El Desemboque
Punta Chueca
Isla Tiburón
Bahía de Kino
Hermosillo
Infiernillo Strait
PACIFIC OCEAN
Gulf of California / Sea of Cortez
BAJA CALIFORNIA SUR
Comcáac Region
0 MILES 100
0 KM 100

precipitation that collectively drained into what is today's Colorado River. As it rushed down from the Rockies, the river reached an uplifted plateau in ancient Arizona that was previously the bed of successive inland seas. Over 5 million years, it carved a vast canyon through sandstone a mile thick, down to metamorphic basement rocks that predate the emergence of complex life on Earth.

The Grand Canyon's worth of nutrient-rich sediment that it excavated was deposited in a delta that is now California's Imperial Valley and Mexico's adjacent Valle de Mexicali. Eventually, accumulating silt pushed the Colorado River eastward to the widening gulf between Baja and mainland Mexico. The Colorado's nutrient bonanza turned the Gulf of California, also called the Sea of Cortez, into one of Earth's most extravagantly profuse seas, home to nearly a third of the world's sea mammal species and a fish census surpassed only by Southeast Asia's Coral Triangle.

Humans first discovered this aquatic bounty over 10,000 years ago. To either side, however, the land lay in a rain shadow between coastal and inland mountain ranges, its desert soils too arid for farming. When Spaniards arrived in 1533, the Indigenous groups they encountered fished and foraged. Before Jesuits could save their souls, however, most in Baja California died from European diseases.

Catholic missionaries had more success across the Gulf, in today's Mexican mainland state of Sonora, where Opata, Yaqui, and Mayo grew corn, beans, and squash along rivers. But one tribe, speaking a language unlike any other, ferociously resisted both pacification and agriculture. Over three centuries, Spanish and Mexican ranchers took nearly all their land and tried to exterminate them. With the government offering a bounty, the ranchers nearly succeeded. Originally estimated at around 8,000, by 1900 fewer than 200 remained. But to this day the tribe that calls itself Comcáac—the People—persists in two villages of about 600 each where the Sonoran Desert meets the sea.

To everyone else, they are the Seri—a name with no translation, given to them by a rival tribe. Their coastal villages, about 35 miles apart, are opposite Isla Tiburón, the largest of the Gulf of California's mountainous midriff islands. From above, those islands resemble giant stepping stones to Baja California—and indeed, giants populate the Comcáac's origin myth. Details vary depending on the telling, but most versions describe a "land maker," Hant Caai, who recruited a turtle to tuck a grain of sand from the sea bottom under its flipper to create terra firma, which he then sowed with trees and plants, both edible and medicinal. The first people to appear were giants, whom the Comcáac don't exactly claim as their ancestors, since one flood changed some into boojum trees—a local columnar succulent—and another transformed the rest into barrel cacti. Except from time to time they reappeared, to impart teachings that amounted to survival skills.

If anyone needed them, it was the Comcáac. The desert was torrid much of the year, filled with things that bit and punctured, but also with several cacti varieties that fruited, and aromatic plants whose roots could be twisted into fibers for weaving baskets tight enough to hold the water they collected from mountain springs. In estuaries near their villages were mudflats filled with clams and crabs, and oysters growing on the roots of the world's northernmost mangroves. The sea swarmed with hundreds of fish species, sea lions, dolphins, and five sea turtle species, but fishing took courage. The 1.6-mile strait between their lower village, Punta Chueca, and Isla Tiburón became known as Infiernillo—Little Hell—Strait for some of the most challenging currents anywhere. Another channel just to the west was named Salsipuedes: get out if you can.

Their boats—*balsas* in Spanish, *hascám* in their clattering tongue—were flat canoes made from reed-grass canes that grew near springs, bundled together with cord spun from chewed mesquite root.* Weighing little, they nevertheless could hold two kneeling turtle harpooners and four green turtles, their favorite prey, both for its rich breast meat and its medicinal fat. From the giants, the Comcáac had learned navigational songs that described the currents and the 15 winds they would encounter in the Infiernillo, and they sang as they paddled.

Often they hunted along the beds of *hataam*: aquatic eelgrass, a favored turtle food that grows between Tiburón and the mainland. Because it propagates by seed, eelgrass nourished the Comcáac as well. When it ripened, its roots withered and detached, forming great cylindrical marine grass tumbleweeds that men in balsas herded toward shore, where the women waited in the surf to catch them. After drying the plants on the beach, the grains, called *xnois*, were threshed with wooden clubs onto deerskin hides, then winnowed by tossing them into the air, like terrestrial grain elsewhere in the world. The seeds were then parched, toasted, and stored in a basket to later be pounded into flour. Mixed with turtle oil or honey, then rolled into balls, xnois was the perfect concentrated food for fishermen to carry on their balsa voyages.

ii. Chef

August 2018, El Puerto de Santa María, Andalusia, Spain: Juan Martín Bermúdez was not altogether surprised to see Chef Ángel León in his doorway so early, or so disheveled, his square face unshaven. Hyperactive since childhood, a trait that regularly got him

* Except for their shallower keels, Seri balsas looked nearly identical to reed boats Sumerians used in Iraq's marshes.

ejected from school, León often stayed up all night imagining unpredictable recipes. A stocky man of medium height in his mid-40s, he had barged into Salarte, the nonprofit Juan directs from an office alongside León's Laboratorio de Investigación Gastronómica, upstairs from his restaurant, Aponiente. He was clutching an article. "This is absolutely inspirational!"

It was a printout from the journal *Science*, including the edition's cover dated July 27, 1973, showing a crude doll made of two crossed bundles of dried grass, bound by a strip of fabric. Where, Juan wondered, had Ángel come up with it? He was further puzzled by the article's title: "Eelgrass (*Zostera marina L.*) in the Gulf of California: Discovery of Its Nutritional Value by the Seri Indians." Eelgrass nutritional?

Juan, a tall, dark-haired, scuba-diving marine ecologist who knew little about gastronomy, had previously directed Spain's Parque Nacional de Doñana, 30 kilometers northwest of them near the mouth of the Guadalquivir River, and later El Parque Metropolitano Marisma de Los Toruños, a wetlands sanctuary along the shore of the Bahía de Cádiz, visible now from Salarte's second-story window. At Los Toruños, Juan had started a program to recover the bay's eelgrass beds, which covered Cádiz Bay back when Christopher Columbus sailed from here for what became America. In the 20th century, dredging, bottom trawling, and coastal overdevelopment had reduced the beds to just a few square meters. Chef Ángel had recruited Juan to revive the aquatic ecosystem that was the source of everything on his world-famous menu.

Ángel León was born in 1977 here in El Puerto de Santa María, a 20-minute ferry ride across the bay from Europe's oldest continuously inhabited city, Cádiz, founded by the seafaring Phoenicians 3,000 years earlier. Fifty miles west of the Strait of Gibraltar, Cádiz and El Puerto were once wealthy ports, the first to receive gold and other treasures from Spain's American colonies, until they were eclipsed by Seville's more defensible harbor 50 miles up the Guadalquivir. Although Cádiz's ancient beauty and beaches attracted tourism, El Puerto became a backwater. Today it has western Europe's highest unemployment rate.

As a disruptive child, fishing the bay with his father was the only thing that kept Ángel calm. His father also taught him to prepare what they caught, and Ángel learned that he had a flair for it. In his teens, he got a job in Seville's La Taberna del Alabardero, and he did so well that within three years, he was working at Bordeaux's Le Chapon Fin, one of the world's first three-Michelin-star restaurants.

His journey to chefdom took him from France to Catalonia, Miami, Buenos Aires, then back to Spain, where he studied Sephardic and Arabic cooking in Toledo. His fusions began drawing gastronomic attention, earning him a lectureship in 2006 at New York's Culinary Institute. By then, the hyperactive boy who never made it through high school could have earned top wages cooking anywhere, but the flavors he'd mastered had been simmering in his imagination for years, and in 2007 he made a radical leap.

Haute cuisine naturally coincides with where the world's money flows—where else could anyone afford meals costing hundreds of dollars, even before the wine pairings? It surely didn't coincide with where Ángel longed to be: home on the Bahía de Cádiz, fishing the tide every morning. His dream scheme was to open a restaurant so extraordinary that wealthy gourmands would make pilgrimages to Europe's poorest city, to inject lavish sums of money into the local economy in exchange for tasting food unlike anywhere else.

The ingredients, he was sure, were readily obtained right where no one else thought to seek them. While working on a trawler one year, he saw nearly half of what nets hauled in discarded as undesirable bycatch. Across the planet's oceans, industrial fishing boats sought coveted species like hake, tuna, salmon, or sea bass, and threw the rest overboard.

In a world where a billion were undernourished, this was criminal, he declared. For gastronomes, it was also the loss of distinct taste opportunities for the sake of a few predictable money flavors.

"Haven't you tasted enough caviar already?" Ángel challenged his potential investors. The waste was unforgivable, he argued. Any fish that gave its life to our taste buds was more than just the choice fillets that made up less than half its weight. Everything that less daring chefs left on the cutting board, he would use.

For that matter, he resolved to use everything aquatic that was routinely overlooked, from what was on the ocean's floor to chlorophyll-rich plankton floating on the surface. In an old stone mill that in previous centuries ground salt and flour by tidal action, he opened Aponiente—"in the direction of the setting sun"—in 2007. In the beginning, some diners, food critics among them, were livid to learn they were paying hundreds of euros for trash fish, or discarded fish parts marinated in sauces made from mackerel eyes, or sea worms, jellyfish, or fish bones, accompanied by savory wafers made from sardine scales. His plankton recipes, one accused him, stole food from whales.

He nearly went bankrupt, but kept on. In 2010, after he'd perfected algae spaghetti, marsh-tasselweed honey, crisped moray eel skin, tuna-head osso buco, and delectable mélanges of unrecognizable sea vegetables, everything changed when Aponiente was awarded a Michelin star.

By the time the second star came, in 2014, he had created sea charcuterie: blood sausages, mortadella, bacon, pigskins, and hams, all confected from unwanted parts of fish, mussels, sea bass, squid—whatever fishermen brought him. That was also the year he offered Juan Martín free headquarters alongside his food laboratory for Salarte, Juan's nonprofit research institute dedicated to restoring the Bahía de Cádiz's fecund salt marshes.

In 2016 Ángel dazzled a culinary conference at Harvard with

bioluminescent soup. The following year, 2017, his plankton pizza with calamar cheese took top honors at Madrid Fusion, where he was named Chef of the Year. Aponiente won its third Michelin star.

Ángel had grabbed back the *Science* article and was brandishing it. "Did you know there was a sea plant that produces grain?"

"Sure I know *Zostera marina*," said Juan. "It grows in intertidal zones between marshes and coastal wetlands. It's critical habitat for all kinds of marine species. But I had no fucking idea about the grain," he confessed.

The reason, he eventually realized, was that in the Bay of Cádiz and nearly anywhere else it grows, eelgrass is a perennial plant that sprouts new shoots each spring from rhizomes—the horizontal roots it sends out. But possibly to accommodate variations in salinity, it also evolved the ability to produce seeds, encased in long slender pods that resemble the grass stems themselves.

The variety of *Z. marina* described in the article clearly behaved differently: detaching from its roots as the seeds ripened, needing to be resown each year. The paper's authors were Richard Felger, a University of Arizona botanist, and Mary Beck Moser from something Juan didn't recognize, the Summer Institute of Linguistics. He wondered if they were still alive.

Ángel was aquiver. If there were an edible marine grain, he needed to know about it. Potentially, this was more than another surprise on Aponiente's menu, although he was dying to taste it. Southern Spain was becoming unbearably hot in the summer, and fast-growing seagrasses can sequester CO_2 35 times faster than rainforests. The world was running out of where to farm without decapitating more forests or dispossessing more wildlife. It was also force-feeding its grain supply with chemistry. Right across from Doñana National Park, Guadalquivir estuaries had been converted into Spain's biggest rice-growing region. Before flooding their paddies, growers soaked the germinating seeds with synthetic ammonia and phosphorus. When the plants emerged from the water, crop dusters doused them with herbicide.

Ángel León now realized that what as a boy he thought to be rice growing along the edges of the Bahía de Cádiz was actually eelgrass, thriving with no fertilizers, handling the competition from other aquatic plants without the aid of human-made poisons.

He directed Juan and researchers from the Universidad de Cádiz to gather *Zostera* seed and plant three small test gardens. Eighteen months later, the grains they stripped from the slender pods resembled brown rice, so Angel cooked it that way. It tasted appealingly like a cross between rice and quinoa. Although its yield was less than rice's, they calculated that they might get as much as 3.5 metric tons per hectare.*

Word got out that this celebrity chef had successfully transplanted and prepared grain from seagrass, something no one had tried before. Journalists appeared.

"If nature gifts you with 3,500kg without doing anything—no antibiotics, no fertiliser, just seawater and movement—then we have a project that suggests one can cultivate marine grain," León told *The Guardian*. "I believe it's a new way to feed ourselves."

iii. Ethnobotany

Four days after that *Guardian* article appeared, Juan Martín received an email from an Arizona ethnobotanist, Gary Paul Nabhan. As a graduate student decades earlier, Nabhan explained, he'd worked

* 1 hectare = 2.47 acres.

under Richard Felger, who wrote the paper Chef Ángel had seen, and had subsequently collaborated with him and his coauthor, Mary Beck Moser, on studies of other traditional Seri food plants. "I'm still closely involved with the Comcáac, as the Seri call themselves," he added.

Minutes later, a reply appeared in Nabhan's inbox. "Ángel is really anxious to learn if that was just a historic thing, or if the Seri still eat the sea grain," Juan wrote.

"Not just historic," Nabhan replied. "They're currently doing eelgrass restoration. I'm working with them."

"When can I come?" Juan wrote back.

Like Ángel León, Gary Nabhan, a first-generation Lebanese American with thick gray hair and pale blue-gray eyes, was another overachieving high school dropout. After a circuitous route that included Washington, DC, for the first Earth Day in 1970, he enrolled in Arizona's Prescott College for its undergraduate field studies. Accompanying his professors to the Sea of Cortez, he sea-kayaked alongside Comcáac through Infiernillo riptides to Isla Tiburón and swam their mangrove estuaries. Later he returned to research his senior thesis, comparing Sonoran and Galápagos coastal desert mangrove ecology.

After graduation, he'd stayed in Arizona, contracting as a field researcher. Someone recommended him to Richard Felger, who was applying to the National Science Foundation to fund a complete Seri Indian ethnobotany that would catalog all the desert and marine plants they used for food, medicine, basketry, and building. Felger's coauthor again would be Mary Beck Moser of the Summer Institute of Linguistics, a missionary organization devoted to translating the New Testament into every language on Earth (purportedly because Christ's return is delayed until everyone has a chance to receive the Gospel). As SIL linguists in their early 20s, Becky Moser and her husband, Ed, had moved to El Desemboque, the upper Seri village, in 1951. The Seri tongue, *cmiique iitom*, being an isolate, few save Comcáac who were born to it had ever tried to master it.

When Nabhan met the couple in the late 1970s, they were still translating the Bible, and still living with no running water or electricity in the adobe house they'd built. Ultimately it would take them nearly 30 years; by then, most Seris could read Spanish, so their translation task of literally biblical proportions was arguably moot. But as Felger's interpreter, Becky Moser had become so versed in Comcáac nutritional and ritual use of plants they joked that she'd become more Seri than the Seri had become Christians.

Nabhan helped write Felger's National Science Foundation grant proposal, and subsequently coauthored several papers with him en route to earning a doctorate in arid lands agriculture from the University of Arizona. He reworked his dissertation—on the ethnobotany of the transboundary Tohono O'odham—into *The Desert Smells Like Rain*, a lyrical 1982 meditation on how Native peoples and native plants could be mutually nourishing.

While working with the Tohono O'odham to revive the traditional foods they ate before Western processed fare brought them diabetes and heart attacks, Nabhan cofounded Native Seeds/SEARCH, a nursery and dispensary of 2,000 varieties of native and farmer-selected landrace seeds, to preserve and promote the crop diversity of North America's desert peoples; it continues to this day. His next book, *Gathering the Desert*, about the delicious edible bounty in their arid region, won the John Burroughs Medal for nature writing, and soon after came a MacArthur Genius Grant.

When Juan Martín's excited response to his email arrived in 2019, Nabhan held the W. K. Kellogg Endowed Chair at the University of Arizona's Southwest Center. Researching how to conserve both nature and culture through traditional farming had taken him to Europe, Africa, Asia, and the Middle East, especially Lebanon, where he'd found his ancestral culinary roots still thriving among his cousins (who showed him a date variety named the *nabhan*).

But his deepest love remained the Sonoran Desert, from Tucson—whose designation as a UNESCO City of Gastronomy he'd spearheaded—down to Comcáac country, where he returned repeatedly.

"My colleague Laura and I would be happy to take you there," he wrote to Juan. "It would be a great interchange. You know things they don't, and they know a lot that could benefit you."

Having read in *The Guardian* that Salarte was attempting to restore eelgrass in its own bay—and that it was networking with other European coastal regions to see which eelgrass variety might produce the most grain—Nabhan hoped that Juan's efforts might help alleviate some problems that, two decades earlier, had brought Laura into the lives of the Comcáac—and his own.

Laurie Monti, an American nurse practitioner, grew up in Bogotá, where her father worked in international development. After becoming a community health specialist at St. Louis University, during the 1980s she served as a war medic in Guatemala, Nicaragua, and El Salvador. Long interested in native plant medicine, in 1997 she entered the same University of Arizona doctoral program Gary Nabhan had graduated from years earlier. They met when he spoke at one of her classes. Besides getting her Ph.D., Gary learned, she was also faculty in the university's public health program.

"The Seri tribe is worried about diabetes," he told her. "Are you interested?"

The drive from Tucson was four hours south to Hermosillo, Sonora, where temperatures frequently exceeded Phoenix's, and then an hour west to the Sea of Cortez, passing through vast acreage that for decades had been pump-irrigated for wine grapes. As the aquifer depleted, fields had been abandoned; in the few remaining, Gary explained, rising heat was causing grapes to mature so fast that they were all sugar and no flavor, suitable only for brandy.

They passed a few attempts at lower-water substitutes: pecans, and agaves for bacanora mezcal. "All failing. Modern conventional agriculture is no longer a viable use of land here."

"The Seri don't cultivate at all?"

"Nope. They are the last remaining hunter-gathers in North America south of the Yukon. All their names for European crops are pejorative. They say they don't taste as good as wild crops."

Where the pavement ended at the coastal town of Bahía de Kino, they turned north. After 16 kilometers, the washboard road rose to where she could see the hooked sandbar peninsula that gave the lower Seri village its name, Punta Chueca—Crooked Point—jutting into swirling turquoise and blue Infiernillo currents. Beyond them rose the mountains of Isla Tiburón. The seascape was staggeringly gorgeous, a jarring contrast with the ramshackle village. Traditionally, the semi-nomadic Comcáac had never erected permanent dwellings. Hoops of ocotillo, a spiny cactus whose name derives from its thin tines that resemble octopus tentacles, might be draped with hides—or later oilcloth or plastic. The prefab slabs that Mexico's housing authority had brought in the mid-1980s for concrete construction seemed even shabbier.

At a shack near the shore, they bought canned juices. A few fishermen, their skin deep umber, lolled by aluminum-and-fiberglass pangas equipped with outboard motors, which had replaced the reed balsas during the 20th century. The last balsa, built in 1990 by Juan José Moreno and blind Alberto Molina, was commissioned by a public television crew working on a documentary about peoples of the Pacific. They were the last two Comcáac who knew how, but both were now gone, their balsa stored in the village cultural center.

Gary took her to El Desemboque, 60 kilometers north. Unlike treeless Punta Chueca, it was shaded by large tamarisks—an Asian drylands species introduced to stabilize western US riverbanks, which now chokes the Grand Canyon's side canyons; seed from there had floated all the way here. He then disappeared for several days—Laurie's cue, she figured, to begin an epidemiological study. Gathering medical histories door-to-door, ingratiating herself in fluent Spanish with women who giggled at her wavy blond hair, she soon realized that half the adult population was prediabetic. The traditional Seri diet of fish, xnois, cactus fruits, and game they hunted had succumbed to the convenience of markets. Packaged white rice was so much easier than all the labor eelgrass grain entailed. Insidiously, the market worked both ways: it was too tempting to sell the core protein in their diet—crab, giant scallops, green turtle, manta ray, and corvina—for cash to spend on junk food. Even their taste for occasional terrestrial game yielded to what they could earn guiding hunters from Mexico City or the US to bag bighorn sheep or mule deer on Tiburón.

Gary kept inviting her back. By the late 1990s she was helping to run Seri clinics and encouraging them to reintroduce Native foods to their meals, though she knew that reverting to diets subverted by the cash economy required time. She became so taken by the Comcáac that she decided to write her dissertation on them, combining cultural ecology with medical anthropology. She accompanied fishermen to learn how their navigational songs were steeped in ecological knowledge. With Seri women, she foraged plants, learning how much their ethnobotany depended on the ironwood tree, which shaded entire communities of other species from the desert sun, and documented how quickly diabetes attacks when native lore is abandoned.

iv. Harvest

"Eventually Gary and I realized that we were swimming in the same waters," Laurie told Juan Martín on his first trip to Punta Chueca in April 2021, "so we got married."

Early that morning, before the sun cranked the temperature, they'd been out with fishermen to see the eelgrass beds. "When do they harvest the xnois?" The following year he planned to return with Chef Ángel and a film crew.

The end of May was when the hataam loosened its grip on the seabed and began rolling in on the waves, but the Comcáac hadn't seriously harvested it for years. "But they will if we pay them," Gary said. The Seri had survived in the cash economy by learning to charge when someone wanted something from them.

The brightly garbed women who, despite the heat, wore long, hand-sewn skirts and long-sleeved blouses, ribboned with colors, now painted their faces only for festivals, they told Juan.

"When is the next festival?"

"When do you want one?"

They went to see Erika Barnett, a lithe Comcáac woman in her early 30s with nearly waist-length shining black hair, who worked with Gary on mangrove restoration. Her grandfather, Miguel Barnett, had been a shaman, as was her uncle Chapo, who paddled the last Seri balsa from Tiburón for the public TV documentary. The production company's budget had been nearly drained by daily demands for more money to pay the boatbuilders, the women weaving a huge ceremonial basket, the ironwood figurine carvers, and Chapo Barnett's holy-man fee, but the film's harried director, a full-blooded Choctaw, said he understood exactly why Natives who'd nearly been exterminated squeezed outsiders for all they could get.

Erika showed Juan the small greenhouse she and her husband,

Alberto, had built with funds Laurie had raised for them, hung with pots made from plastic Coke bottles cut in half, each containing a red or black mangrove seedling, the two predominate local species.

Since 1980, more than a third of the world's mangroves have been cleared—or lately, lost to salt water overwhelming their brackish environments. It was no different here. "Our mangroves used to flood four or five times a year," Erika said. "Now it's 20 to 30 times. We keep planting behind them to stay ahead of the rising sea." She pointed to a pile of construction wreckage a hundred meters closer to shore, collapsed by a storm surge that flooded a third of the village. "I was born in that house."

Seven more homes were in the risk zone 20 meters from the shore. Punta Chueca's namesake sandbar was now bisected, its curved point now an islet. In a few more years, it would be submerged.

"We think by 2080 the village may have to retreat a mile farther back," said Erika. By then, they'd be backed up against the jagged volcanic Sierra Seri, with nowhere left to go.

When Gary and Laurie broached Juan's interest in a xnois festival, with women hauling in the harvest from the surf and pounding it to free grains they'd then prepare, she was dubious. "Nobody does that anymore. It's so much easier to just buy packaged beans and rice. I think the last in our family who did was my great-grandmother."

But they were already collecting grains, she acknowledged, because of a blue carbon grant that Gary got them after drone photos revealed dark gaps torn by hurricanes in the Gulf of California's largest remaining eelgrass bed. Erika was directing a team of Comcáac fishermen paid to drop biodegradable burlap bags containing mudballs mixed with xnois seed from their pangas to replenish the carbon-sequestering seagrass.

Whoever worked on the xnois festival would also be paid, Laurie said. "And you'll be rescuing knowledge for your kids."

Erika couldn't refuse. They would all do anything for Laurie. In the teeth of the pandemic, she and Gary had driven down from Arizona with a truckload of oxygen machines, medicines, masks, disinfectant, water filters, and hundreds of sheets and pillowcases. The Seri had believed themselves so isolated that they'd escaped contagion until Laurie, listening to a Desemboque woman's labored breathing, diagnosed her over the phone.

She and Gary had arrived on the day of a COVID funeral, the first of seven. The village health clinic had no stethoscope, the ambulance's tires were flat, and the state's electric company had cut their power because it hadn't been paid, so there was no running water or ventilation. Laurie, Becky Moser's linguist daughter Cathy, and director Lorayne Meltzer of the coastal ecology center that Gary's alma mater Prescott College had founded in nearby Bahía de Kino, overnight had raised $5,000 for the first load of medical supplies. Before the pandemic ended they'd raised $250,000 more to revamp both village clinics and place off-grid solar panels and water storage tanks on every house. Laurie didn't return home for weeks. Between her early intervention and the Comcáac repertoire of anti-inflammatory herbs, there were no more COVID deaths.

The following May, Juan Martín returned with a film crew. A last-minute conflict had prevented Chef Ángel from joining them for the Fiesta de Xnois, but Gary had also invited another culinary star, James Beard Award–winning baker Don Guerra, founder of Tucson's Barrio Bread, where lines form daily, even in blazing summer. In the kitchen at Prescott College's Kino Bay Center, the Spaniards filmed him trading fiesta recipes with a Comcáac herbalist, Laura Molina, whose grandfather taught her which healing plants to collect and the songs for asking permission to take their leaves.

She made xnois tortillas and cookies. Guerra shared bread he'd baked with eelgrass grains Gary had provided. He'd brought an electric mill to show how he ground the flour, mixing the xnois with wheat, which, he told her, "Spaniards introduced to my Yaqui ancestors in the 1650s."

Molina, in a satiny purple blouse with red and white trim for the filming, frowned at the mill. "We don't have those."

"You do now," said Guerra, eliciting a smile that stretched the blue and red face paint across her broad nose and cheeks.

They didn't have ovens, either, but that was no problem. "We bake in pots," she said. The flour mill would speed preparation of the weekly traditional and medicinal meals Molina had begun offering, with no packaged junk. Afterward, she tested everyone's glucose levels; those who'd cut back on Coke and potato chips were already showing improvement.

Comcáac men took the Spaniards out on boats, singing of *Xnois cacáaso*, the redhead duck that first discovered *Zostera* here. "Every stand of eelgrass is dedicated to a different ancestor," explained a barefoot fisherman in an orange slicker. "When April's full moon brings the xnois, we say it's a gift from them."

That morning was windy, with a significant Infiernillo chop, and they barely found enough floating eelgrass to film. But they'd already been harvesting for a week, and the women had held back a pickup truckload's worth of hataam for the filmmakers. They waited on a beach just north of Punta Chueca, where in recent years rental campsites had appeared to accommodate the latest injection to the local economy: seekers of a psychedelic rush from smoking the secretions of the Sonoran Desert toad. Never mind that the bug-eyed amphibian, whose neck glands exude a hallucinogenic toxin when stressed, lives far north of Seri country, near the US border. A self-described toad doctor had appeared to broach a business deal with an Indigenous shaman. If young gringos were willing to pay hundreds of dollars for a guided spiritual high, Chapo Barnett was happy to comply.

After filming the pounding and threshing, the Spaniards stopped by Erika and Alberto's greenhouse, now expanded to grow squash, radishes, lettuce, beets, tomatoes, chilies, onions, chard, and cantaloupe. "Comcáac are hunter-gatherers of the desert, not farmers," said Erika, her cheeks decorated with concentric red, blue, and white triangles. "But during COVID, food was scarce, so we bought seeds and tried this."

Some elders had criticized them, which Gary Nabhan kept in mind when proposing his latest project, which, after some discussion, the Comcáac agreed to. For the past three years, he and a geographer colleague had experimented with growing crops under photovoltaic panels north of Tucson at Biosphere 2, a privately funded 1990s experiment to re-create Earthly ecosystems under glass domes, robust enough for humans to survive someday on Mars. Although a two-year test failed—within months, eight Biospherians sealed inside were driven out by runaway CO_2—it was an instructive reminder that believing human technology might ever replicate nature is the ultimate hubris. It was now a University of Arizona learning laboratory.

They found that kale, carrots, lettuce, fava beans, sorrel, and spinach did well in the shade of solar panels mounted high enough to walk beneath. The photovoltaics themselves, whose peak optimum temperature is 77°F, actually performed better, because the plants' aspiration kept them cooler.

Europe already produced tractors small enough so that even mechanized farming could happen in solar fields, making use of soil that otherwise went wasted below the panels. While it might take a while for America to retool, Nabhan wanted to try agrivoltaics where there were no tractors because there was no farming—yet. One hundred and seventy-two solar panels now stood at the entrance to Punta Chueca, waiting to be connected.

"The Comcáac love going out to out to forage together," Gary told Juan. "But they also want to access plants closer to their village.

Under these agrivoltaics, we'll plant things they've already used for centuries: chiltepines, wild oregano, wild onion, prickly pear—stuff that can stand the desert heat."

"Will they accept that?"

"It's a concept that even their ancestors understood. Desert plants do best in the shade of nurse plants like ironwood and mesquite," he said, mopping his brow with his T-shirt. "This is another kind of nurse plant—in fact, it's just another kind of solar collector, because every plant is a solar collector."

XNOIS ES XNOSOTROS

EN LO ANCESTRAL HAY NUESTRO FUTURO!

"Eelgrass grain is us, our heritage is our future!" read the hybridized Spanish–*cmiique iitom* banner in Punta Chueca's sandy central square. Under a shaded ramada on its south edge, face-painted Comcáac women in a rainbow of long skirts and blouses, heads swaddled in colored cottons against the fiery sun, had set out tables. Out came bowls of fish and corn ceviche; platters of mussels, shrimp, and fried mullet; plastic cups of tomatillo salsa; slices of prickly pear fruit, Laura Molina's xnois cookies, xnois energy bars mixed with oats and honey, and hunks of Don Guerra's fresh xnois bread. As young girls circulated with shallow baskets heaped with toasted xnois to sample and the Spaniards filmed, Erika Barnett carried out the main dish: a vat of shrimp xnois pilaf.

Digitally adept young Comcáac had put out the word on Facebook, Instagram, and TikTok to Indigenous and sustainability groups, and nearly a hundred visitors had come, some from as far as Mexico City. On a bandstand near the shore, the colorful women sang, and René Montaño, a Comcáac elder with a wispy white chin beard, told a version of the redhead duck tale.

"He spread xnois seeds in the Canal de Infiernillo for our ancestors to use. He gave us permission to collect it and told us to care for it well."

Overhead, brown pelicans flapped; higher, a pair of frigate birds hovered motionless on thermals rising from the pale sands. Seri men who had competed to see who could gather the most hataam stood near their boats, waiting for the winner to be announced. Twenty pickup-loads of eelgrass had been gathered, but after all was threshed and winnowed, the total harvest amounted to a little over six kilos of grain.

"A week of harvesting for only six kilos?" asked Juan Martín.

"I told you," said Erika Barnett. "It's a lot of work."

v. La Bahía de Cádiz

For two years, Chef Ángel León had been telling people that eelgrass grain was a superfood, that their studies found it contained more high-quality protein than wheat, barley, oats, or rice, and more high-quality carbohydrates and fiber than lentils, wheat, or millet. Not only did it have more vitamins A, B, and E than any other cereal grain; it was the only one with essential fatty acids omega-3 and -6. "We have discovered something more transcendent than anything else we have done before. We have opened a door to the future," he told the trade magazine *Foodservice Consultant.*

Learning to propagate eelgrass was the centerpiece of their research. "They only got six kilos from a thousand hectares of *Zostera* beds?" he was now asking Gary and Laurie. "That's a little surreal, no?"

"The Comcáac weren't trying to get as much as possible," Laurie replied. "'This is how we do it,' they said to us. 'You don't go and take it. You wait until the sea offers it to you.' They treat it as a sacred being. It was all done the old traditional way, consulting their elders."

A week after the Fiesta de Xnois they were in Spain, in Chef Ángel's Aponiente office. Mounted head-outward on the wall behind him were two dozen multihued fish species that he served. Two hours before he'd be starting dinner, Ángel wore sweatpants and a black T-shirt that featured a cuttlefish.

"Would the annual harvest have been more? Say, 600 kilos? Four hundred? There are no records?"

"The only amounts that mattered to them were enough or not enough."

León glanced at Juan Martín, who shrugged. So far, his own eelgrass tests had yielded nothing near their early predictions of 3.5 tons per hectare.

The day before, Juan had taken Gary and Laurie behind the old mill that housed the restaurant to see ancient saltpans that Salarte was converting into experimental ponds to test various *Zostera* varieties he collected. Under a cloudless sky, they walked along earthen dikes where Neolithic humans, then Phoenicians, Romans, Visigoths, Muslims, Sephardic Jews, and finally Spanish Catholics had produced salt. For centuries it was a strategic industry controlled by dukes and marquis, because it was needed to preserve meat—and also to control people: merely threatening to raise the price of salt kept them sufficiently subdued.

The Bahía de Cádiz was lined with more than 10,000 hectares of hand-dug ponds and channels like these. As late as 1970, before Westinghouse and General Electric brought commercial refrigeration to Spain, there were 170 active saltworks here.

"Today, there are four," said Juan. Along with an ancient tradition, thousands of jobs were lost.

When Juan was first named a Bahía park director, most surrounding *salinas* were abandoned and trashed. As he set about bringing back the natural marshlands, he'd discovered that until the 20th century, humans actually benefited wildlife here. Fifteen-foot tides filled rows of saltpans that drained by gravity to shallow pools, where sun and wind evaporated what remained to white crystalline salt. Each time the sluice gates opened, organisms entered with the tide, where they found shelter from waves and predators in the calm pond waters. Micro-ecosystems developed. Although Juan's mission was to nurture remaining virgin marshland, he was amazed to find even greater biodiversity in the saltpans: 127 species of microalgae, fish, and crustaceans "in the water alone," Juan said. "That's not counting all the shorebirds and ospreys that feed or nest."

A bull-necked man in a straw hat and chest-high waders named Ricardo Ariza, an Aponiente employee assigned to the eelgrass project, was mucking a pond. "Ricardo is one of 12 children," said Juan, introducing them. "They grew up in these marshes, foraging for shellfish like their great-great-great-grandparents. He knows far more about how this ecosystem works than I do. He's the one deciding where to plant."

"We need to keep the water flowing through these tanks," Ricardo told them. "You need plenty of current to stir up the nutrients from the bottom for eelgrass to grow."

"That's why the Infiernillo Strait, with its currents, has the most eelgrass," said Gary.

"How do you plant the seed?" asked Laurie, looking at a pond swirled with olive-green seagrass.

"We don't," Juan said. "The seeds we collect go to a lab that germinates them in a petri dish, then sends us seedlings that we plant in the mud. There's maybe only four square meters of eelgrass left in the bay, scientists say. Our seed reproduction success rate is only 7 or 8 percent."

"That's normal in the wild," said Gary.

"But to restore eelgrass, we need every plant to germinate we can get," said Juan. He showed them tanks where they were trying cold-water varieties from northern Spain. "We're trying to identify which plants tolerate more acidification, more salinity, or more heat to see if we can select for those traits. We talk with Spain's top oceanographers, with Dutch scientists replanting eelgrass in the Waddenzee, and with seagrass people in Bangladesh, Venezuela, Canada, Cornwall, Wales, Sweden, and Japan. We consult the University of Virginia's Environmental Institute in Chesapeake Bay, which knows more about *Zostera marina* than anyone. We visit the Comcáac." He removed his dark aviator glasses, his handsome sunburned face crinkled against the noon heat. "There's a lot to do."

They drove to Salina Balbanera, another site a half hour east that Salarte was restoring on the edge of Parque Natural Bahía de Cádiz, where Ricardo and his brothers were raising flounder and more *Zostera.* The flat landmass filling the horizon across from Balbanera was Isla Trocadero, where the Arizas hailed from, increasingly flooded at high tide. Next to a two-story structure covered in peeling white stucco that once housed salt workers, one of Ricardo's many sisters-in-law, Isabel, sat them under an awning and brought iced beer, shrimp pancakes, and omelets made from salt-tolerant pickleweed and their own chickens' eggs. Then, fresh oysters, clams in garlic olive oil, a bowl of green olives, and dishes of fresh sardines.

"This is a perfect lunch," said Laurie.

"These are just appetizers." Isabel laughed, heading inside. Presently she emerged with platters of fat mullet and *dorado*—gilthead bream.

"Everything except the beer comes from here," said Ricardo, beaming.

"We want people like these foragers to have a future," said Juan, toasting him.

"Thank you for bringing us," said Gary, who'd been watching flamingos and spoonbills hunt shellfish in the salt flats. "This is about much more than reviving eelgrass. You see *Zostera* as part of a mosaic of habitat and landscape restoration, including human livelihoods. So many people go after a silver-bullet miracle crop but forget its context, and it fails. I love that people have a useful place in an eelgrass ecosystem."

"I credit Chef Ángel," replied Juan, hefting the last oyster. "He has personally invested €400,000 in a project that we don't know will work. Sure, he's hoping to cultivate it someday on a reasonable scale. But he understands that, realistically, this isn't a commercial venture, it's a long-term experiment that is at least healing the habitat. We see fish and seahorses returning. We're growing *Zostera* in captivity from our own seed. No one's ever done that before. But there's so much research and testing left."

"Aponiente can't top this meal," Laurie said, finally pushing away her plate.

"Just you wait," said Juan.

The following evening's 15-course dinner began with hors d'oeuvres on a glass-enclosed patio at Aponiente's entrance, overlooking an estuary gilded by sunset. A server in a black coat—more than 20 would attend them that evening—poured a sparkling rosé made from blue grapes grown in the nearby town of Rota. Then, calamar tartare, drained for three days, topped with a strip of yellow koji dotted with sculpted squid flowers, bathed in vinegar infused with oregano, paprika, and garlic, and served on a thick clear-glass lens. The melded flavors and texture were both subtle and overwhelming.

Next, on a satiny white ceramic pillow: glistening strips of bacon—not pork, but sea bass belly—matched with a moist date confection. Then a foamed sea urchin atop a weightless, flourless wafer made from fish scales. More rosé, then a pan-fired snowflake pastry in an emulsion of parsley and baby onions, flecked with tiny shrimp and wasabi buttons on a bed of salt grasses and marsh flowers.

Again, those were just appetizers. "*¿Todo bien?*" asked the server who guided them inside the saltwork-turned-restaurant, its limestone walls hewn from the ancient underlying seabed. A final hors d'oeuvre, he said, would precede the meal: sardines marinated overnight in smoked olive oil, seared in salt with smoked eggplant and served on thin, crisp Spanish glass bread.

"*Mejor que bien,*" said Laurie. This was beyond eating. More like dreaming.

Earlier, she'd been upset to learn she couldn't ask Juan to join them for dinner.

"It's their protocol. Each dinner costs hundreds of euros," said Gary.

"I'm ethically against paying that much at a restaurant, and having our colleague excluded because of price. The whole thing is ridiculously expensive. It's just food. They're just making a profit."

But now, as the dream deepened, she surrendered. The warm staff doting on them, their pride in the provenance of each dish, attested to a happy workplace. "Between servers, cooks, kitchen help, and fishermen, your dinner employs 70," Juan had told them. Twenty-five more worked in La Taberna del Chef del Mar, the casual bar run by Ángel's wife, Marta, which offered his culinary creations to the community at affordable prices.

In the hour they'd spent in his office before he changed into his long-sleeved, collarless chef attire and they descended to dine, Ángel said that despite the hype about him discovering another delicious marine delicacy, his goal with *Zostera* grain wasn't to serve it at Aponiente.

"Give me the choice between finding new gourmet recipes or discovering new ingredients, discoveries excite me much more than hearing how great Aponiente's food is. A discovery could be transcendental for humanity. We're obsessed with finding a food that anyone can buy, costing maybe €3.50 to €6.50 per kilo. If in two years we know how to produce a truly sustainable cereal grain for people, I'd be thrilled."

"Well," Gary replied, rubbing his goatee, "of all grains, *Zostera* certainly requires the least energy and freshwater inputs."

"Exactly. We're under no illusions that it'll be easy. Maybe we have to invent a harvesting machine to get enough kilos per hectare to make it viable. Most important for me is that this project is a way of looking at the sea in the future. It's the chance for planting marine gardens, having different diversified foods that require people to maintain the environment and create a circular economy."

He quoted what later they found printed in their menus: "Why is our planet called Earth when three-fourths of it is covered with water?"

He added, "What I want now is that you enter into our dream. I will generate sensations, textures, flavors, new elements in your mouth that you don't need soil to tell you. If land didn't exist, we'd still have enough protein. The dream we offer is that the sea can give you everything. Nothing needs to come from the land."

"Two-thirds of us is water," the evening's menu continued, "practically the same mineral composition as the sea, as was the water that harbored us for nine months before we're born. As are our tears, both of sorrow and joy."

Moving sentiments, but Ángel's assertion that land was superfluous was belied by various delectable fruits of the vine. El Puerto de Santa María being a vertex of Andalusia's famed Sherry Triangle, the pairing for the next course was a dry white fino that whiffed of the sea, pressed from local palomino grapes, then fermented and fortified expressly for Aponiente. It accompanied the charcuterie cart wheeled up to their table, hung with a dozen different sausages of varying color and diameters up to a large baloney, two small hams and a large slab skewered on a rotisserie, from which a masked server sliced prosciutto—all made from red tuna.

Except for the bread crisps, everything came from the surrounding wetlands and the continental shelf beyond—although the main ingredient in the plankton butter was now farmed faster in Aponiente's laboratory than harvested from the sea in fine nets. "Remember that half our oxygen comes from plankton," said Ángel as he circulated among the white-linen-draped tables. "We wouldn't be here without it."

The other 18 diners, among them the manager of one of Spain's oldest football clubs, had been so hushed in the presence of such immaculately presented food that the American scientists had barely registered their presence. Each course, only a few bites, was just enough, its flavor so unexpected and engrossing. They kept coming: A vichyssoise of slow-cooked sea vegetables with a chilled white muscat of Alexandria from an ancient local vine. Braised sea snails in red wine and halophyte-garlic oil with a chilled white from a seaside vineyard. Razor clam sashimi floating in a yellow clam stew, paired with aged manzanilla sherry.

The next sherry, which accompanied a plump local oyster swimming in velvety green salsa, was an oak-aged Raya Cortada from the same winery as the previous, and came with an explanation from the sommelier, something about yeast and reverse oxidation.

"This many glasses into the meal, it's too complicated to follow," said Laurie.

"Exactly," agreed the sommelier. "The notes are very complex."

It was all ambrosian, they assured her. They'd entered an oenological dimension where they'd probably never be again, where liquid tasted so unique its memory would be irretrievable.

That also described the iced cuttlefish kakigori in warm sea broth, as deliciously mysterious as a koan, with a deep-cellar amontillado that imparted strangely apt hints of iodine. Then came whiting in an onion reduction, marbled with plankton, with golden oloroso sherry. And then, what could qualify as the main course: slices from a two-foot porchetta log on a plank, four inches thick, its center cut from an Atlantic tuna's nape, coated in moray eel roe and sea lettuce, then wrapped in moray eel skin and served with icy sparkling fino.

Dazed, amazed, they looked at each other, Laurie in silver bracelets and a pastel rainbow silk scarf, her dress pale lavender; Gary in a long-sleeved blue linen guayabera. As a native-plants gastronome, Gary had eaten and written about food all over the world, but this was completely original. Reflection would come later, but first, desserts: a macedonia, somehow made of shellfish and fruit gel; a caramelized kelp-apple tarte Tatin pastry; and chocolates too lovely to eat, but they did. The sweet wines were two muscats and a 15-year-old blend of olorosa and Pedro Ximénez, Spain's emphatic answer to port.

They walked back to their hotel under starlight, still tasting. A month later, in June 2022, Aponiente would be named Earth's most sustainable restaurant by the World's 50 Best Restaurants for its maritime take on snout-to-tail cooking: wasting no part of every fish, finding something savory to do with everything in the net, and eating lower on the marine food chain than anyone had ever imagined.

Whether Chef Ángel could conjure a sea grain that could feed the masses was uncertain, but in the 21st century, uncertainty was the rule. Warming seas, acidifying seas, overfished seas, overpopulated and over-stressed civilizations—it was a lot to ask.

"We loved meeting Ricardo and Isabel," Laurie told Ángel. "It was moving to realize that just like the traditional knowledge of the Comcáac, the knowledge here goes back thousands of years."

"Now we're bringing it forward," Ángel said. "With the land over-exploited, the sea is the future of food. All your work, all the work we'll keep doing—we're opening windows into that future. Sooner or later, the world will have to look through them."

CHAPTER TWELVE

UMAMI

i. Manifesto

Vincent Doumeizel, the senior advisor on oceans to the United Nations Global Compact, has an unusual theory of how western civilization developed. A lean, fiftyish Frenchman who also fronts a rock band, Doumeizel isn't a historian but an economist. His expertise is food security and sustainability; he also directs the Food Programme at the Lloyd's Register Foundation.

His idea doesn't involve our Mesopotamian predecessors' advances, but rather, something they lacked. That was edible seaweed—not rooted seagrass, but marine macroalgae from which land plants evolved. The deeply tidal upper Arabian-Persian Gulf that birthed Sumerian culture, he says, wasn't ideal seaweed habitat. Instead, hunter-gatherers learned to cultivate onshore grains growing along delta rivers—and the rest is human history as many of us think of it.

Elsewhere, however, human history began independently on paths that never strayed far from the sea, from which all life springs (we and seaweed, in fact, share common biological ancestry). More than 2,000 years before Mesopotamian agriculture, Doumeizel writes in his 2021 book, *The Seaweed Revolution,* humans entered the Americas. This earliest incursion, he suspects, was not over the Bering land bridge that only became passable later, but by following what he calls the "kelp highway" along the Pacific coast—forests of bull kelp off Alaska and giant bladder kelp off California, which resumed again off Ecuador, Peru, and Chile. In between, from Mexico to Panama, these nomads or navigators—it's unclear if they walked or used watercraft—would have found abundant tropical red and green algae. Like kelp, those seaweeds were not only nourishing; they were keystone species that hosted bountiful fish and crustaceans.

A site in southern Chile's lake district, Monte Verde, evidences human habitation at least 14,500 years ago. Discovered in 1976, Monte Verde provoked controversy among archaeologists because it predates the Beringia migrations that led to widespread permanent human presence in the Americas. Gradually, however, it's become accepted that at least one band of humans who likely picked their way down the Pacific coast lived there for a time.

What fascinates Doumeizel is that several species of seaweed were found at inland Monte Verde, some from the Pacific, 36 kilometers distant—meaning the settlers valued it enough to travel or possibly trade for it. Some were in cooking hearths; others were formed into quids evidently to be chewed, suggesting that these earliest Americans used marine algae medicinally as well as nutritionally.

"All nine seaweed species recovered at Monte Verde are excellent sources of iodine, iron, zinc, protein, hormones, and a wide range of trace elements, particularly cobalt, copper, boron, and manganese," wrote the site's discoverer, archaeologist Tom Dillehay, in the journal *Science.* The Mapuche and Yagán of southern Chile, he noted, still employ those species medicinally. Along coasts of every inhabited continent and island, Indigenous use of marine vegetation suggests that seaweed was once integral to the human diet and pharmacology—except, Doumeizel believes, at the top of the narrow Persian-Arabian Gulf. Thousands of years later, he says, Mesopotamia's dependence on land-based crops has consequences.

"Mesopotamia directly influenced Greece and Rome. In the *Ae-*

neid, Virgil says nothing is more worthless than seaweed. Aristotle placed it lowest on his Great Chain of Being. The Vikings ate dry seaweed to avoid scurvy while sailing to Greenland and Iceland, but as Greco-Roman influence pervaded Europe it was forgotten. Before they were colonized, New Zealand's Māori and Australia's Aborigines ate seaweed. The same in America, where many First Nations seaweed traditions have disappeared. Only Japan, Korea, and China managed to maintain an appetite for these foods."

Gradually, Doumeizel lost his own appetite—for land-based agriculture. "I was working in social compliance," he says. "I realized sugar, cocoa, and tea essentially depend on modern slavery: 40 million people kept in misery, three times more than America imported from Africa. Hundreds of millions are polluted and starving. Our supply chains drive climate change and biodiversity loss. Intensive agriculture destroys the soil and uses up the water. I realized there is no solution."

On land, that is. In 2020 Doumeizel's UN initiative published the "Seaweed Manifesto," endorsed by the secretary-general, to convince us to make an enormous dietary leap for the planet's health and our own. Seaweed, it says, is an important source of vitamin B_{12} and polyunsaturated omega-3 and -6 fatty acids. It boosts immune systems—"both human and animal, meaning less antibiotics in animal feed," Doumeizel explains. A 2023 World Bank study agreed that both livestock and pets represent huge market opportunities for seaweed. There are also much-touted indications that adding red algae to feed reduces bovine methane eruptions by up to 95 percent.

A vegetarian, Doumeizel would prefer that sea vegetables feed humans, rather than the bovids, swine, and poultry that humans consume. "But it's still better to feed salmon or sheep seaweed rich in protein than GMO soy meal that deforests Amazonia and must be transported around the world."

With no underground roots, every bit of seaweed photosynthesizes. Kelp, the biggest marine alga of all and one of the fastest-growing organisms on Earth, absorbs five times as much carbon dioxide as land-based plants. What eventually happens to that captured carbon is unclear, but when kelp decomposes, at least some stays in the ocean depths. Seaweed's nutritional value and prodigious appetite for excess agricultural runoff is uncontested. As land-based crops increasingly suffer parching heat, water scarcity, and soil salinity, our destiny, its proponents contend, may hang on whether we can domesticate and raise vast amounts of it.

Humans have long cultivated marine algae, but mainly just six of the 12,000 total seaweed species, all believed to be edible. Three that we farm are sushi components—nori, kombu, and wakame—and we use three others for making noodles or thickeners like agar and carrageenan.* Ninety-eight percent are raised in Asia. The UN's Global Seaweed Coalition, which Vincent Doumeizel heads, is charged with expanding seaweed farming on Earth's five other continents and their inhabited islands.

Whether we'll ever realistically crave umami-flavored, savory sea vegetables enough to ease pressure on the land and let forests grow back is uncertain, but to avoid calamitous famines and torrents of hungry migrants, we still must produce as much food by 2050 as we've grown in all human history. So we'll have to do something.

During talks, Doumeizel often produces a compostable cup made from seaweed, which he says can actually be eaten. Already, several

---

* Carrageenan, extracted from red algae and widely used to stabilize foods like ice cream, peanut butter, and processed meats, has been suspected of causing intestinal inflammation and even bowel cancer, especially in countries with high sugar and additives in their diets. The EU prohibits carrageenan in baby formula.

American and European startups produce plasticware and plastic wraps made from seaweed that swiftly biodegrade once they've served their purpose.

The challenge is that petroleum-based plastics that end up in the ocean are still far cheaper. (That is, to their manufacturers. The rest of us pay the oil industry's heavy subsidies and, worse, the environmental consequences.) In the adolescent 21st century alone, the amount of plastic has more than doubled. It's no longer just about octopi and sea turtles dying after mistaking plastic bags for jellyfish; it's about us. Single-use forks and coffee cup lids—even those with the comforting recycle triangle—break into micro-bits that end up in our bloodstream, breast milk, placentas, and our brains.

Unless our heads are too stuffed with polystyrene to calculate the actual price of petro-plastic, if we factor in all the damage it does, seaweed-based cutlery that dissolves in the water that bore it starts becoming economically competitive.

"Seaweed is the greatest untapped resource on this planet," insists Doumeizel. "If we get this right, we're starting a great journey: the first generation to sustainably harvest the potential of our largest common good, the ocean."

A self-described "optimistic citizen of the world," he leaves unspoken that if we don't, we could be the last.

ii. Water Farm

Bren Smith, wearing a wool plaid shirt over several layers, plus a polar-fleece scarf, chest waders, red rubber gloves, and a green whaling cap over his smooth pate, heads out on a February morning to inspect kelp and shellfish lines and do a biweekly water-quality check. His Thimble Island Ocean Farm, which floats and bobs on 20 acres of Long Island Sound, is 10 miles east of his home on New Haven's Quinnipiac River, an 1870s Gothic Victorian built by the inventor of the deep oyster dredge. His upstairs windows overlook some of the best oyster beds left on the northeast US coast.

Until the early 1990s, waters here were filled with lobster boats, but those crustaceans have crawled north to Maine and Canada. "A good, easy day to farm," Bren says to Ron Gautreaux, who's piloting: overcast and hazy, 35°F, flat and calm, little wind.

Ron, a rangy biologist with an inverted U-shaped mustache, weaves them through the granite outcroppings that make up Connecticut's Thimble Islands, some big enough for only one home. Former residents include President Taft, Tom Thumb, Orson Welles, Jack London, and the pirate Captain Kidd. The Statue of Liberty's pink base was quarried here.

Through the early 20th century, these were some of the planet's cleanest waters, filtered by billions of oysters that bred in inlets where rivers ran to the Sound. An acre of oyster reef can filter 40 million gallons of water in a day, protect shorelines from surging storms, and provide habitats fish love—but it took only that dredge, and untreated sewage from burgeoning cities, to kill 99 percent of the crop, sabotaging Connecticut's oyster industry.

All that's visible of Bren's farm is a grid of small black and white buoys, spaced 50 feet apart, the white ones at the corners. Hooking one, they haul up a line heavy with dripping brown blades of sugar kelp, eight feet long and ruffled like lasagna. Pulling a fillet knife from his belt, Bren cuts a piece, nibbles it raw, then blanches the excess brine and iodine by dipping the rest in a thermos of hot water, turning it bright green. He tastes again.

"Mouthfeel like noodles. Rich umami." Come April, it will be harvested, unfurled, trimmed, sorted, blanched, dried, and sent to be turned into plant-based meat substitutes, bioplastics, and soil biostimulants.

During COVID, when they couldn't risk doing any of that in their processing plant, it had occurred to Bren's marketer, Samantha Garwin, that Connecticut, once the world's top producer of fine cigar wrappers, had at least a hundred mostly idle tobacco barns. The farmer she contacted was more than willing: hanging brown sugar kelp from horizontal lathes between the beams took the same skills that four centuries' of tobacco workers have used. The plants even look similar.

With a global respiratory pandemic driving even more people to quit smoking, the unexpected rental income from diversifying with a crop that wasn't a cancer delivery system was heaven-sent—until Connecticut's Department of Consumer Protection, concerned about flies landing on the kelp, shut them down. "Meaning it's okay for smoking, but not for eating?" protested one of their seaweed buyers, to no avail.

"It completely gutted an idea for leveraging existing infrastructure," says Bren. "In South Korea, they dry it on land every day."

A deep breath stretches the suspenders of his chest waders, then expels as a sigh. "It's a climate crisis! To get lots of kelp and a regenerative economy growing, maybe a few flies are a lower priority?"

During the 2010s, Bren Smith became America's poster boy for seaweed, starring in dozens of national articles, broadcasts, and YouTube videos that featured him, telegenic in his trademark red-and-black-checked shirt, extolling the promise of capturing carbon, cleaning up excess nitrate pollution, and producing chemical input-free, healthy food from the oceans. Unlike on land, where family farms had largely given way to agro-industrial behemoths, he asserted, with less than a $30,000 investment small-plot sea farming could produce huge amounts of food and provide a viable living.

Folksy but articulate, self-effacing (he never learned to swim, he admitted, because fishermen believe it prolongs a shipwreck's agony), the press loved him. "We don't have to feed, fertilize, or water our crops," he'd say, lifting ropes draped with curtains of kelp alongside mussel socks and oyster cages. "When storms wash over our farm, it just sinks and pops back up. If tides rise, people can build walls or flee, or we can embrace the ocean as a solution."

People might be ravenous for hopeful stories, but would they be hungry for kelp? Was seaweed a savory tale too good to be true? Could shellfish really become staple proteins, when so many people are allergic, including Bren himself—and his wife, who, when he sent her a box of his farm-raised clams the day after they met, broke out in hives?

Born to American parents who fled north during the Vietnam War, Bren Smith grew up playing on a Newfoundland fishing village's stony beach. After draft dodgers were pardoned, they returned for better schools when he was 11. That worked for his sister, who went on to Harvard, but Bren ached for the sea. He dropped out, hauled traps on a Massachusetts lobster boat, and got caught peddling marijuana. A few nights in jail later, he hitched to Alaska, where he worked in a salmon cannery, on crab boats, and baiting cod lines on a Bering Sea trawler.

After two years he went back to New England, got his high school degree, and enrolled in the University of Vermont. Each summer, he went back to Alaska to earn tuition by chasing black cod destined for McDonald's. He was there in 1993 when the cod fishery and 37,000 livelihoods vanished overnight in his beloved Newfoundland. Bren could see where industrial fishing in the Aleutians was also headed: to maximize catch, they'd run 40,000-hook longlines that illegally encroached Russian waters. No romance, only waste, with thousands of pounds of dead bycatch tossed overboard with each haul.

Before his senior year, he returned to Newfoundland, where new salmon farms promised an alternative to hunting marine wildlife.

What he found was an aquatic feedlot, with an industrial pig farm's worth of salmon shit fouling the sea bottom. Cattle notoriously eat six pounds of feed per pound of beef: a reason why 70 percent of US grain is consumed by animals, not people. Salmon are even worse: their conversion ratio is 10-to-1, and they're carnivores. Along with daily antibiotics to prevent epidemics from decimating a monoculture, the pellets Bren shoveled into their packed cages were exhausting South America's anchovy and herring stocks. He fled.

A Vermont politician hired him to do community organizing. The farmers, miners, and service workers reminded him of fishermen he missed. He learned he was good at public speaking. He campaigned for Congressman Bernie Sanders and followed him to Washington as a labor policy analyst. He attended, and hated, Cornell Law School. In the early 2000s he was living off-grid in a used Airstream parked in the woods across the road from his artist girlfriend's parents in suburban Connecticut. He wrote for the leftist press—*The Nation, Common Dreams, Z Magazine*—cofounded a labor think tank, and coauthored books on globalization and American war crimes in Iraq. None of that paid, so on weekends, he and his girlfriend sold street art in New York City parks.

Then he read in a local paper that Branford, the coastal town next to where he had his trailer, was leasing tracts of open water for $50 a year per acre in hopes of resurrecting Connecticut's oystering tradition. A thousand dollars got him 20 acres. Another $1,500 bought him a questionable boat. The rest of his money purchased cages and mesh bags, buoy markers, and oyster seed. After learning it's possible to drown oysters, his skills improved. He added clams, tried scallops, and eventually was growing enough to start a CSF—a community-supported fishery, whose members bought shares to front his annual expenses in return for a monthly crate of shellfish.

In 2011, Hurricane Irene buried his oysters and nearly all his gear in mud. A year later, Superstorm Sandy all but wiped him out. He was wondering what, if anything, might be less vulnerable when the same local newspaper featured a University of Connecticut professor, Charles Yarish, who was teaching vocational high schoolers to grow kelp at an aquaculture lab in Bridgeport, just 20 miles down the coast.

Dwindling fish stocks had led Yarish, an evolutionary marine biologist, to investigate culturing seaweed. He became a world expert—except no one in New England where he lived farmed aquatic plants, so in 2006 he helped two Maine entrepreneurs start a company, Ocean Approved. Their products were frozen kelp cubes to give smoothies a nutritional boost, kelp extract iodine supplements, canned seaweed salad, and pizza sauce.

Yarish's Bridgeport laboratory farm consisted of polypropylene lines suspended below the surface between buoys—which, when pulled up, were hung with a wall of sugar kelp, known as kombu in Japan, where it's pickled, dried for soup stock, and used to flavor sushi. Thriving on nutrients in cold ocean waters that upwell to the surface, its growing season is November to April.

His vocational students showed Bren how to strip spore tissue from the dark, central nutrient-transporting stipe of kelp blades into buckets during breeding season, to then pour into aquariums containing spools of string, which the kelp seed adhered to. In the fall, they outplanted by winding the seeded string along the length of the poly lines. By December, they'd be heavy with long blades of kelp.

If teens could grow kelp, so could he, Bren figured. He sketched a working design for a three-dimensional, floating scaffold of horizontal lines hung with vertical ropes. At the four corners, anchor lines would attach to buoys light enough to simply go under in high seas, then resurface as the surge subsided. Yarish, pleased to see his research applied, taught him multitrophic aquaculture: using the whole water column to grow food. At the bottom, clam beds, dug into the mud. Above them, oysters, started in cages and then hung in multi-tiered mesh nets resembling Japanese lanterns that could also hold scallops, alongside long socks filled with mussel seed, tied like sausages. (The mussel spores that Bren normally had scraped off his oyster gear, he realized, were the sea offering another free crop to raise.)

Along horizontal rows spaced 50 feet apart grew the kelp, whose appetite for carbon creates what Yarish called a "halo effect" that reduced shell-dissolving acidification in surrounding waters. By 2013, Bren Smith had the second-largest kelp operation in the US, after Maine's Ocean Approved. The local public radio station aired a piece about his farm, which was picked up by NPR's *Morning Edition,* which prompted a *New Yorker* Talk of the Town feature. Next, *The Wall Street Journal* had New York chefs try his kelp. They became some of his biggest customers.

Leveraging the attention, Bren launched a Kickstarter campaign to scale up. "Imagine a farm designed to restore rather than deplete our oceans—a farm growing local food but also biofuel and organic fertilizer," went his pitch. His 3D farm's kelp, he said, could absorb 160 tons of carbon annually and, with its shellfish, filter the water and absorb nitrogen runoff before it could feed harmful cyanobacteria blooms.* Like a coral reef, his dense rows of kelp and hanging nets would provide a habitat for 150 species of fish and crustaceans, his entire flexible setup acting as a shock absorber for storm surges—and producing nutritious food.

* Unlike seaweed, cyanobacteria, the oldest organism on Earth—often called blue-green algae due to the water color when it accumulates—isn't algae. Its mats can grow so fast on agricultural nitrogen effluent that it cuts off light to other waterborne organisms. When it dies en masse, its decomposition depletes oxygen, contributing to vast dead zones in lakes, deltas, and seas.

"Our seaweeds contain more protein than soybeans and more calcium than milk"—plus all those omega-3s, he'd add, and more vitamin C than orange juice. His goal was $30,000 to expand fivefold—but if he raised more, he would create a network of local ocean farms and train the next generation of 3D ocean farmers. "Join us to jump start the blue revolution!"

His project successfully kick-started, his image blossomed on the internet. He and a rural community developer friend, Emily Stengel, cofounded a teaching nonprofit, GreenWave. In 2015 his 3D farm won the Buckminster Fuller Institute's $100,000 Challenge prize and was honored by the Clinton Global Initiative. It was named one of *Time* magazine's 25 Best Inventions of 2017. His 2019 book, *Eat Like a Fish*—part memoir, part manual, part manifesto, part cookbook—won a James Beard Award. Besides boutique restaurants that garnished salads with kelp noodles, his customers included the plant-based burger industry and Loliware, a startup turning seaweed into compostable drinking straws and flatware. Even Purina showed up, having done nutritional analysis on 200 different seaweeds to see which might be suitable for the pet-food industry.

Why, he wondered, wasn't the people-food industry doing the same?

In 2020 Bren Smith was named one of *Rolling Stone*'s "25 People Shaping the Future." The following year brought GreenWave the Curt Bergfors Food Planet Prize, worth $2 million. Then came $2.5 million more from the US government's Build Back Better Plan, plus rolling grants from the Walton Family Foundation and other private philanthropies, and from the National Oceanic and Atmospheric Administration. Thimble Island Ocean Farm was now a training facility for GreenWave, which had a $5 million annual budget for research, training, and subsidies that included two years' worth of free seed to jump-start new growers.

iii. Umami?

Kodiak, Alaska, has six canneries, mainly for salmon, halibut, flatfish, and the backbone of their extractive fishery, pollock. Some years they collectively process nearly 400 million pounds. Most of what's left after canning gets turned into fishmeal, earning fishermen another two cents a pound. The remaining waste, nitrogen-rich fish sludge, flows into the Gulf of Alaska—so much, that satellites passing overhead can see Kodiak's entire harbor turn white.

To Kodiak fishermen Nick Mangini and Alf Pryor, it's a perfect opportunity awaiting: to recapture all that nitrogen with seaweed. Kelp—especially a species known as winged kelp, or wakame—is a veritable nitrate sponge. But thus far Alaska offers only carbon credits, not nitrate credits—and the reality with carbon, says Alf, "is you get maybe a hundred dollars a ton. That equals fractions of a penny."

So until the world got serious about paying for the collateral-damage costs of CO_2 and pollution, the kelp market was mainly for food—except it wasn't much of a market. The only sizable Alaskan company specializing in edible seaweed, Juneau's Barnacle Foods, makes kelp-based hot sauces, condiments, and pickles, not main course items. A few years ago, Nick, a commercial fisherman for three decades, contracted with a company owned by a Getty heir for high-quality food-grade kelp to market in the Bay Area as sea vegetables. He invited Alf, a fourth-generational Kodiak gillnetter, to partner with him. They soon saw that, unlike raising crops on land, growing kelp doesn't displace habitats—it creates them. Nick hadn't seen side-stripe shrimp for decades. Now they were living among the dense foliage on their lines. That year, they sold 400,000 pounds of kelp. The next year, the same.

With Alaska's salmon catch getting shaky and snow crab numbers tanking, they couldn't believe their good fortune. Then the pandemic hit. Purchases were cut in half. The next year they dropped to 60,000 pounds, and never rebounded. "It turned out the kelp was for foodie cafeterias at tech companies like Google," says Nick. "Since COVID, everyone works from home."

"We stripped hundreds of thousands of pounds back into the water. A waste," grunts Alf.

Scrambling for income, they got small grants from the US Department of Energy's Advanced Research Projects Agency–Energy, which was interested in growing massive amounts of kelp for biofuel. ARPA-E wanted an innovative harvesting design, so Alf rigged the "Kelp Buddy"—a 16-foot barge, shallow-hulled to float atop the kelp lines.

After reports that Australian tropical red seaweed cut methane in cow belches by 95 percent, a Faroe Islands company, Ocean Rainforest, bought 80,000 pounds from Nick to send to South Dakota for a livestock-feed test pilot. The trial showed that cold-water sugar kelp doesn't reduce bovine methane, but produced other benefits.

"Initial findings show that adding fermented kelp to both pig and dairy cow feed is really good for gut health," they told Nick. "Some cows are averaging 10 percent more milk. The sows are producing more piglets. But these are early trials, so we'll see."

In the meantime, GreenWave kept them going. GreenWave had begun the Kelp Climate Fund to support licensed farmers with subsidies of a dollar per linear foot of seeded kelp line, up to $25,000 annually per farm for three years. As the number of participating farms from New England and Nova Scotia to Alaska went from nine to 50, and the tonnage of absorbed carbon and nitrogen grew accordingly, the awards hit $600,000 and would keep rising as more farmers were incentivized.

"You're giving me free money—what's the catch?" growers would ask GreenWave's director of training, Lindsay Olsen, daughter of an Alaskan fishing family.

"There isn't one," she'd tell them. "We're just rewarding work being done instead of continuing to talk about kelp's potential to capture carbon."

"That saved me the last two years," says Alf.

"I worry about people getting too reliant on subsidies," says Nick.

"The goal isn't subsidies forever," says Bren Smith. "But the idea that free-market dynamics work when we're facing climate change—the largest market failure in history? It's insane to think we can build this just by selling kelp noodles. When blue carbon credits"—offsets for protecting or planting carbon-absorbing marine species—"are finally real because the market decides to reward businesses that help mitigate climate change, then subsidies won't be necessary. But to smooth out the bumpy ride while innovation, processing, and marketing develop, subsidies are key."

In Europe, Japan, and the US, subsidies are routine for land-based farms, although the source is government taxes, not grants from a nonprofit. In the US, because the more acres planted, the higher the subsidy, they mostly reward those who least need them: corporate owners of massive industrial farms. (Energy subsidies, reports the International Monetary Fund, are even more skewed; including the free pass they get for damaging health and the environment, fossil fuel companies worldwide receive more than $7 *trillion* in government handouts annually, at least 35 times more than for renewable energy development.)

In return for GreenWave's largess, farmers photographically sample their crop monthly and anonymously share growth and harvest data via a My Kelp app. In collaboration with PlantVillage, a two-time Xprize winner for using AI to aid drought-stricken African farmers, images are analyzed to estimate yields and quality, then summarized so ocean farmers can see regional trends and how much carbon and nitrogen they're collectively absorbing. After the waiting list for their in-person training swelled to 8,000,

GreenWave switched to a web-based ocean farming hub with how-to videos, interactive designs, gear lists, budgeting plans, and networking tools to connect growers with markets. Their first online six-week Kelp 101 course drew 1,300.

They flew kelp farmers from Maine, the biggest seaweed-harvesting state, to Alaska to learn the speed-harvesting techniques Nick and Alf had developed, and flew Alaskan farmers to see how their Maine counterparts process their kelp.

"I worried about giving trade secrets to our competitors," Nick said afterward, "but it turned out to be great networking. I still talk to those people."

GreenWave also brought a vice president of Organic Valley, the world's largest organic family-farmer dairy cooperative, to Alaska to broach the idea of forming a seaweed co-op, so growers could set standards and prices as the industry developed.

"Between five or six farmers," said Nick, "we could easily put together a couple million pounds and guarantee large processors the volume of healthy Kodiak kelp that they need." Nike had contacted him about trying kelp for packaging, and possibly shoe soles—both potential uses for nonfood-grade kelp that gets bio-fouled by crustaceans and other organisms at the end of harvest as waters warm. "With a cooperative sharing the revenue among its owners, we could handle those large-scale markets."

"*Co-op* is kind of a bad word in Alaska," said Alf. "But I think that's where we'll end up."

In 2017, a Maine coastal development nonprofit, Island Institute, researched kelp's potential to augment the state's lobster fishery's dependence on a single species. The lone buyer then was the small kelp-cube purveyor that UConn professor Charles Yarish had helped start, Ocean Approved. In 2018, an author of that study, Briana

Warner, a former US foreign service officer in Africa who'd followed her husband to Maine, became Ocean Approved's CEO. Renaming it Atlantic Sea Farms, she started courting lobstermen and investors.

Five years later, 27 farmers were landing a million pounds of kelp raised during lobster's offseason, which Atlantic Sea Farms purchased and sold frozen—packaged whole or in cubes, including berry-flavored—or in its line of kelp burgers, salsa, and kimchi. The company provided farmers free seed, technical assistance, and processing in a 27,000-square-foot facility. Some grossed over $100,000.

But like other commercial seaweed pioneers, such as Cape Cod's AKUA, which featured Nickelodeon-sponsored SpongeBob SquarePants kiddie-kelp burgers—Atlantic Sea Farms still operated on venture capital, not profits. (The exception, Alaska's Barnacle Foods, had achieved break-even by targeting the specialty gourmet niche.) All promoted kelp as rich in umami—the "fifth flavor" after sweet, sour, salty, and bitter. Yet although umami is the deliciousness our tongues savor in beef, mushrooms, anchovies, and parmesan cheese, nutrient-packed seaweed remains far from being a staple on American and European plates. So like the Faroe Islands' Ocean Rainforest, Warner started looking to animal feed as her crop's future.

On Penobscot Bay, with so many rocky inlets and spruce islands that Maine's total shoreline surpasses California's, Atlantic Sea Farms' top grower, ruddy-jowled Keith Miller, has been lobstering for more than a half century, since he was six. He's not sure how much longer he can—not that he's going anywhere, but the lobsters are.

"We used to get hard-shells from December through May. In June, they'd go into the rocks to shed. But the last few years, we're seeing soft-shells in March. Hard-shell lobster is now a very short season. They're hard to come by. The climate is changing greatly, so we're pushing more towards kelp."

He and his son run over six miles of kelp lines—32,000 feet. But with the Gulf of Maine warming by 1 degree Fahrenheit per decade since the '80s, even kelp is now vulnerable. After two straight 170,000-pound years, in 2022 they got less than half that.

"Strap kelp"—what Mainers call wakame, often used in seaweed salad—"killed us. We grew maybe 20 percent as much as before. More climate change, I guess."

iv. Burials and Polymers

"We're building an army of climate adapters and mitigators," Briana Warner says, but a race is on between seaweed's capacity to inhale atmospheric carbon and the still-rising emissions that threaten it. It's also uncertain what happens to that carbon. One theory is when kelp dies or breaks apart in storms, its buoyant pieces can float thousands of miles. When they finally sink, the carbon stays sequestered in the ocean depths a century or more. A century, however, only postpones the problem. There's also no way to measure how much carbon from decomposing kelp simply dissolves into seawater. Nevertheless, millions in venture capital have backed schemes to accelerate that process, by funding startups to grow kelp from floats in the open ocean, then sink it—and then sell carbon credits to companies seeking to offset their fossil-fueled exhaust.

The stated goal of one, Running Tide Technologies, was to remove a billion or more tons of carbon every year. "What we have to do is run the oil industry in reverse," its CEO told NPR. "The kelp will sink and essentially become part of the ocean floor. That gets you millions of years of sequestration, making oil"—meaning it will be buried and, over eons, be compressed into liquid hydrocarbon again. Its carbon credits customers included Shopify, Stripe, and the Chan Zuckerberg Initiative.

A team of experts that Running Tide hired, some from the Woods Hole Oceanographic Institution and the Monterey Bay Aquarium Research Institute, cautioned that sequestration timescales were likely far less and recommended more research. Then, in 2022, amid problems raising kelp in the nutrient-poor open ocean, Running Tide suffered a mass exit of scientists after it filed to patent a controversial process to speed kelp growth by artificially fertilizing the ocean with iron filings. Other scientists questioned how its vision of millions of floating kelp microfarms might compete for the same carbon with phytoplankton, the base of the world's food chains, or turn the ocean into an obstacle course for migrating whales—let alone the unknown biochemical impact of massive amounts of kelp on the seafloor.

In 2023, Running Tide abandoned kelp and switched to chipped Canadian lumber scrap. After dumping 25,000 tons off western Iceland, in 2024 it closed, blaming lack of governmental financing for carbon control, and leaving behind a 10-meter-high mountain of woodchips on the Icelandic shore.

Far less controversial is the chance for seaweed to remedy one of technology's most tragic unintended consequences: failing to foresee the dangers of creating an inexpensive, practically indestructible material. Atlantic Sea Farms is also now the main American supplier to Silicon Valley–based Loliware, one of the first companies to make plastic from seaweed, not from oil.

Loliware founder Chelsea ("Sea") Fawn Briganti, parlaying winsome looks and model-worthy long blond hair combined with keen entrepreneurial acumen and heartfelt convictions, has become a convincing spokesperson for alternatives to senseless, yet still profitable, trashing of land and sea. Briganti grew up on Oahu, collecting garbage flotsam on beaches where she swam. At the Parsons School of Design, where she graduated in 2010, she visualized her company's first product: an edible cup made from fruits, vegetables, and seaweed, in citrus, vanilla, and cherry flavors. That segued to seaweed-based disposable knives, forks, spoons, and Blue Carbon drinking straws—neither paper nor plastic—all "Designed to Disappear" in compost piles or the ocean.

Then Loliware pivoted again, into making seaweed-based injection-molding resins, compatible with existing machinery for producing the plastic feedstock pellets sometimes called nurdles. With regulatory pressure against single-use plastics tightening in Europe, it partnered with plastic giant Montachem International to license EU and UK manufacturers to use its resins, whose proprietary formulas combine 12 red and brown seaweeds from around the globe, including Maine kelp.

A life cycle analysis, says Briganti, showed that an early version of Loliware's resin netted about half the carbon footprint of other bioplastics derived from corn, sugarcane, wood, or fermented microbes—and an infinitesimal fraction of what's emitted in producing conventional plastics, especially given fossil-fuel subsidies.

Redirecting those subsidies is part of Loliware's and Montachem's strategy, by entering into agreements with national governments. "Because we're creating jobs and capturing carbon, we'll start seeing biomaterial become subsidized."

But it can feel like a race against a runaway train. As the world migrates from combustion to electrification, the fossil-fuel industry is intensifying plastic production to maintain profits. With oceanic plastic now doubling every six years and predicted to outweigh all marine fish by midcentury, more companies are alchemizing seaweed into 100 percent biodegradable plastic. In 2022, UK-based

Notpla won Prince William's £1 million Earthshot Prize for its seaweed packaging for take-out meals, edible bubbles for liquids such as salad dressing, and plastic-free films to wrap food—and then disintegrate. Hamburg-based one•five GmbH touts a seaweed substitute for the ubiquitous plastic coating in packaged food wrappers.

All these products work. Making them as cheap as polypropylene, polyethylene, and all their 150 petroleum-based cousins is the trick.

"We must keep building until we have an unlimited seaweed supply, essentially funded by governments," insists Loliware's Sea Briganti in her white office in the San Jose biotech incubator BioCube, her trademark golden locks coiled into a formidable braid. "With that, we can come down the cost curve in the next five years and get very close to plastic pricing."

v. Sea + Soil

In British author Thomas Lawson's 1907 novel *Friday, the Thirteenth,* the stock market crashes on that date.* In 1989, that happened for real on Black Friday, when an airline's stock collapse, due to a strike, also dragged down the global junk bond market. Among the casualties was a New Zealand investor, Keith Atwood. Suddenly broke, he found work substitute teaching in Auckland at a school for troubled kids. When the literacy teacher he'd replaced, a redheaded single mother named Jill Bradley, returned, she found her class in good hands. They became friends. After encouragement from Tane, her

* The superstition surrounding the number 13, considered unlucky in many countries, may date to Christ's Friday crucifixion after being betrayed by Judas, the 13th to arrive at the Last Supper.

adolescent son from a past relationship with a Māori man, they started dating.

Keith, hearty and square-jawed, treating bankruptcy as an excuse to go fishing more, proposed they spend summer break working on North Island organic farms in exchange for meals. With Jill driving because he'd lost his car, they volunteered at several. All were fighting fungal invasions during an unusually wet summer—until the last one. It was a dairy with rich, rolling pastures, fruit and nut orchards, herbs, medicinals—and no disease.

The difference was what the farmers, a German couple, were adding to their soil: seaweed.

The experience cemented their relationship, and their resolve. They bought an acre on the edge of a mangrove estuary and started gathering seaweed from the beach. After three years of testing various fermented slurries on kiwifruit orchards, they determined that brown spiny kelp, *Ecklonia radiata*, was easy to work with—and cut fungal rot in half.

Nearly three decades later, it's still the main ingredient that their company, AgriSea, uses in soil additives and animal supplements, as well as a foliage spray used in Italian vineyards. They've expanded from an old barn that stored fertilizer to a 58,000-square-foot factory surrounded by eight hectares of test fields, with more than 30 employees. All family-owned, no venture capital investors: Tane and his wife, Clare, now run it, with Keith and Jill until her death in 2023 serving as active founding directors, and Tane's Māori father, Taonui Campbell, heading their special projects team.

Their success owes to seaweed's capacity for storing minerals and trace elements, and for producing vitamins and amino acids. Its disease resistance apparently derives from chemical defenses that free-floating seaweeds expel because they're completely exposed. Since seawater immediately dilutes them, the compounds must be especially potent.

The result is that sea vegetation promotes growth in land dwellers. From ruminants to insects, they probiotically boost gut health. "A beekeeper thought our amino acids and vitamins would help his apiary," Tane told Bren Smith when they met. "He mixed our product in water, sprayed the bees, then watched them feeding off each other. He reported amazing growth in vigor and colony strength. It's now a steady part of our business. We export it."

For plants, their additives are biostimulants, coaxing roots to exude photosynthesized energy for soil microbes and fungi, which in turn fix nitrogen and release growth-promoting enzymes. Repeated tests show they reactivate soils whose natural microbial activity has succumbed to relentless applications of synthetic fertilizer.

Biostimulants—enhancing germination, lending vigor and resistance to both heat and drought—are a growing, multibillion-dollar global industry. But although AgriSea's deliver NPK—nitrogen, phosphorus, and potassium: the three key nutrients whose percentage ratios are displayed on every bag of conventional fertilizer—biostimulants are still not considered fertilizer replacements.

"They're wonderful, but immeasurable," says Karlotta Rieve, author of a World Bank study on new uses for seaweed. Topping the list were biostimulants, just above animal feed and pet food. "They're not really NPK alternatives. They work differently. They improve root growth and nutrient uptake from traditional fertilizers, so they're often used in combination."

Only 2 percent of AgriSea's customers are organic farmers; the rest still use artificial fertilizer, albeit a third less than they used to. Unless someone figures out how to measure what natural biostimulants do, that may not change. Or, it simply confirms that the yields we demand from fields are unnatural, and biostimulant companies will merely reduce, not replace, their customers' synthetic nitrogen runoff, as AgriSea was doing—until recently.

"We weren't in a rush to grow any faster than the environment," says Clare Bradley. They still don't farm seaweed, relying only on wild spiny kelp that washes up during storms onto beaches of Māori communities who gather it for them, leaving 30 percent for the tidal ecosystem, whose sustenance requires it. "We haven't grown beyond that. We've been too committed to regenerative, Indigenous Māori practices to restore New Zealand food systems. We haven't raced to scale up business. But the race now is about the environment."

To that end, they've formed a partnership with GreenWave. After reading Bren Smith's book, they invited him to speak in New Zealand. He invited their whole family to visit Connecticut. With warming seas now slowing wild kelp reproduction, they saw how by using red lights and refrigeration, GreenWave keeps a seed bank of microscopic sugar kelp gametophytes in suspended animation in their hatchery, which allows them to outplant early and get double the yield of naturally fertilized plants. In effect, they artificially inseminate their kelp—increasingly, a necessity.

Imagining large-scale monocultures in the ocean, the Bradleys were wary of seaweed farming. But with GreenWave they felt an alignment of vision and values, and watching their children connect with Bren's daughter was reassuring, so they agreed to share their fermentation process to test how American sugar kelp compares with their spiny kelp—and if favorably, to consider what purpose it might serve.

"It's a big leap of faith for us," Tane said. "You know, family secrets."

Bren knew. "But with planetary stakes rising," he said, "resilience against coming storms depends on collaboration, not competition."

Tane glanced at Clare. She is fair; he has a grizzled beard, but their smiles mirrored each other's.

"It's a trust relationship," said Clare. "Not one plus one equals two, but one plus one equals 15. What can we create together?"

Every July, Bren Smith takes his family to Newfoundland, where his toddler, Willa, picks blueberries and chanterelles, laughs at moose, and plays on her father's boyhood beaches. Together they watch lobstermen with icebergs floating behind them pull cages filled with crustaceans displaced from hot New England, amid robust kelp.

"You just doubled my carbon footprint," he told Willa when she was born. "Daddy needs to grow a shitload more kelp." But now, turning 50, he realizes that's not going to work.

Within 10 years, he knows he will be growing a different crop in Long Island Sound.

"It's important," he tells his team, "not to be obsessed with kelp. Every fisherman has had to diversify over time. We have to be willows, not oaks." Already California's giant kelp forests that Katie Arkema studied are nearly gone, devoured by an explosion of purple urchins whose main predators, sea stars, since 2014 have succumbed by the billions to possibly the largest marine disease outbreak ever, suspected to be temperature-triggered. Because of kelp, GreenWave's outreach has always been in temperate zones, from Alaska to New Zealand. Now they've begun working in Puerto Rico with the Woods Hole Oceanographic Institution on farming warm-water red algae.

Shellfish is another matter. In 2016, bay scallops were a $2 billion industry from the mid-Atlantic to Maine. Then climate change surged (few dared to utter the unspeakable: *tipped*). With sea temperature rise accelerating threefold within two decades, by 2023 scallops were all but gone, and everyone feared oysters would be next. Fish can swim north, but immobile shellfish can only broadcast fragile spawn. Overnight, the thousands of blue mussels encrusting GreenWave's ropes fell off. Barracuda were spotted off the tip of Long Island, tarpon off Cape Cod. Around his kelp, Bren was finding seahorses.

"Lobsters leave, blue crabs come in. We used to say 'invasive species,' but they're climate refugees. Forget the whole 'native species' thing. We're going to have new ones."

But while they awaited tropical shellfish to arrive, biostimulants were already real. New European restrictions on toxic or carbon-intensive soil amendments had food giant Nestlé moving into seaweed biostimulants. What excited Bren was how AgriSea's fermentation explodes kelp cells so that everything from grass to tomatoes to spinach could use their growth hormones and compounds. And so simply—fermentation barrels get stirred, sugars get eaten.

"And so cheaply makes them scalable," Bren told his staff. "You don't need a ton of kelp to impact a plant's heat tolerance. A hundred pounds covers 45 acres. This isn't new. They spray biostimulants on apple orchards from planes. It's a multibillion-dollar industry. We can weave kelp into it."

Yet if sea farming's main application becomes soil farming enhancement, that doesn't help solve the growing scarcity of arable land—a virtue that seaweed champions like Bren have long touted. But he was confident seaweed would find its rightful place.

"Things are moving fast," he kept reminding his team. They were working with a company that raises soldier flies for the chicken-feed and lobster-bait industries. The collapse of herring- and mackerel-bait fisheries meant kelp to the rescue, adding ocean nutrients to soldier-fly protein replacement in bait bricks.

His staff itself, 15 of 19 of them women, continually reminded him that big change is possible. Seven tribal women were now running a GreenWave-supported farm on Long Island's Shinnecock Nation. A GreenWave advisor and former CEO of one of several seaweed enterprises founded by women, AKUA's Courtney Boyd Myers was the daughter of the corporate vice president who invented the Burger King Kids Club. Her own company sold SpongeBob kelp burgers.

Her burgers were also soy-free. "It's not just how much carbon that kelp absorbs stays sequestered, but how much carbon gets *avoided*," says Bren. One of seaweed's biggest selling points, he stresses, is that there's no need to raze Brazil's Amazon or plow Argentina's pampas to sow it.

"Our official scientific position," he tells funders, "is that kelp does good shit. It's fine to try to quantify this, but what we really need now is action. Are we going to research and debate this stuff to death—which leads to real death? What's the minimal amount of information we need to move forward?"

The rising, warming waters around the Thimble Islands are proof enough of the need. If humanity can find the courage to decarbonize in time, with enough kelp still in the water, its halo effect might be enough to counter acidification and keep coastal ecosystems alive—even though sea levels will inexorably climb.

"It'll just mean more ocean farmland," surmises Bren. "Someday, I could be farming in between the skyscrapers of Wall Street."

CHAPTER THIRTEEN

A TALE OF THREE CITIES

In 1950, two-thirds of humanity was rural,
one-third urban—a ratio expected to reverse by 2050.
What hope, if any, does that portend?

i. Innovate

Carlos Delgado grew up hearing bombs, and watching their smoke waft over Medellín from his classroom windows when Pablo Escobar's drug cartel practically ruled the city. One of his friends had his house blown up. Things had calmed somewhat by 1995 when he entered the University of Antioquia, which had seen five faculty chairmen assassinated during the previous decade. Escobar was finally dead, but Colombia's long-standing civil war raged on.

Carlos's initial interest, animal husbandry, ended when he learned that most industrially bred livestock spent their lives in pens or cages. Instead, he became a vegetarian and a wildlife biologist. With Colombia first globally in bird species, second in plants and amphibians, and third in reptiles, he wanted to see it all, even though leftist guerrillas controlled access to most of Colombia's 46 national parks.*

He somehow dodged trouble until one morning just outside his

* Today, 60.

hometown. He'd hiked at dawn into a nature preserve in the forested mountains and gorges south of the city to collect stool samples. Scatology had become his passion; so much could be learned by poking through an animal's feces, he told his horrified mother when she found jars of bat poop in her refrigerator. From 9,500 feet, he'd gazed down on Medellín, filling the bowl of the Valle de Aburrá and climbing the mountainsides. Its brick architecture glowing scarlet in early morning light, its distant traffic quenched by nearby trills of Andean solitaire thrushes and the warbling of chestnut wood quails, Medellín looked worthy of its soubriquet, City of the Eternal Spring . . .

. . . until he turned, and two Kalashnikovs aimed at his chest reminded him that it was also still known as the murder capital of the world. "What are you doing here?" the guerrillas demanded of this lanky, curly-headed backpacker clutching a trowel.

If he couldn't explain it to his mother, how was he going to make the FARC—las Fuerzas Armadas Revolucionarias de Colombia—understand why he was collecting wood fox excrement? Not quickly, it turned out. As they marched their captive through the forest, he'd name the species they passed: laurel, oak, yolombo, wax palm, orchid, and giant tree fern. With trowel, gloves, and a bottle of bleach, he showed them how bones found in a puma's poop revealed that marsupials formed part of its diet. "See these tiny red teeth?" he said, teasing apart turds of a *tigrillo*, the world's smallest spotted cat. "They eat shrews!" After five days, they believed him and let him go.

Two decades later, the foxes he studied are showing up in Medellín. "They're our leading roadkill," mourns Delgado, who now heads the ecology department at Medellín's CES University.* The reason they're in the city is a rat explosion. In his neighborhood, he's also seen four raptor species preying on them. But without enough

* Formerly Centros de Estudios en Salud—Institute for Health Science Studies.

natural predators to keep them in check, people are using rat poison, which then kills the wildlife that eats them.

Something, Delgado worries, is slipping out of control in this city, which in the early part of the 21st century abruptly went from murder capital to—according to *The Wall Street Journal* and the Urban Land Institute—being the world's most innovative city: winner of Harvard's Veronica Rudge Green Award in Urban Design and Singapore's Lee Kuan Yew World City Prize. Suddenly, instead of war councils of narco-chieftains, Medellín was hosting the World Urban Forum and a global Intergovernmental Science-Policy Platform on Biodiversity and Ecosystem Services conference—IPBES: biodiversity's analogue to the Intergovernmental Panel on Climate Change, IPCC. Instead of US embassy travel advisories warning citizens to stay away, Medellín was now a top retirement destination for Americans.

Medellín's accomplishments, Carlos Delgado acknowledges, have been extraordinary. But he objects when it's portrayed as a paragon of urban sustainability.

"Today, everything is sustainable—sustainable transport, sustainable food, sustainable clothing. But if you really look, where's the sustainability?"

In a world where most *Homo sapiens* are now city dwellers, importing nearly all their food with limited space for their wastes, can any city truly be sustainable? Medellín is rat-infested because trash pickups have been overwhelmed by a spreading global phenomenon: in less than a decade, an influx of refugees—in this case, fleeing Venezuelans—has swollen the city's population of 2.5 million beyond 4 million. As in cities elsewhere, their influx is unsettling but, despite grandstanding politicians vowing to deport them, largely tolerated, because refugees typically work more cheaply than natives, further distorting Medellín's top-heavy class disparities.

"*Sustainability* is a neutered word," says Delgado. "If we keep hiding the truth in flowery speeches, we'll never act. We have to get real."

More than 400 ravines drain the mountains surrounding Medellín, often separating poor hillside neighborhoods within eye- and earshot of each other, but a chasm apart. During the bad years, barrios were further divided by invisible borders demarcating the territory of rival gangs. Wandering into the wrong sector was often lethal.

Among the few neutral spaces were neighborhood reading rooms, founded during the 1980s and '90s by residents themselves in response to the constant violence. Sometimes these tiny public libraries were situated directly between combating sectors, with a separate door for each. During frequent eruptions of barrio warfare, they were where parents could safely leave their kids, who could throw themselves under tables if bullets flew.

In 2003, Medellín elected a new mayor. A booklover, Sergio Fajardo visited all the city's libraries. He was charmed by a program in poor barrios called Endless Reading: every year they chose a book that children would read aloud, each voicing a different character. Proclaiming Medellín *una ciudad educadora*, during his term he founded five new libraries, in depressed areas or situated conveniently near Metro stops. His aide and successor, sociologist Alonso Salazar, founded four more.

During the bloody 1990s, Medellín's Metro had been its one hopeful achievement: Colombia's first subway, running on quiet rubber tires north and south, with one transversal line. Still, it didn't serve a huge portion of the populace who lived in hillside slums so precipitous that not even taxis could enter. To reach jobs in town, people walked at least an hour, then later trudged up two hours or more to their homes via steep, crumbling staircases, paying gangs who controlled separate sections for permission to pass. Fajardo had been to Caracas's fancy Hotel Humboldt atop Mt. Ávila, reached only by cable car. Why couldn't they do that for Medellín's poor?

The first Metrocable line was inaugurated in 2004 to serve high barrios in northeastern Comunas 1 and 2. Within a week, it carried 30,000 low-income passengers a day. Then something unforeseen happened. People from the valley below realized that the Metrocable's glass-enclosed, polyhedral gondolas must offer spectacular views. As each new line went in—there are now six—even residents of Medellín's posh barrio El Poblado made sightseeing excursions to areas of the city they'd previously dreaded.

To accommodate them, cafés opened in plazas around the cable car stops, and they kept coming. For the first time, Medellín was becoming connected.

In 1991, Angela González had followed her brother José to Medellín from their red-dirt pueblo 100 kilometers to the south. Five years earlier, he'd squatted on the city's western outskirts just above Comuna 13, called the most dangerous barrio in Medellín. His corrugated tin shack was alongside a ravine called La Escombrera, where feral pigs rooted through a dump filled with construction rubble. The pigs, Angela learned, fed on bodies the *paramilitares* tossed there.

For 120,000 pesos, about $30 US, Angela bought her own parcel one ravine over and started a daycare center for children whose mothers trekked daily to clean houses or sew in Medellín's textile factories. In the afternoon she sold flavored ices to kids playing *fútbol* in the dusty street. As Comuna 13 kept growing with shacks perched on every bluff, she started selling arepas.

Then, in October 2002, came Operación Orión, a joint military-paramilitary operation to roust the FARC and other guerrilla groups from Comuna 13. More than a hundred died or disappeared. Hundreds more were jailed. The week of carnage destroyed most of the barrio's pirated utilities.

Instead of buying inflated black-market propane to bake her arepas, Angela started selling tanks of gas herself. She still does—except her store is much bigger, and much more fully stocked, due to another miracle of public transportation that no one had envisioned in those dark days. "Since *las escaleras* came," she shouts over reggaeton booming from a nearby choreographed break dance, "Comuna 13 is at full employment!"

The civil engineer who built the Metrocables, César Augusto Hernández, had convinced Mayor Salazar to try something unprecedented for sinuous Comuna 13, and far cheaper: six flights of covered outdoor escalators that would rise the equivalent of 28 stories in six minutes instead of 60, all the way to Angela González's lofty bodega.

"Any mayor can build schools, hospitals, or parks," Hernández had said, "but you could do something completely new that could make the community feel something they've maybe never felt before: pride."

As residents rallied behind the project, the gang warfare that regularly interrupted construction gradually subsided. Many Comuna 13 residents were so poor they'd never seen an escalator; the mayor bused them, goggle-eyed, to Poblado shopping malls to try them. Houses bordering the electric staircases were offered bright-colored paint. Local graffiti artists were invited to adorn walls. When the escalators opened, not only Medellín's upper classes arrived to ride them, but soon young international tourists flocked, galvanized by viral social media.

"They keep coming," Angela says, grinning beneath a mass of dark curls. On every flight, visitors are greeted by busking musicians, rappers, and street performers. German girls in cutoff jeans take selfies with internationally celebrated graffiti artists they follow on Instagram. Each level has cafés with outdoor tables, and shops selling ice cream and souvenirs—except for Angela's.

"By the time they get this high, tourists have already made their purchases. I sell cooking oil, eggs, and rice to the people who live here. Everyone now can afford staples."

For a dollar apiece, her elder son guides tourists to the roof of the three-story brick house he built next door, the highest overlook in this now inviting, music-drenched barrio. Her younger son works the lower levels with a hot dog cart. Before *las escaleras eléctricas,* they lugged propane tanks up and down the treacherous staircases for tips.

But in this shaky, weather-beaten century, is basing an economy on tourists sustainable anymore—or anywhere? In equatorial Colombia, will tourism wither in rising global heat that's already making it difficult to be outdoors—Comuna 13's main attraction?

Angela González isn't ready to worry about that. "Our lives have been changed by the escalators. Imagine trying to get a stove up here before them."

The resurrection of Medellín included lining the banks of its ravines, gorges, and the Río Medellín that bisects its valley with 25 new public parks. A 19th-century botanical garden was revived next to a futuristic museum, Parque Explora, part of a complex including the Universidad de Antioquia and Ruta N: three connected, vine-draped modern towers made from nontoxic building materials, designed to save water, utilize natural lighting, and incubate 200 tech startups.

Throughout town, helmeted riders on blue EnCicla bicycles, a public bike-sharing program, zip down leafy median strips, planter-lined sidewalks, and 11 new pedestrian promenades lush with vegetation. The forester with flowing gray hair in charge of these transformed green corridors, Mauricio Jaramillo, explains, "We plant to lower the temperature. We start with tall trees—silk tassels and flowering acacias with wide, shady crowns. Under them come palms, and then shrubs to increase the niches for birds and pollinating insects."

His challenge is that 80 percent of Medellín's trees were introduced from other continents. "We try to use native species, but often we don't know what they were, because this Aburrá Valley has been cultivated for 5,000 years"—including by his family, who settled in El Poblado 300 years ago, when it was still *ranchitos.* Daily, he walks the medians to see what is surviving, as Medellín segues from being a city of eternal spring to what feels like eternal summer.

"Over the past century, 4°C warmer." So far, philodendrons, heliconias, and spurges are doing okay, as is the verbena, its purple flowers adapting to new Africanized pollinators. The one that worries him is Medellín's signature tree, the yellow flowering guayacán—because of not just temperature, but airborne chemistry.

Since indulgent tax laws enable the nation's wealthy to shelter much of its GDP offshore, to finance its budget, Colombia relies on selling its coal and oil. As the latter, a high-sulfur crude, is expensive to refine, sulfur particulates trapped in Medellín's bowl often measure 10 times the limits acceptable in the US or Europe. Also, with Medellín's made-over reputation—prosperous, innovative, safely inland from hurricanes and rising seas—attracting everyone from refugees to entrepreneurs to American retirees, an explosion of cars has immobilized traffic, sometimes actually shutting it down when the air gets too dangerous.

To stave off environmental emergency, the city asks Mauricio to plant even more trees. He's visited nurseries all over Colombia, looking for species that might survive the heat and toxicity. Standing under a ceiba he planted, from which pink blossoms flutter into a sizzling, gridlocked midtown intersection, he doffs his visor to mop his forehead. He's survived so much. So many of his forestry classmates were killed by guerrillas. Guerrillas kidnapped his wife; the ransom all but bankrupted them.

"But now I'm tired, watching this decline," he says. "Every day, more pollution. Medellín claims we're sustainable, but sometimes I fear nothing's sustainable unless we have a world war. Overpopulation is decimating so much. Sometimes I see no future for the human race."

Then he glances up into the ceiba's branches, crowded like a refugee camp with squabbling parrots and turtle doves, and softens.

"From necessity, solutions must appear. My hope is that coming generations commit to the fact that we must change. A new civilization could still blossom."

ii. Retrofit

Luis Aguirre-Torres and his wife, Courtney Ann Roby, both in blue jeans and black parkas, stand at the construction site of a new conference center next to Ithaca, New York's brick city hall. It's a November 2021 afternoon; Luis wears an Adidas stocking cap pulled down over his gray widow's peak; Courtney's apple cheeks are framed by straight, chin-length light hair. Although little about the pandemic was lucky, the delay on this project meant that Luis was hired just in time to prevent the conference center's builders from perpetuating the same mistake we've all been making for more than a century.

"This whole building will be fully electric, except for"—his eyes roll heavenward—"the kitchen. Their chefs claim electric induction stoves will exceed load capacity when they cook for 300 people."

"Induction stoves are low-power," objects Courtney.

"Right. But the kitchen isn't the problem. The entire building will exceed peak load, so the architects are worried they'll have to change the transformers, which run up to $150,000. Fortunately, I am good at finding money."

She grins. It's true.

"So I did. Then they didn't have a reason for not doing it. Now it's down to the chefs. But we're making progress."

Luis Aguirre-Torres isn't merely good at finding money; he conjures financing where no one thought to look for it. That wasn't what Ithaca's Common Council expected when they hired him eight months earlier as the city's director of sustainability. In many cities, his duties would be recycling and composting programs, installing LED streetlights, planting vegetables in vacant lots. But in 2019, after the UN warned that civilization had only 11 years to cut global CO_2 emissions nearly in half, or else, progressive Ithaca, home to Cornell University and Ithaca College, had approved its own Green New Deal to achieve town-wide carbon neutrality by 2030.

Then COVID arrived. Everything stopped—except global temperatures, which kept rising. Normally, Luis would have been away for months at a time, advising South American governments on green technology. Instead, he was grounded in Ithaca with Courtney, a Cornell classics professor, when he saw the job advertised. The challenge of weaning an entire city from fossil fuels intrigued him, and getting to stay home with his wife appealed, so he applied. His qualifications were peerless: Ph.D. in electrical engineering from University College London; former presidential fellow during Obama's term for promoting Cleantech industries in Mexico, where he'd midwifed a constitutional reform to advance solar energy; and State Department envoy to develop climate change legislation in Latin America.

Plus, he checked the Hispanic diversity box. So hiring him in

March 2021 seemed a no-brainer—until council members realized Luis meant to take them at their word. Before getting his doctorate, he'd earned honors in computer science at the Universidad Nacional Autónoma de México, so he knew how to build a model to calculate what was needed to fully electrify heating, cooling, and cooking in 14,000 single- and multifamily Ithaca residences and 600 commercial buildings.

Retrofitting the city for energy efficiency alone would reduce half its emissions, but 40 percent of Ithaca's gabled Victorians, Gothic revivals, and brick apartments were built before 1920. Every wall hid frayed, ancient wires and desiccated insulation. Prying off clapboards to weatherize plaster walls, rewiring, and converting from gas or oil . . .

"It'll take $2 billion," Luis told Ithaca mayor Svante Myrick.

Myrick, elected councilman in 2007 when he was still a Cornell student, then mayor in a landslide four years later, looked stricken. "Luis, our entire budget is only $80 million."

"There may be a way." Luis led him through the figures. Ithaca wasn't all highly educated, well-paid Cornell professors. It had a substantial low-income population, many of them minorities or immigrants. To change windows, insulate, and fully electrify run-down single-family homes could cost $50,000 each.

"Then most wires between houses and utility poles will need replacing. Their copper won't handle the increased load, 50 percent at least. We'll have to change transformers, poles, and cables. Substations will need energy storage to compensate for differences in voltage and frequency. It's super complex."

Ithaca couldn't possibly finance all that. If it took $2 billion for its small population of 32,000 plus 25,000 students during the school term to go clean, decarbonizing the world's cities—at least 500 with over a million inhabitants; 34 with over 10 million—would cost more money than existed.* Was humanity irrevocably doomed to keep using fossil fuel?

"There's no way to issue bonds or count on state or federal incentives to cover all those," said Mayor Myrick.

"We need to tap into the private sector," said Luis—but he didn't mean for donations. "We need to change the role of government to be a catalyst for investment."

But who would invest in Ithaca instead of Wall Street?

"I have an idea," Luis said.

"All my life, I've been obsessed with really wicked problems that seem unsolvable until you start dissecting them." Luis's doctoral dissertation on patterns in entropy, for which he modeled the chaotically expanding internet, had so impressed AT&T Laboratories that it offered him more money than this son of a Mexico City electrician had ever seen. So he moved to the US, where the first tricky problem he encountered was named Courtney.

It was 2003; she was on a joint master's fellowship with AT&T Labs and the Department of Energy. Love of mathematics had led her to electrical engineering, but when they met she was having a dark night of the soul, realizing she might never control what her research was being used for. "Everything's about how can this be militarized," she protested. A robotic eye she'd worked on was probably being used to spy. The nanoscale optical filter she helped design

* Two billion dollars divided by Ithaca's population is roughly $35,000 per resident. According to the World Bank, in 2023, 56 percent of humans—4.4 billion—were urban. At $35,000 apiece, electrifying the world's cities to Ithaca's standards would cost $154 trillion. The total global money supply is approximately $83 trillion. (Presumably, economies of scale and simpler residential construction elsewhere would substantially reduce per capita electrification costs.)

for the fastest, most energy-intensive laser would allow it to precisely target and destroy the interior of anything.

"Why do we do science and engineering the way we do? How did history lead us here?" she'd challenge Luis, who was simply trying to get her to marry him. After five proposals, she agreed, but by then she'd tracked how we think of science back to ancient Greece and Rome, so for her doctorate she switched to studying classics at Stanford.

Following her to California, Luis worked with Governor Arnold Schwarzenegger on green legislation. Newly steeped in the urgency of climate change, Luis founded a Cleantech think tank and soon was doing the same for the governments of Mexico, Brazil, Colombia, Chile, and Argentina.

Having dealt with big federal budgets, he now saw advantages to the scale of Ithaca's challenge. "Decarbonizing an entire city at once, you create economies of scale. You can bulk-purchase in industries like heat pumps that are never bought in bulk."

He started pitching a creative financing scheme to equity capital firms that offered investors a chance to make money and minimize their risk. The profit would come from the difference between what people currently spent on energy and what they would save by maximizing their dwelling's efficiency and going all-electric. That savings, plus state and federal energy incentives, would make low-interest loans affordable to nearly anyone, and the interest would accrue to the investor.

Although the city needed a sum that only banks or wealthy private equity firms could provide, it also had smaller sources of income: government renewable-energy incentives and even manufacturers' rebates that, when buying in bulk, add up. Those sources, Luis realized, could significantly reduce risk for both city and investors.

"Say investors pool $100 million and want a 10 percent return to cover administrative costs, risk, and profits to their limited partners," he'd explain. "We put that pool of money in a managed bank account. Say we then get $5 million in sustainability grants from the state and federal governments. Instead of spending that on heat pumps, we put it into the same pool. Now we have more leverage to bulk-buy heat pumps at scale, and we relieve the pressure to generate a 10 percent return, because now we only need $5 million more."

It might sound like slick accounting, but after a decade of investment loans in Latin America, Luis had learned that, unlike commercial banks, private equity capital was pricey but more flexible. Still, to get investment bankers to work with Ithaca, they'd have to manage the risk of low-income borrowers with no credit history—23 percent of the population, his model calculated—because the mission was not just decarbonization, but also social and environmental justice. For that, Luis said, the state could create a loan-loss reserve—a common credit inducement used by governments to assure partial risk coverage to lenders.

"That brings the probability of default practically to zero," he'd conclude, "because we have a diversified portfolio of 6,000 buildings and protection for the highest risk bracket. Once you do all that, the cost of capital is lower, the risk is lower, and you have an affordable program."

They'd also generate at least a thousand jobs: substantial for a city of 30,000. "It will be a massive workforce development plan that will help not just the city but the entire region. We'll start slow as we get everything figured out; maybe two projects a month for the first three months, then ramp it up. In a year, probably 50 per month, and then 100."

Recognizing that government would be too clunky and bureaucratic to manage the huge amounts of money and work, he contacted BlocPower, a Black-owned company known for deftly financing low-income retrofits and electrification in Brooklyn. He kept calling private investors. He called the United Nations energy program and arranged for Cornell scientists to present Ithaca's Green New Deal at the COP26 meeting in Glasgow.

BlocPower's CEO, Donnel Baird, was also in Glasgow, discussing heat pumps and electric vehicles with Secretary of Energy Jennifer Granholm and Vice President Kamala Harris. When Harris asked about specific BlocPower projects, Baird mentioned the city of Ithaca. Soon thereafter, the director of the White House's Climate Change Advisory Group called Ithaca to talk to Luis.

Which was when the trouble hit. He'd already seen vicious comments that the same few obnoxious readers posted to articles about him: saying he was a foreigner with an inferior education from a Mexican university who had no business telling an Ivy League city what to do; he and the Black CEO of the company he hired have a deal to get rich from this project; you know how those people are.

"I quit," he told the mayor.

"You can't let a few jerks get to you," said Mayor Myrick. Half Black himself, Myrick had seen plenty of the same but had flummoxed the haters by being reelected twice, once with 89 percent of the vote.

"They're not getting to me. They're getting to the council."

The hate mail was being copied to every alderperson on Ithaca's Common Council. One of them had actually forwarded him a racist email calling him incompetent, with a note suggesting that the writer had a point: Luis needed to better explain what he was doing.

Another openly objected to how Luis was racing ahead of the council's normal pace, making executive decisions to meet the 2030 decarbonization goal. The council still hadn't approved his financing plan.

Meanwhile, climate change wasn't waiting. "I don't need to take this," Luis told the mayor. "I can do this anywhere. I think people will be very happy to hire me and the $100 million I just raised."

"What?"

It was actually $105 million: $50 million from a private equity fund, Alturus, specializing in industrial-scale decarbonization and eager to back an entire city. The rest was from Bank of America, Goldman Sachs, and Microsoft.

"Let me call them," said the mayor. The next week, Luis's plan was approved unanimously.

That was in October 2021. But three months later, Mayor Myrick accepted an offer to direct People for the American Way, the advocacy group founded by television producer Norman Lear to defend democracy. An acting mayor took over. Luis didn't begrudge Myrick, who had served for 10 years and now had a chance to make a national impact, but he sorely missed him. "He wanted to do things differently. He told me to just run with it."

So he had. But with Myrick gone, suddenly everything needed to be approved by internal committees, lest decarbonization programs impact the workloads of assorted city departments. Meetings would often end before he had a chance to explain adequately. The resistance he felt became clearer when he was invited to present to the World Economic Forum, then to the Bloomberg Green Summit, and then to the United Nations and the Federal Reserve.

"I kept hearing 'You are not empowered to represent the city at that level.' Then, when I was invited to the White House for the party after the Inflation Reduction Act was signed, I was told that I was no longer allowed to meet with anybody outside the city, or talk to the press anymore. I was just trying to make a name and a reputation, they said."

He explained to them how much Ithaca had already influenced federal policy and state policy and what that meant.

"They look at us and realize that this could work, and we get funded. Same for the state government. In the governor's State of the State address he mentioned things they've learned from Ithaca. That puts us first in line for both federal and state funding. Instead, I heard 'That wasn't the reason we hired you.' When they said that, I was like, okay, then that's not the reason for me to stay here. So," he told them in October 2022, "I quit."

"Common Council didn't really understand what they were signing on to when they passed that resolution. It's turned out to be much more complicated than any of us realized. And it's rolling out much more slowly."

Peter Bardaglio, the former provost of Ithaca College, heads both the Tompkins County Climate Protection Initiative and Ithaca's branch of the 2030 Districts Network, composed of cities nationwide that aim to cut energy and water consumption and transportation emissions in half by 2030. It may be coincidental, but since Luis quit, Peter's back has been aching badly. "Luis was a great example of what's called charismatic leadership. He generated energy. Since he left, the council just hasn't demonstrated strong support for electrification."

Even BlocPower got off to a slow start, he says. Despite Luis's insistence on linking sustainability and social justice, instead of retrofitting low-income residences they began by focusing mainly on commercial buildings. He suspects their investors were pushing for a faster return. Bardaglio has praise for the deputy director who replaced Luis, who completed a town greenhouse gas inventory and laid out a plan for moving forward.

"But without Ithaca's Common Council taking ownership, it'll be difficult. I'm hoping in forthcoming elections we'll get a new generation in there—folks who really understand the crisis that we're facing, and why we need to move as quickly as possible."

A grimace, as he waits for the pain pill to kick in. Just 70, still with a full head of hair, Bardaglio wasn't expecting his body to suddenly betray him. "It was sad," he adds, "to watch Luis encounter such active opposition, especially around his ethnic identity. But I think the resistance to him was more fear that Ithaca wasn't going to be able to deliver, and all that attention would backfire."

If fear of failure stalls action, failure is guaranteed. "A classic instance of a self-fulfilling prophecy," says Bardaglio, sighing, especially troubling when it involves a failure to face how to survive the climate we've inadvertently broken. Still, a year after Luis's departure, they'd made some headway. Nearly a dozen buildings were ready for the fuel-free future: An ice cream store with geothermal heat pumps, LED lighting, new windows, and solar photovoltaic panels on the roof. An old auto parts store, stripped to the cinder blocks, insulated, solarized, and heat-pumped to net zero. A historical center with an adjacent retail outlet selling architectural salvage, each retrofitted with easy financing.

"But it's a tricky time. Interest rates are high, supply chains are gummed up, inflation's giving a lot of people pause. So it's rolling out a lot slower than any of us hoped."

Despite the racial bile spewed at Luis Aguirre-Torres, prosperous, livable-sized Ithaca is among the most educated and progressive cities anywhere. If Ithaca can't reach its emission goals by the United Nations deadline of 2030, can any city?

"Just because things are moving more slowly," says Bardaglio, "doesn't mean that it's failing. We're showing that it can be done."

Especially since Luis reemerged as the director of financing solutions for the New York State Energy Research and Development Authority. Suddenly, he was in charge of a statewide version of Ithaca's program, just as the national Inflation Reduction Act was releasing state-sized block grants for decarbonizing as much of America as possible. Already, New York had passed the US's first all-electric building act, prohibiting the use of fossil fuel in new structures* (gas stoves excepted).

Between IRA money, surcharges on ratepayers' electric and gas bills, state bonds, and a greenhouse gas initiative in which multiple states collect money from utilities and other highly polluting entities, Luis could now access billions in electrification incentives. His job, he told Bardaglio, was "to stretch every dollar as a catalyst for private investment or philanthropic investment." No longer was he seen as a guy moving too fast and breaking things. His mandate now was to disrupt.

He and Courtney still lived in Ithaca, and a small group, including Bardaglio and the deputy who replaced him, still met with Luis regularly. But unlike when he began, Ithaca's Green New Deal faced inflation, foreign wars, and rising electricity costs—partly because as natural gas usage declines, the cost of maintaining its remaining infrastructure doesn't.

"Transitions are difficult, but they happen," he told them. "Killing whales for whale oil still continued for 50 years after we replaced it with petroleum."

He'd then describe finance mechanisms that could "leverage $100 million into $300 million worth of affordable loans to decarbonize..."

"Great. But what happens to the Inflation Reduction Act if a Republican gets elected?"

Deep breath. "Then the whole thing gets canceled. Or it still exists, but it's unfunded. So we can't say with absolute certainty that this is going to work."

Silence.

"Look, nobody expected this to be easy. I'm exhausted. Every morning at 5:00 a.m. I'm glued to the Bloomberg channel, trying to

* Those under seven stories beginning in 2026, and all new buildings after 2029.

understand the latest dynamics. The risk of a change of administration probably adds 2 percent to the cost of capital. Every day there's a new yield or mandate that prevents you from doing something. It's very complex."

Something else: since he left his job with Ithaca, the climate had effectively tipped. Within a decade, instead of 20 days over 95°F, the state of New York expects 45. "We'll have fewer rainy days, but more rain when it does. Shorter winters, but with much colder days. All those jobs built around winter, from recreation to snowmelt management for hydropower—people will need to find something else to do several months a year . . ."

For a rare moment, Luis had left even himself in the dust, seeing the future unspool. "Okay. It wasn't until reality hit us head-on that we're understanding the intricacies of what we were trying to do. We're clearer now. A ton of people are trying to do the right thing. I'm optimistic we can make this happen. Otherwise, I wouldn't be able to sleep."

iii. Rise Above

In 2007, when Marc Collins Chen became French Polynesia's minister of tourism, he inventoried the 48 airports distributed among its 121 islands: all at sea level. Then all the hotels: same. Back then, there wasn't yet much worldwide alarm about sea level rise, but in the Pacific, they knew. Not too many decades in the future, Chen saw no solutions beyond planned retreat.

"It's tough to tell people they'll have to abandon where their great-grandmother is buried. That's their cultural identity." Being a government official, he looked into hard defenses, like in the Netherlands, but concluded that eventually they don't stand up to nature. New

Orleans, he saw, spends billions on levees that later the city has to redo. That led him to wonder: Instead of perpetually fortifying cities against rising seas, could they possibly build cities to rise with them?

The idea of living on water, Chen learned, was as old as civilization. Fisherfolk in the Sumerian marshes inhabited human-made, unmoored islands; Uros built them out on Lake Titicaca to elude Incan raiders; Mexico's Xochimilcas used them to farm crops. In the 1960s, American futurist Buckminster Fuller and Japanese architect Shoji Sadao designed a tsunami-proof tetrahedronal floating city to house 1 million in Tokyo Bay—an idea that, blessed by the US Navy as seaworthy, Baltimore entertained on a smaller scale for Chesapeake Bay. Fuller saw it as a solution to the high cost of waterfront real estate.

"Floating cities pay no rent to landlords," he noted—possibly why none had been built.

PayPal founder Peter Thiel, some libertarian billionaire pals, and free-market economist Milton Friedman's software engineer grandson saw something else a floating city might avoid paying: taxes. In 2008, they started the Seasteading Institute to explore siting one in international waters. Interested in technologies they might develop, Chen invited them to Tahiti, but their "aquapreneur" manifestos—to "leave behind archaic governments which assert control over their citizens"—wasn't what he felt was realistically needed.

Buckminster Fuller's ideas, he believed, were ready for the 21st century. "Around the world," he told the UN's Itai Madamombe when they met, "3 million people a week move into cities. Nine of the world's 10 megacities, where they mainly go, are coastal. That means in the next 30 years we'll need to double the amount of real estate: two-and-a-half-trillion square feet of built environment. With shorelines being swallowed, there's only one way to do that."

The occasion was the 2017 United Nations Ocean Conference in

Fiji, which Madamombe, who specialized in multinational water partnerships, had helped organize. Born in Zimbabwe, she studied public health at Harvard on a Pan American Health Organization scholarship, then worked at the UN during the AIDS crisis. For years she was an aide to Secretary-General Ban Ki-moon, helping to plan the UN's first climate summit.

At the Ocean Conference, following the opening kava ceremonies and exchanges of hibiscus-bark wreathes, she and Marc Collins Chen met on a climate panel. A tall Pacific islander with long, dark hair worn tied back, Chen presented floating cities as a priority for the not-so-distant future. Itai, also tall, with a cascade of curls and memorable cheekbones, was captivated by his alternative to dumping rocks in advancing seas to fortify shorelines, which ravaged coastal ecosystems and was ultimately futile.

They were a striking pair, fluent in six languages between them. Marc, raised partly in Mexico City, with an engineering degree from the University of Texas and an entrepreneurial stint in Tahiti before entering government; Itai, a UN insider who grasped the nexus between the environment and public health—it was a love match, and soon a business partnership.

They called their company Oceanix. "Imagine a city rising with the water," said Itai as she introduced Marc to people at the UN. "It's anchored to the seabed, but it floats. It can't be flooded."

There were precedents, he'd explain. "The 1.5-mile-long Evergreen Point Floating Bridge linking Seattle to its eastern suburbs—and, right here in New York City, Pier 57. It's been in the water for 65 years. Most people don't realize it's floating. It's the same technology we'll use for floating cities."

They could begin with modular floating neighborhoods that could be linked together into villages and more. "In theory, you can keep on multiplying," Itai said.

They developed a cadence. Marc: "Unlike the Seasteaders, these wouldn't be little independent republics, but extensions of existing coastal and port cities, with mixed-income housing."

Itai: "Imagine trying to turn New York into a sustainable city. You face so many hurdles, trying to undo and retrofit. But this is a clean-slate, sustainable solution for both sea level rise and the housing crunch."

The key to sustainability, said Marc, was water recycling. "There's precedent for that, too. Namibia's capital, Windhoek, a city of 400,000, has been recycling 100 percent of their wastewater into drinking water for 50 years."

Windhoek's treatment plant accounted for a quarter of its drinking water. A floating city could be designed from scratch to produce all of it, he said. "The entire city becomes one big rain collector. Every raindrop gets trapped and treated. We don't waste anything."

They requested an analysis by the Global Footprint Network, which calculates the number of global hectares needed to provide a person's needs and absorb their wastes. Given Earth's limitations, the optimal figure was 1.6 hectares. The average US citizen uses 8.1. Worst is the United Arab Emirates: nearly 12. The GFN ran Oceanix's numbers: half a hectare per resident.

People would generate their own solar energy and commute to mainland jobs on solar-powered ferries. "Like the ones already operating in Kerala, India," said Marc.

"People will grow their own food," said Itai.

"Think seaweed, fish, shellfish, and at the bottom things like sea cucumbers to clean up," said Marc. "Multitrophic aquaculture, each layer feeding the next. A program called GreenWave is already doing this."

"We don't want to leave the oceans poorer than we found them," said Itai. "We want floating cities to actually rehabilitate the ocean. Sustainable floating cities."

"I hate the word *sustainable*," said architect Daniel Sundlin, a partner at BIG, the Bjarke Ingels Group, known for some of the planet's most stunning new buildings. "It implies a status quo, but actually, we're heading downhill. We need to find a way to get back up. I think *regenerative* is the only way to move forward."

"Exactly," said Marc. "Regenerative."

To find the right architectural firm to design their vision, they'd started at the top. Most recently, BIG had finished the 279-foot CopenHill, a trash-to-energy generator in its home base, Copenhagen, which incinerates 95 percent of the city's solid waste so cleanly that they topped its half-kilometer-long, slanting roof with a ski slope. In the same city, to affordably house college students in one of the world's priciest real estate markets, they designed Urban Rigger: apartments in stacked shipping containers, floating in the harbor.

In late 2018 Marc and Itai met Sundlin at the Bjarke Ingels Group's office in Dumbo, Brooklyn, a block from the East River. Among BIG's New York collaborations were the Lower East Side's BIG U and the Billion Oyster Project with Henk Ovink. Now in his mid-30s, Swedish-born Sundlin, sandy-haired with a stubble beard, had spent his whole career with BIG. After hearing Bjarke Ingels talk about "a building's narrative" and how BIG involved people from different professions to ferment their designs, it was the only place he applied.

To Sundlin, Oceanix's own narrative of affordable, scalable, floating infrastructure was instantly compelling, especially with coastal real estate becoming even more expensive as encroaching seas make it scarcer. "The future is not about iconic architecture but iconic infrastructure," he told them. "Focusing on how buildings perform—and in this case, how cities perform."

The chemistry was obvious; BIG loved working with people trying to change the system. "We aren't interested in building a prettier mousetrap. We're interested in really switching the dial," said Sundlin. Marc and Itai were on a journey with major cultural, political, technical, financial, and environmental challenges. The technical part, BIG could handle.

A few months later, they were back in BIG's 55,000-square-foot gleaming white Brooklyn workspace, illuminated by fluorescent halos encircling white concrete pillars, with blond plywood scoop chairs around blond wooden tables, floor cushions, and a succession of glass-walled workshops to convey the sense of a huge team collaborating, including the lab where they 3D-print mockups.

One of these was now a model of their floating city: a hexagonal layout resembling an aquatic snowflake, formed by clusters of triangular lots that fit together into neighborhoods linked by walkway bridges—it was all modular, and could be repeated on any or all sides without compromising the underlying floating infrastructure.

"We involved about 10 different consultants," Daniel Sundlin later recalled. "Engineering and mechanical systems, ecological systems, food systems—every facet of making society run."

Among the first was MIT's Center for Ocean Engineering. "Is there anyone here who thinks this is ridiculous and cannot be done?" Itai had asked.

Department chair Nicholas Makris had some concerns. "These floating barges need a very sheltered location to survive Category 5 hurricanes." Makris knew; he grew up in Glen Cove, Long Island, sailing with his father. "You should start near the coast, in a relatively benign environment," he said. "Farther offshore, sea-states get increasingly difficult, with high waves constantly hitting you. You would need to design something similar to an offshore oil rig."

They assured him their plan was to be anchored firmly to the continental shelf, close to land. A naval architect Makris consulted had mentioned the challenge of having many platforms in close proximity without destroying each other. "You'll need to reinforce your concrete with steel, otherwise it'll fracture too easily. Finding long-lasting materials will be critical."

"The Romans made concrete that's lasted a very long time," said Marc.

"Yeah, and nobody knows exactly what they did," said Makris. "The problem is weathering wave action. You get a bending moment from any kind of wave."

"We know we have to build it in harmony with the sea."

"Everyone wants to be in harmony with nature, but nature is sometimes the enemy." Makris held up his hand. "See this big scar on my thumb? That was a wave. But, yeah," he acknowledged, his thick fingers combing his receding mane. "This isn't like you're trying to invent a new energy source."

In April 2019, the United Nations Human Settlements Programme—aka UN-Habitat—along with Oceanix, MIT, and the Explorers Club, held the "High-Level Roundtable on Sustainable Floating Cities" at UN headquarters. Bjarke Ingels introduced BIG's design: six prefabricated, 4.5-acre hexagonal islands, each designed to house 300 people, modularly clustered into six villages of 1,800, which connected together to form a city of over 10,000. Each module would produce its own power, purify its own water, and grow much of its own food: in rooftop and vertical hydroponic and aeroponic gardens; beneath the platforms in shellfish cages and seaweed farms; and in greenhouses using recycled water and composted wastes.

The 70 participants included engineers, urban planners, energy experts, marine biologists, sociologists, conservationists, a dystopian philosopher, and, attending remotely, Nigerian students whose three-story, Dutch-designed, timber-framed school since 2013 has floated atop plastic barrels in hyper-crowded Lagos's lagoon. Among the economists present was Nobel laureate Joseph Stiglitz. BIG's designs for waste and water recycling would make excellent sense anywhere, Stiglitz opined. "Some of the most important benefits won't be in floating cities but on land."

The hollow steel and green-concrete hulls of each floating caisson could be used for growing mushrooms and the edible insects expected to meet part of future human protein needs.

"They'll also house infrastructure that's usually underground," added Daniel Sundlin. In land-based cities, upgrading or repairing sewer lines, water or gas pipes, and electric or fiber-optic cable means digging them up and reburying them. "In a floating neighborhood, you'd go belowdecks to fix or replace them without disrupting life."

For a day, they brainstormed possibilities. Unlike coastal development that damages the planet twice by excavating fill somewhere and dumping it, BIG's contingent explained, floating cities would benefit maritime habitats in populous coastal areas by acting as living reefs to attract and reestablish aquatic ecosystems. To rustproof their steel anchors, low-level current from solar panels would react with dissolved CO_2 in seawater to form a coating of limestone—"regenerative biorock," Oceanix called it, because it created habitat for water-filtering mollusks.

"A lot of new habitat, because modular systems can be added indefinitely," said Itai Madamombe. "We could scale Oceanix City up to 2.7 million people."

"And do it affordably," added Marc Collins Chen. "Because the water is a commons, we don't realize how cheap it is. When wind farm companies first wanted to lease space off the eastern seaboard, they discovered it cost $3 per acre per year. In Manhattan, that would be $43,000 per acre."

By the end of the day, a believable vision had emerged from multiple perspectives of a functioning, floating city that could generate jobs and be a living innovation lab, a social research space, and a mixed-income community—a requisite for the UN's blessing—interspersed with shops, businesses, and green spaces, both walkable and navigable via canals and bridges linking the modules. MIT's Nick Makris was impressed.

"I learned a lot here," he said. "If engineers tried to do this on their own, they'd design something that nobody wants. It takes everyone. Just start small and stay near the coast. Anywhere a boat can safely be moored."

The roundtable's consensus was for Oceanix to find a suitable city partner to develop the world's first prototype. That required seed money. The UN's imprimatur brought them credibility. Soon, venture capitalists from Prime Movers Lab, which specializes in breakthrough scientific startups, were in for $2 million.

Apart from BIG's sweat equity in exchange for 5 percent of the company, Oceanix until then had been totally funded by Marc's and Itai's personal savings. Their relationship had weathered a try at marriage, then improved when it ended—so much that they kept living in the same house as they kept working together.

They contacted coastal cities that seemed like prospective partners. Marc, Oceanix's chief technical officer, explained they could build a floating concrete slab for $1,500 per square meter, compared to $16,000 for waterfront property in Monaco or $100,000 in Hong Kong. They entered discussions with Abu Dhabi; building their prototype in a country that could easily afford it, they reasoned, would finance valuable lessons for designing floating cities in the developing world's Dar es Salaams and Chittagongs.

While talks with Abu Dhabi stretched on through the pandemic, they spoke with laboratories working on green substitutes for CO_2-intensive Portland cement, including concrete made from desalination's briny effluent. They explored mining the ocean for construction minerals and building zero-waste systems. To assure that floating city platforms wouldn't fall apart in a few decades, they pushed construction companies for 150-year guarantees. They recruited Tom Goreau, whose solar-trickle-charged Biorock was rebuilding reefs from Indonesia to Panama, and Bren Smith to advise them on 3D ocean farming. ("Sorry," Bren told them, "no seaweed underneath your city. It needs sunlight to photosynthesize.")

In 2021, BIG's scale model of Oceanix City was displayed in the Smithsonian's *Futures* exhibition as an example of how blockbuster technology could radically change the world. Other examples included flying taxis, DNA designer babies, hyperloop trains approaching the speed of sound, a solar-powered water harvester to pull vapor from desert air, and a biodegradable, tree-sprouting funeral urn. It was nice to be included, but with the Abu Dhabi deal languishing, it also was frustrating to merely be a futuristic museum exhibit while climate news reminded them daily that the need was now, in the present.

But that November, all the answers they'd prepared over months of questions from the Emirates' sheikhs paid off when they signed a memorandum of understanding with Busan, South Korea, the port from which Samsung, Hyundai, and LG products flow to the world. The initial 15.5-acre prototype in Busan's protected harbor would be three triangular platforms. One would be a research center, including a hydroponic farm. The second would accommodate visitors, with an eco-lodge hotel, eco-retail outlets, and an organic restaurant at the waterfront's edge—possible, they said, because unlike buildings along the shore, it could never flood.

The third platform, with six multistory hexagonal residential buildings of varying heights, would house 12,000, and be scalable to 100,000. Inspired by the efficient use of space in beehives, the hexagons would mesh around a garden courtyard. Each was designed for a low center of gravity to evenly distribute weight—unlike a new floating neighborhood in Amsterdam's River IJ, where a stove and dish cupboard can't be on the same side of the kitchen, lest the floor begins to list. Power would come from photovoltaic panels, both rooftop and floating.

With the Busan contract signed, Prime Movers Lab doubled its $2 million stake. Bjarke Ingels Group officially became the design architect, and SAMOO, a division of Samsung, the architect of record. Icelandic Danish artist Olafur Eliasson signed on to design the public spaces. Oceanix CEO Itai Madamombe relocated to South Korea to begin discussing hull construction with Samsung and Hyundai and final design permits with Busan's government.

That process typically would take a year, and inevitably, there would be tweaking. Harbor contamination, for example, could scuttle plans for 3D farming mollusks beneath the platform, although Marc was discussing filtration strategies with seagrass experts in Busan's universities to gradually restore marine ecology, as with New York's Billion Oyster Project.

The projected cost, $627 million, "is a lot cheaper than land reclamation," Marc said. With construction estimated at two to three years, and with Busan a finalist to host the world's 2030 Expo and anxious for Oceanix City to be its centerpiece, all was expected to be underway by early 2025.*

There was another reason to move briskly: new evidence that since the 1990s, the seas had absorbed 91 percent of excess greenhouse-effect heat and were warming all the way to the bottom of the water column. Warmer water expands; along with newly discovered

* Losing eventually to Riyadh, Saudi Arabia, didn't sink Busan's floating neighborhood. It's slated for completion in 2028.

melting on the underside of Antarctic ice shelves, that meant sea level rise is accelerating.

"The planet is 70 percent water. The future of humanity is on the water. It's our next frontier," says Itai Madamombe. "Oceanix will be very busy in the coming decades."

CHAPTER FOURTEEN

THE ATOLLS' LEGACY

i. Father and Son

When Jelton Anjain was tiny, his father left Ebeye Island, so while growing up he barely knew him. In 1985, when Jelton was in fifth grade at the Jesuit school and everyone started calling his father Moses for having led his people to safety, it was like hearing about a stranger. Not until they reconnected years later would he come to understand the national atrocity that led to his father's absence, and his heroism.

Now, having reached the same age his father was then, it's plain that more heroics are required.

Ebeye, Jelton's birthplace, is one of 94 islands in Kwajalein Atoll—the largest of 29 rings of coral isles surrounding the submerged rims of ancient volcanoes, which together make up the Marshall Islands, midway between Hawai'i and Australia. Spread over 750,000 square miles in two parallel chains, the Marshallese atolls total just 70 square miles. Ebeye, a mile long and an eighth of a mile wide, with roughly 13,000 people, is among Earth's most densely populated places. With no high-rises, yet three times more crowded than Manhattan, the island is often called the biggest slum in the Pacific.

At the end of World War II, barely 100 lived here. Until then, Japan had occupied the Marshalls since 1914. Natives lived under thatch made from coco palms brought by their Austronesian ancestors 3,000 years earlier. They ate pandanus, taro, breadfruit, and coconut, and gathered fish, clams, crab, and turtle eggs from atoll lagoons. In early 1944, US armed forces won a critical battle for Kwajalein Atoll's strategically positioned, namesake island, and never left. Following Japan's defeat, the Marshalls became a United Nations trust territory administered by the US, which converted Kwajalein Island into a military base and expelled hundreds who'd always lived there to Ebeye, five miles away.

With base groundskeepers, mess hall cooks, and housekeepers earning US minimum wage, Ebeye's proximity attracted hundreds more—today, nearly a thousand Ebeye residents board a ferry each morning to work at US Army Garrison–Kwajalein Atoll, the nation's biggest employer after the government. In 1979, the Republic of Marshall Islands declared independence, which it finally gained in 1986 after entering a Compact of Free Association with the US: a semicolonial arrangement in which the currency is the American dollar, the postal system is the USPS, Marshallese waters are controlled by the US military, and most of Kwajalein Atoll is a US missile test site. Ebeye children grow up watching rockets streak overhead.

Compared with what ultimately martyred Jelton's father, however, antiballistic missiles are mere fireworks displays. Beginning in 1946, Ebeye's population swelled further when refugees from two northern atolls, Bikini and Enewetak, arrived after the Americans began exploding atomic bombs on their ancestral islands.

The translators had spoken like missionaries, promising that for "temporarily" leaving their homes, God would thank and bless them for bringing world peace by ending wars forever. Instead, the aboveground thermonuclear testing seemed to go on forever. Over 22 years, the US bombarded Bikini and Enewetak 67 times, but wars have failed to end.

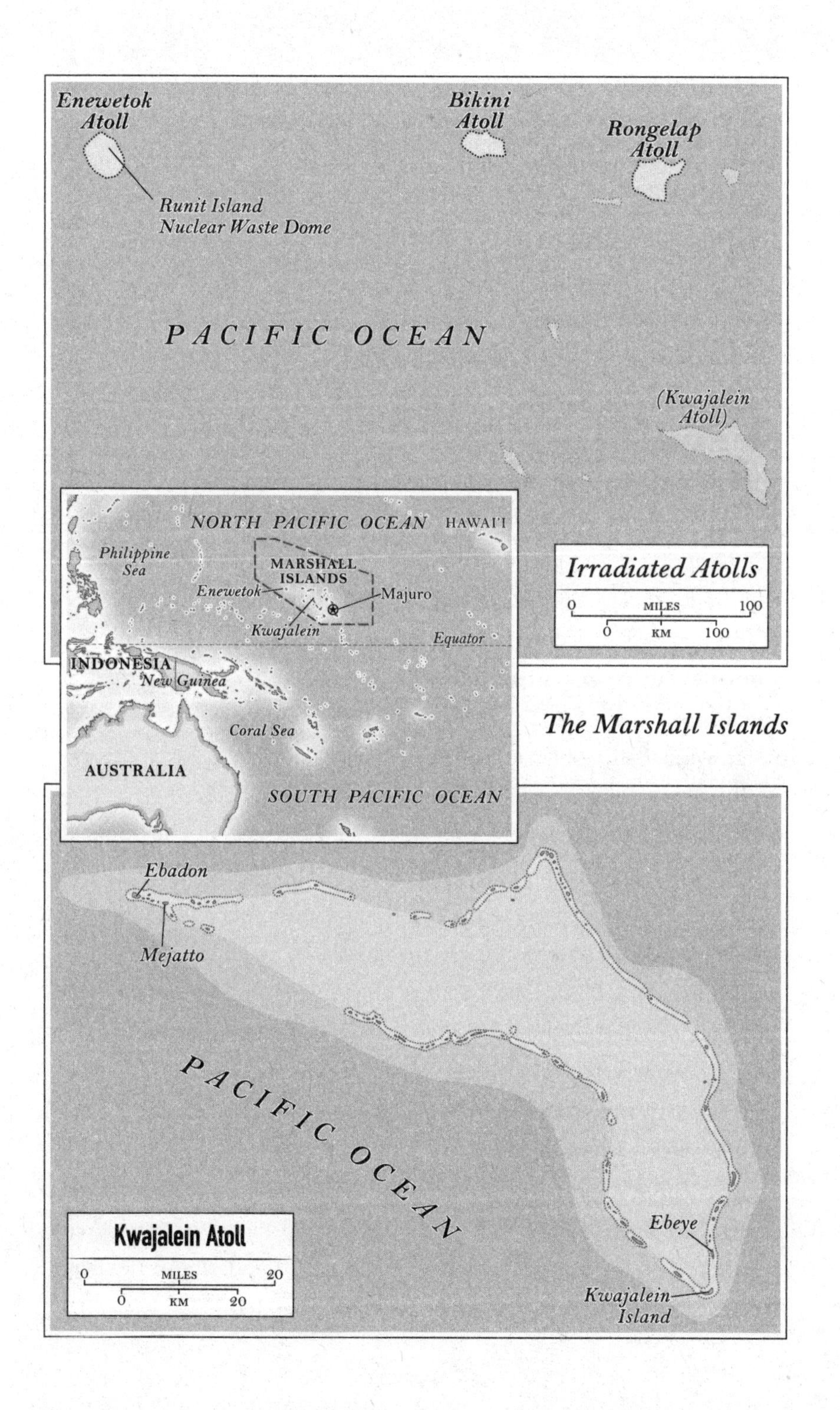

The Marshall Islands

The refugees were first taken to distant, uninhabited islands—which, they discovered, lacked lagoons for safe fishing or enough fruit and coconut trees. Malnourished, realizing they weren't returning home "within months," many drifted to Ebeye. However cruel their dispossession, even worse was coming: On March 1, 1954, a hydrogen bomb nicknamed Bravo blasted a mile-wide crater off now-vacated Bikini. Due to a miscalculation of its lithium deuteride fuel's volatility, the expected 6-megaton explosion measured 15, the biggest thermonuclear explosion ever—a thousand times bigger than Hiroshima.

Also miscalculated was the weather. The day before, the US Atomic Energy Commission had dismissed concerns when northerly winds above 20,000 feet shifted eastward. The following morning, John Anjain, the magistrate of Rongelap Atoll, 120 miles east of Bikini, was drinking coffee when he saw what resembled a sunrise—but to the west, followed by roaring. As he watched, a cloud mushroomed 130,000 feet into the stratosphere. That afternoon, white ash from 300 million tons of vaporized sand and coral dropped two inches deep on Rongelap; children played in it like snow. It fell on their food and in their drinking water, which tasted bitter, and people began vomiting.

Two days would pass before 82 Rongelapese were evacuated to Kwajalein Island. They were given showers; by then radiation burns were appearing. Children's hair and fingernails fell out. Once their skin lesions were treated and their hair and platelet loss measured, they were resettled on an islet near the Marshallese capital, Majuro, where women had serial miscarriages or bore deformed dead babies, some without bones.

After three years, they were told they could go home. "Even though radioactive contamination of Rongelap Island is considered perfectly safe for human habitation," the AEC's radiological expert wrote, "the levels are higher than in other inhabited locations in the world. The habitation of these people on the island," he added, "will afford most valuable ecological radiation data on human beings."

Ten years later, 1967, the magistrate's son, an infant when fallout filled his cradle, was among the first Rongelapese to develop thyroid nodules, and later the first radiation victim to die, at 19. In the late 1970s, when the magistrate discovered his own nodules, he was rejoined by his younger brother, Jeton Anjain, who years earlier had left to study dentistry at the hospital in Ebeye, where he married and had three children, including his near namesake (Jelton never learned why an *l* appeared in his name). What Jeton found on Rongelap after his marriage ended transformed him into a public health crusader. In 1981 he was named the Republic of the Marshall Islands' minister of health, but by 1983 his entire family had thyroid cancer, so he resigned to seek justice for Rongelapese.

They elected him senator. His 1984 resolution to relocate his people was unanimously affirmed by the RMI's parliament, but the US Department of Energy's project manager for the Marshalls refused, insisting that Rongelap's radiation levels were lower than places in America—despite a DOE report that Rongelap's soil was even more contaminated than Bikini's.

Taking matters into his own hands, Jelton's father contacted Greenpeace, which had refurbished a trawler for campaigns against whale hunting and nuclear waste dumping. In May 1985, Greenpeace's *Rainbow Warrior* moved 300 Rongelapese to Mejatto Island in the far northwestern reaches of Kwajalein Atoll, in four 110-mile trips, 11 hours each way. Greenpeace then sailed to New Zealand to lead a yacht convoy to French Polynesia, to protest nuclear testing in Moruroa Atoll. While moored in Auckland, French underwater commandos attached bombs to the *Rainbow Warrior*'s hull, sinking it and killing a crew member.

For leading his flock out of their irradiated homeland, Jeton Anjain received the 1991 Right Livelihood Award and the 1992 Goldman

Environmental Prize. By then, he and his son Jelton, now a teen, were trying to forge a relationship. There was so much Jelton intended to ask his father, but in 1993 Jeton died of cancer, likely due to radiation exposure.

February 2023: Kwajalein's deputy education commissioner, Jelton Anjain, now nearing 50, heavyset, with graying forelocks above his broad, coppery face, bounces in his white government SUV down the five-mile unpaved causeway that connects the north end of Ebeye to six islets. On the first, Lojjairok, is the walled royal family compound of Mike Kabua, *iroij*—chief and landowner—of Ebeye and also *iroojlaplap*: paramount king of Ralik, the Marshall Islands' western chain,* to whom the Americans pay $21 million in annual rent for their army base and missile range. On the second-to-last islet, Guegeegue, is Kwajalein Atoll High School and Jelton's house. He'd raced up there to get his family back to Ebeye, lest a looming king tide strand them. The last time, the overwash yanked boulders from the oceanside seawall, blocking the causeway.

He needs to be on Ebeye himself, to meet two Americans, Eric Rasmussen and Gregg Nakano, who've recently returned with a plan that could determine the fate of the islanders. But as he sprays through the sheet of water already covering the causeway, he meets another white government vehicle heading right into it. As it passes, he recognizes the big sunburnt face and bald pate of the man in the front passenger seat.

Of course, Jelton realizes, Eric wants to see what for years he's been warning would happen.

---

* Ralik ("toward sunset") also includes Enewetak, Bikini, and Rongelap. The eastern chain, Ratak ("toward sunrise"), includes Majuro, the Republic of the Marshall Island's capital. Although land inheritance is matrilineal, male *iroij* wield the power.

ii. Disaster Relief

After World War II, Eric Rasmussen's father, a Norwegian resistance fighter who spent years in a Nazi prison camp, had wandered to California, where he married Eric's mother. Eventually, PTSD scuttled both his marriage and, as there was no money, his son's acceptance to Southern Cal, so Eric joined the US Navy. Testing high enough for nuclear submarine duty, for the next few years he roamed the Atlantic and Arctic. Mustering out before his GI benefits expired, he studied the Great Books curriculum at St. John's College in New Mexico, learned molecular biology on the job at Los Alamos National Laboratory, and applied to Stanford medical school.

Graduating Stanford summa cum laude afforded him his pick of futures, but he missed the camaraderie of manning a nuclear sub with fellow smart misfits. The Navy welcomed him back as a medical officer. He served on everything from submarines to aircraft carriers and became a naval commander and fleet surgeon to the Third Fleet. During three war deployments to Bosnia during the 1990s, he noted the appalling lack of coordination between military and civilian humanitarian relief agencies, resulting in risky, redundant missions. He proposed a DARPA project, Strong Angel, which pioneered satellite-based field networks allowing troops, disaster relief teams, NGOs, and UN refugee services to share alerts, live interactive maps, translation services, and supply-line tracking for food and medicine.

To learn if solar photovoltaics could be deployed everywhere from soldiers' packs to battleships, he visited the Rocky Mountain Institute in Snowmass, Colorado, to meet its founder, green energy guru Amory Lovins, who lived in an all-solar home warm enough to grow a banana tree in its atrium. Then came September 11, 2001.

Between genocide in Kosovo, an American invasion of Afghanistan, and another looming in Iraq, multiple refugee crises were imminent. Refugee camps strain resources and patience in host countries; years often pass before residents can safely return home (if ever), so he and Lovins convened a brainstorming summit to design a near oxymoron: a sustainable refugee camp.

They gathered personnel from the UN, its World Food Programme, and the US State, Energy, and Defense departments, plus scientists and Afghani human rights workers, to a retreat outside Santa Barbara. Among the most notable ideas that emerged was mycologist Paul Stamets's proposal to impregnate cardboard care-packaging with mushroom spores, to be buried and watered for nourishment within 10 days, and a commercial crop within a month. Mostly, though, it was apparent how unsustainable the refugee problem was.

Two months later, Eric Rasmussen was in Iraq, leading humanitarian aid efforts in the wake of the 2003 American invasion. There he met Gregg Nakano, an infantry commander during the previous Gulf War, following Saddam Hussein's 1991 seizure of Kuwaiti oil fields. A 1988 Naval Academy graduate and son of a US Air Force attorney, Gregg had been raised on Sun Tzu's dictum from *The Art of War*: "Know your enemy and know yourself, and you can fight a thousand battles without disaster."

"But if you keep reading," he later told Eric, "Sun Tzu says the ultimate generalship isn't winning battles. It's accomplishing an objective without ever having to take the battlefield."

To better know his country's potential enemies, after leaving the Marines Nakano spent four years in China and Iran, learning Mandarin and Farsi, then got a master's degree in security studies at Tufts University. To pay off his student loans, he'd joined USAID as a civilian-military liaison officer, just days before 9/11.

When the US again invaded Iraq, he returned with USAID's Disaster Assistance Response Team, DART, to assess rebuilding needs once Saddam Hussein was overthrown. At the humanitarian operations center across the border in Kuwait, he met Eric Rasmussen. For the next 10 months they shared intel as each made forays into shattered towns like Basra and Nasiriyah.

A year later, in 2005, they'd meet again in Banda Aceh, Sumatra, epicenter of a 9.3 earthquake and tsunami that killed 228,000. Gregg was USAID's ground coordinator, trying unsuccessfully to wring information from the aircraft carrier USS *Abraham Lincoln*, which had collected digital imagery of how the tsunami had pummeled the coast—until Eric helicoptered in, brandishing a letter from the secretary of defense.

"Guys," Eric explained, "you have permission to share any unclassified images with people doing humanitarian coordination." He left the ship with six CDs of GPS imaging that showed what was broken and where people huddled in trees, which he turned over to Gregg's team.

Following his second tour in Afghanistan, where a bomb blasted his eardrums, Eric retired from the Navy as a disabled veteran. After serving for three years as the first CEO of InSTEDD, a disaster and disease response NGO founded by Google.org's Larry Brilliant, he started Infinitum Humanitarian Systems, IHS, the global disaster response team for the Roddenberry Foundation, the philanthropic arm of the Star Trek franchise.

At a post–Hurricane María FEMA conference in Puerto Rico, where IHS installed water-purification systems, Eric presented mapped data gathered by local students he'd hired to survey villages. Using a shared app, they'd identified pharmacies still shuttered, nonfunctioning ATMs, bus stations without buses, and everything else still unrepaired. Impressed, the Office of Naval Research asked Eric where he'd like to do it again.

"Yemen," he replied. Its war imperiled 11 million children, but

Navy lawyers demurred at entering a Middle East quagmire involving the Saudis.

They asked where else. He recalled a recent conversation with Gregg Nakano about the Marshall Islands.

Over 14 years, they'd stayed in touch, comparing disaster notes. Between them, during the new century they'd worked earthquakes in Iran, Haiti, Nepal, and Turkey; a typhoon in the Philippines; and hurricanes in Guatemala, New Orleans, and the Caribbean. In 2016, while getting his doctorate at the University of Hawai'i, Gregg had attended a public health conference in Majuro, Marshall Islands. At night, he drank beer with a local Asian Development Bank analyst who explained that although the Marshalls were in the prime tuna belt 4 degrees north of the equator, little profit from the catch ended up in the country. The RMI's main income was US lease payments and registering ocean vessels under the Marshallese flag, to take advantage of low taxes and lax inspections. Most people were poor and suffered gastrointestinal ailments from nitrate-contaminated wells.

After the conference, Gregg impulsively caught the weekly turboprop island hopper from Majuro to Kwajalein. Discovering there was nowhere to stay on base, he took the free ferry to Ebeye, but found he couldn't check into the sole hotel or buy food because no one took credit cards. He ended up staying with Mormon missionaries.

Wandering the islet, he was stunned by the contrast between the immaculate army base, resembling a leafy Honolulu suburb complete with fiber optics, McDonald's, Burger King, and Subway—"Almost Heaven" read a popular base T-shirt—and trashed, nearly treeless Ebeye, just five miles away: so packed that hundreds of children roamed at night because families had to take turns with beds. There were no parks, just ubiquitous, never-empty, half-sized concrete basketball courts. Houses of peeling plywood and corrugated sheet-metal rotted into the sea. Drinking water was so dubious that people employed on the base took as many jugs as they could carry every day to fill there. The lagoon was so saturated with heavy metals from the missile base that local grouper and snapper were too toxic to eat.

Ebeye's typical diet—Spam, packaged white rice, junk food, sodas—correlated directly with its diabetes rates. Fish from far across the lagoon or outside the atoll was still safe, but gasoline was nearly $10 a gallon, and there was no sailing culture anymore. On an island packed with descendants of history's greatest mariners, who'd navigate thousands of miles by stars and stick charts, Gregg saw not a single outrigger canoe.

"Kwajalein is a microcosm of our military-industrial complex's brutal impact on Indigenous cultures and the developing world," he told Eric. "There's no better example. If we can do something here, it'll prove that we probably can do it anywhere."

A Tufts program called ALLIES—Alliance Linking Leaders in Education and the Services—paired civilian undergrads with peers from military academies on trips abroad to get acquainted for when they later became policymakers and officers. It inspired Gregg to start Pacific ALLIES. A year later he was back on Ebeye with service cadets and Tufts students, looking to meet young Marshallese. "See Jelton Anjain," the city manager told him.

As a lost, fatherless kid on a lost island, Jelton had hung with other toughs, smoking cigarettes and worse, when at age 10 a priest convinced him to try school. Embarrassed to be the oldest in his class, he stuck through and earned a scholarship to Loyola College in Baltimore. Under the Compact of Free Association, Marshallese are free to migrate to the US, but upon graduating in political science, Jelton returned to Ebeye. Although several government departments and the Catholic church wanted him, his passion was education.

Yet after years of working in both public and parochial schools,

Ebeye's kids seemed more aimless to him than motivated. In 2017, Jelton tried an experiment: a monthlong summer field school he called Spartan Camp, pairing top high schoolers, both boys and girls, with students at risk like he'd been. Mindful of Ebeye's high teen pregnancy rate, he and his wife chaperoned. The curriculum was STEM—science, technology, engineering, and math—which to Jelton hearkened to the way their ancestors learned: hands-on, and by observation.

"Get them doing math without realizing it," he told parents.

He had students imagine a bridge between Ebeye and Kwajalein Island so they wouldn't need the ferry. He gave them Popsicle sticks and glue and told them to build one, a foot long, six inches high. No instructions. "However you want to do it."

An hour later, they'd made a sturdy model. "I'd never known how many inches there were in a foot," one admitted. They wanted more.

STEM, Jelton believed, might also mean their survival. If waters kept advancing and their generation had to leave, they would need skills. The program would identify the sharpest kids and help prepare them for leadership someday, even if it meant leading Marshallese away from their homeland, like his father had done: Moses in reverse. There was so much to teach them. Then suddenly, this buzz-cut ex–US Marine with ramrod posture earning his doctorate in education appears with a clutch of American undergrads, offering to peer-mentor Marshallese students in computer literacy, marine science, English, climate science . . .

"Truly a match made in heaven," Gregg later told Eric when he introduced them.

"Like getting married," agreed Jelton.

In 2018, Jelton, Gregg, Annapolis midshipmen, West Point and US Coast Guard Academy cadets, and a Whitman College student taught Marshallese kids beginning robotics. To study geology, oceanography, and marine ecology, they surveyed and cataloged Ebeye's shores, the STEM students showing their mentors places where their island was narrowing. The previous fall, Ebeye's school reading level had ranked 42nd out of all Marshallese public and parochial schools. Jelton and colleagues took STEM-style problem-solving into their classrooms. By spring, they'd jumped to number 1.

Then in 2019, Eric Rasmussen came with Office of Naval Research seed funding—to propose exactly what, he wasn't sure, but Spartan Camp felt like where to start. He arrived with 21 new Pixel phones loaded with two free programs. The first was the same software, translated into Marshallese, that students in Puerto Rico used to survey needs after Hurricane María. Distributing phones, Spartan Camp T-shirts, and green reflective vests, he, Gregg, and Jelton divided Ebeye into five sections and sent students to every household with a questionnaire:

Did they have clean drinking water that day? Enough food? A working toilet? Lights after sundown? A phone? Any income from work? In the past decade, had anyone in this home moved to the US or to another RMI island? Was anyone sick or injured? Did children need vaccinations? Glasses? Dental care? Were they in school?

Using a 360-degree camera Eric provided, they photographed every street, alley, and house on the island—images he'd arranged to upload to Google Street View. "Before," Eric told his Navy funders, "they never saw themselves reflected anywhere on the internet. Now they can."

A third of Ebeye households, they discovered, had no clean drinking water. A quarter had ill or injured people. The government had no idea how many people lived on Ebeye—in one household, they found 44. For social services, medical teams, or Google Maps to locate anyone required something Ebeye lacked: addresses. Repeated waves of arrivals had cobbled shelters wherever they could.

Designers of the second program on their phones had used Oxford's supercomputer to divide the entire globe into 57 trillion 3-by-3-meter squares, each assigned a combination of three words, separated by periods. What3Words included a GPS to identify every square and could locate any other by its W3W name.* "For example," Eric explained to Ebeye Hospital doctors, "that doorway is *wobbling .vent.divisions.* Mongolia, which has few named streets, has adapted What3Words as its national addressing system."

Although W3W was available in several languages, it was uncertain if it was translatable to Marshallese, because 57 trillion three-word combinations require at minimum a 50,000-word vocabulary. But English was a skill that STEM students were perfecting, and knowing a word's meaning wasn't necessary to pinpoint locations in the event of accidents or disease outbreaks.

They also documented changes on the reef. The plan was to photograph the same high tides on Ebeye's lagoon side and ocean side every year at roughly the same hour—but then the pandemic intervened.

For nearly three years, the isolated Republic of the Marshall Islands was the only COVID-free nation on Earth, because no inbound flights were allowed, but Jelton kept Spartan Camp going with local STEM technology. When Eric and Gregg returned in early 2023, they saw what two master builders from Lae, a minuscule atoll 100 miles west, had taught 11th graders to make: three full-sized outrigger sailing canoes, two blue and white, the other painted orange with leopard spots. Since Ebeye had so few trees, the masters had brought breadfruit trunks from their island for the hulls. The out-

* https://what3words.com.

riggers and booms, attached with coconut-fiber rope to absorb wave shock, were made from driftwood heaped by overwashes on Ebeye's causeway.

"We've entered races," Jelton told them. "No one can believe that high schoolers did this."

Two of his young boatbuilders, Lowa Kabua and Kobie Loeak, sporting sunshades and downy mustaches, were hoeing in the school's adjacent community garden.

"What do you guys intend to do with your STEM education?"

"College in the States," said Kobie. "Then come back."

"Same," said Lowa.

"And then?"

"Start with a low government job, then become a councilman, then a senator. Then I'll try running for president."

"Yes!"

During COVID, Eric and Gregg had plotted an idea for Ebeye. Periodically, Eric lectured at Singularity University, a non-degree-awarding Silicon Valley startup incubator. His audience was budding entrepreneurs with schemes to disrupt the world positively while enriching themselves.

Pulling no punches, Eric would outline the stark political and climate realities of the world in which they hoped to make billions, then challenge them to turn existential problems into opportunities for making both money and a difference. A much-decorated military hero who still rushed to disasters worldwide and a gripping speaker, he'd elicit huge applause and promises that doing well by doing good was their every intention. Now he would give them their chance.

He was proposing that ONR, the Office of Naval Research, fund a Marshallese-directed Kwajalein Atoll Sustainability Laboratory on Ebeye. KASL would invite foreign companies and researchers to test climate adaptation ideas in one of the planet's most difficult places.

The laboratory would employ local STEM graduates, who would earn good wages and participate in decisions affecting them. Should any new sustainable products or technologies succeed, their islands would benefit and earn intellectual property royalties. They'd learn new skills to use on Ebeye—or in the diaspora, if it came to that. They'd also take pride in knowing the world would gain from innovations tested and lessons learned on their atoll.

They'd brainstormed KASL with Eric's network of Washington think tanks. After neglecting the Marshall Islands for decades, the US was newly interested, not only because of Kwajalein's missile range and tracking station for North Korean intercontinental ballistic missiles. China had recently signed a security pact with the neighboring Solomon Islands. Three Chinese groceries had opened on Ebeye. Across Micronesia, drowning atolls were now caught in a geopolitical tug-of-war.

"These are not small island nations," Eric kept reminding people. "They are large ocean nations."

In February 2023, Eric and Gregg walked around Ebeye for five days, presenting KASL to the mayor, to senators, to the woman who directed the local UN International Organization for Migration office, to doctors, to the Catholic priest.

"Average temperatures have already increased 1.5°C here, and you're on track for 3.5°C," Eric said bluntly. "Within 15 years, overwashes are expected annually—the entire ocean sweeping over the islands."

Often occurring with no warning under blue skies, overwashes result when resonant waves from storms far to the north and south meet near the equator, smashing into atolls. Until recently, various scenarios for sea level rise calculated that island nations wouldn't drown before 2100. But modeling by the Netherlands' Deltares and the US Geological Survey that included overwashes now warned that by midcentury, most atolls would be uninhabitable because wave-driven floods would seep into their thin lenses of underground fresh water, leaving it too saline to drink.

Eric Rasmussen's big, 6-foot-1 frame commanded gravitational authority, even in a Hawaiian shirt. So did the braided blue-and-yellow bracelet on his wrist, a gift from a Ukrainian girl when, amid a snowstorm two months prior, his disaster relief team built her family an insulated prefab shelter after Russian missiles destroyed their house. "Since 1993, you're up 210 millimeters—nearly 8 inches," he told everyone. "You're rising 7 millimeters a year. In 15 years, add a storm and you're swamped. Where we're sitting will be underwater."

He'd just delivered a death sentence for their homeland, but no one argued. Everyone knew—yet Marshallese kept living as though it couldn't possibly be true. Their government had hired Deltares, which designed a plan that would make islands smaller but up to 12 feet higher, carving out huge chunks of reef to pile atop places they most wanted to save, to then surround with more seawalls. But who would pay for that? What did Deltares know about islands without any river deltas? Even on Ebeye's lagoon side, new seawalls had cracked within a year.

The islanders' cognitive dissonance—knowing what can't be changed, yet acting as though maybe it still could—augured their oncoming grief over the unfathomable loss of the very ground beneath their feet. KASL's painstaking statement of purpose attempted to compassionately bridge the chasm between their denial and acceptance:

> **Live better now,**
> **Transition more gently,**
> **And then be more welcome.**

When Eric and Gregg presented KASL's prospective Marshallese board members with their intended research objectives—

1. Energy
2. Water
3. Health Care
4. Education
5. Food
6. Shelter
7. Sanitation
8. Communications
9. Transportation
10. Jobs creation
11. Ecosystem Regeneration
12. Cultural Preservation

—they were told to move the final category from last to first. "We've always had problems with one through 11," said one senator. "We can live with some diarrhea, but now that we're looking at leaving our islands after 3,000 years, help me preserve my culture," she pleaded. "That's our highest priority."

But how? Attempts at digital twins of entire ecosystems always prove incomplete, because nature's variables are uncountable. Wouldn't a digital twin of an entire culture also be hopelessly inadequate? Eric and Gregg were talking with University of Hawai'i archivists, but no matter how much history, art, lore, and knowledge is archived, culture ultimately is people.

"You are already navigators like the world has never seen," Gregg told them. "We want you to be the new climate navigators."

iii. The Laboratory

"I hope we don't have to become climate nomads," said Jelton Anjain. Daily, his school principals checked NOAA* weather charts to get kids home before high tides, canceling classes whenever they exceeded five feet. In the Hotel Ebeye's restaurant, overlooking a la-

* National Oceanic and Atmospheric Administration.

goon filled with sunken Japanese war planes and submarines, they were showing him a PowerPoint on Eric's MacBook with three tiers of potential KASL research projects.

MIT and several national laboratories were interested, Eric said, once the Marshalls gave KASL the go-ahead. They'd need to form a Marshallese NGO to work out intellectual property agreements, but Apple and Google already were offering student internships. "You could start Tier One this summer at Spartan Camp."

Tier One possibilities included emulating Roman cement by burning coconut waste to extract lime from seashells; releasing sterile mosquitoes for dengue control; beginning a fleet of modern outrigger canoes; installing water purification systems with technology from the International Space Station; building a prefab climate lab with the same technique Eric's Ukraine team had used; and continuing "Oceans Connect Us" Zoom encounters begun during COVID between middle schoolers in the Marshall Islands, Chile, and New Zealand to discuss living on coasts in the 21st century.

"Already done," said Jelton. "The kids are now running it themselves."

"Terrific. The second-tier stuff we're not quite ready for, but they'll prove very desirable: inter-island transport drones; salt-tolerant crops; floating lagoon homes; bladeless wind power; 3D ocean farming; coral reef seeding; harvesting water from air; and a whole list of things MIT wants to try, including 3D-printing buildings with plant-based polystyrene."

"Floating houses?" said Jelton.

"With vegetables growing under rooftop photovoltaic panels. My construction engineer from Ukraine wants to come for Spartan Camp and put one on floats right here," he said, pointing to the lagoon.

Jelton frowned. "Even in the lagoon, wind and waves can get fierce."

"That's why we test." Eric refrained from describing a wine-induced dream he'd had of ringing inner lagoon shores with self-sustaining floating neighborhoods that, when waters subsume the atolls, would still outline where they once were, entitling Marshallese to keep valuable fishing rights.

"The last tier is stuff that would be fabulous, but the people with the ideas haven't invented them yet." This included DARPA's reef-regeneration project; transmuting oceanic plastic waste into 3D-printing resin; saltwater-powered illumination; wave-energy generators; clearance of unexploded ordnance from out-islands for drone-monitored farming . . .

No harm dreaming, although, with the exception of MIT, which long had a Department of Defense–funded field station on the Kwajalein base for developing national security technology, whether companies could be convinced to test in a place at the farthest end of supply chains, which still had bandwidth issues and was served by only three commercial flights a week, was unknown. The enticements Eric used to tease corporate contacts were 1) capitalizing on goodwill by helping a sinking nation stay afloat, 2) proving their product could work in the hardest place, and 3) "If we don't, China gains a greater foothold."

Their last day on Ebeye, cloudless and sunny, the king tide hit at new moon—a swell waxing as it rolled over submerged reefs that, centuries ago, Marshallese navigators had noted on their stick charts with cowrie shells. Like a flash flood, within three minutes the dirt causeway leading to Guegeegue Island resembled a river. Rollers bounced over the seawall, tumbling its boulders, connecting the ocean with the lagoon. When the road was completed in 1991, it was supposed to withstand king tides for a half century. Spray-soaked, standing ankle-deep, Eric recalled his lunch a month earlier in Portland, Oregon, with Marshallese poet laureate and RMI climate ambassador Kathy Jetñil-Kijiner.

A poem she wrote in 2018 recounted her four-day canoe voyage through 20-foot waves to Runit, an islet in Enewetak Atoll, where a 1958 misfire had scattered unexploded chunks of plutonium. Under a concrete dome wider than a football field, US soldiers had interred 3 million cubic feet of radioactive debris. Decades later, a *Los Angeles Times* exposé revealed that 130 tons of radioactive soil were also airlifted from Nevada's atomic testing site and dumped there.

Jetñil-Kijiner had climbed atop the dome, whose 18-inch-thick concrete had started cracking in the 1980s.

> You were a whole island, once.
> You were breadfruit trees heavy with green globes of fruit
> whispering promises of massive canoes.
> Crabs dusted with white sand scuttled
> through pandanus roots
> Beneath looming coconut trees, beds of ripe watermelon
> slept still, swollen with juice.
> And you were protected by powerful irooj, chiefs birthed
> from women who could swim pregnant for miles beneath a
> full moon. . . .
>
> You became crater, an empty belly.
> Plutonium ground into a concrete slurry filled your hollow
> cavern. You became tomb. You became concrete shell. You
> became solidified history,
> immoveable, unforgettable. . . .
>
> My belly is a crater empty of stories and answers
> only questions, hard as concrete.
>
> Who gave them this power?
>
> Who anointed them with the power to burn?*

* Excerpted from "Dome Poem Part III: 'Anointed,'" by Kathy Jetñil-Kijiner.

As he watched a quarter of the causeway dissolve, Eric thought of the seawater now lapping at Runit's dome, intruding beneath its crumbling edges. He and Gregg had long been civilians, but in these islands, they were still yoked to American military history. They understood that KASL might be perceived as atonement, and that any such offering was wildly inadequate.

But it also was needed, they and their colleagues here knew, because something equally unfixable for which Marshallese were equally blameless was bearing down.

Before catching the thrice-weekly United flight back to Honolulu, they made the rounds in the capital, Majuro, pitching KASL to the Asian Development Bank, to the US ambassador, to the president of the College of the Marshall Islands, to Jo-Jikum: the youth environmental collective Kathy Jetñil-Kijiner founded. All were enthused, but nothing could happen without the royal blessing of Iroojlaplap Michael Kabua, king of the Ralik sunset chain and RMI senator, son of the founding president, uncle of the current president, and the most powerful man in the country. "He's your one-stop shop," the ambassador advised.

Mike Kabua's office walls were hung with war clubs, ceremonial daggers serrated with shark teeth, palm-thatch ceremonial fans and dance sticks, and a row of framed *marmar*—leis—plaited from hibiscus, coconut fiber, and pandanus leaves, then woven with flowers and bone, and encrusted with abalone, snail, limpet, and cowrie shells. His charcoal silk shirt was hatched with gold. A former canoe racer, still barrel-chested in his 70s, with flattened gray hair and a thick nose, he never stopped smiling, even as his strikingly wide dark eyes penetrated.

Eric presented him a 2014 Cabernet Sauvignon from the Chateau Ste. Michelle winery, 10 miles from his home in Seattle.

"Thank you," he whispered.

They'd been allotted 15 minutes. Eric gave him a bound version of the KASL PowerPoint. "The founding advisor to the Kwajalein Atoll Sustainability Laboratory is education minister Kitlang Kabua"—the president's niece. "On my first visit to Ebeye, she challenged us not just to test climate adaptation ideas and then leave, but to seek how Marshallese could benefit."

They now had 77 ideas, he said, that scientists from MIT, Singapore, and US national laboratories had proposed for testing on Ebeye, subject to Marshallese approval.

"Some aren't quite ready, others are ready to try this summer in Spartan Camp, which Jelton Anjain wants to take to Ebadon"—an island on the opposite tip of the crescent-shaped atoll from Ebeye, owned by Mike Kabua's family—"if you find that acceptable."

No reaction. "We want to help Jelton," Eric continued, "so we're bringing an engineer and a STEM teacher from the US. Does this make sense so far?"

"Mmmm."

"Gregg and I are supported by a grant from the US Office of Naval Research. We aren't paid by KASL or by any business associated with it. Anybody bringing an idea to test is on their own dime. They stay on Ebeye, not on the base, to enhance the local economy."

Kabua sat behind his desk, which was draped in red-and-black island fabric, impassive. Eric stood. "Marshallese get first say on whether an idea is valuable or not. Some eventually may become businesses on Ebeye, or there may be a licensing fee for Ebeye, or for Marshallese. Does that make sense?"

"Mmmm."

Eric dropped his hands. "That's what it is. We intend this to keep going. ONR has committed to support us for three years, so Gregg and I will be involved at least that long. Any questions on any of that?"

A mumble. Eric looked to Gregg.

"He said he's surprised to hear this is happening."

"I got blown up in Afghanistan and am mostly deaf," Eric apologized. "I have hearing aids, but you are a soft-spoken man, so I may look to my friend to tell me what you said. We published a paper in an engineering journal encouraging people to bring us ideas. It got us invited to Washington, DC, to speak about KASL. It got us invited to speak at the UN General Assembly. But this cannot be real until you say it can. This rests on you."

"I'm just really surprised this is happening on Ebeye," Kabua repeated, louder.

"I hope 'really surprised' also means really pleased—not really annoyed."

The iroojlaplap laughed, his wide eyes smiling now.

"I'm glad, sir. Do we have your permission to continue?"

"Please do," said the king. "We don't want to leave here."

That summer, United Airlines melted down over July 4, canceling thousands of flights, including their return to the Marshall Islands, and another wasn't available for a week. Their 14-day expedition, months in planning, had to be crammed into seven, including the five-hour boat ride from the Kwajalein base airport to Ebadon, a forested island where a hundred people living in plywood shacks fished transparent waters and gathered pandanus, coconut, and breadfruit. Idyllic—until the sea started eating their white beaches. The swells were bigger, the currents had changed, outrigger sailors said.

Eric's construction engineer, Dan Kenney, had arranged a World War II–era assault landing craft to deliver two dozen sheets of four-inch-thick expanded polystyrene foam made from recycled plastic, the same they'd used in Ukraine. With Ebadon villagers helping to cut them with hot wires and join them with insulating adhesive, under a tropical sun and portable lights by night, in a week they built a three-room, 12-by-24-foot climate security research station, anchored with ground screws. They skinned it with the cement technique they'd wanted to test by roasting seashells and dead corals over a coconut-shell fire—but the villagers already knew how: that was how they extracted lime for chewing betel nut.

The Marshall Islands had no Starlink internet contract, but Eric knew someone who knew Elon Musk. Their solar-powered satellite broadband terminal was connected to separate servers for educational materials and data storage. The station's environmental sensors would monitor coral reefs, ocean temperatures, air quality, barometric pressure, wind, rainfall, dew point, solar radiation, and any heat-index change from the atoll's perpetual low to mid-80's F. Periodic ground observations—beach erosion, biodiversity, mangrove and pandanus mapping—would be done by residents.

Gathered data would upload via a Starlink WiFi cloud 600 yards in diameter, which included the church and the K–8 school. The village's internet allotment was 50 gigabytes a month, but a separate local server contained a complete copy of Wikipedia and nearly all of the online video tutor Khan Academy, which students could access as long as they liked without draining connectivity.

Gregg's latest Pacific ALLIES included Dan Kenney's wife, Veronica, a former Boeing engineer turned STEM teacher; two male future teachers from the nonprofit Educators of America; and two female Coast Guard Academy cadets. One, a Filipina who the Marshallese kids thought resembled them, became an instant star to the island girls, who flocked to watch her build the climate station alongside the men.

"She's why a bunch of Marshallese girls 10 years from now will apply to the Coast Guard Academy," Gregg told Jelton.

Jelton's own son, Ranton, had joined the US Army. In 2019, alongside Swedish activist Greta Thunberg, he'd addressed a press conference at UN headquarters as they filed a formal complaint that inaction on climate violated the Convention on the Rights of the Child. He was now stationed in Virginia; whether his future would be stateside, Jelton couldn't predict. Over the past decade, the Marshall Islands' population had dropped by nearly one-fourth, from 55,000 to 42,000. There were now large Marshallese enclaves in Hawai'i, Oregon, and Springdale, Arkansas, where thousands worked in Tyson Foods poultry-processing plants.

Jelton himself wasn't going anywhere. He was expanding Spartan Camp all around Kwajalein Atoll; this year's was actually on Mejatto, an island three miles from Ebadon with twice the population, where 30 years earlier he'd buried the father he'd barely begun to know when cancer took him.

Five Spartan Camp veterans from Ebeye were here as peer mentors. The other 32 participants were descendants of the irradiated survivors Jelton's father had rescued from Rongelap.

Unlike verdant Ebadon, Mejatto, the only place the US had allowed nuclear refugees to settle en masse, had no aquifer. People collected rain in cisterns that were cauldrons for chronic intestinal disease. Students here knew their families received compensation money, but most didn't know why. One day, through tears, Jelton explained their history to them.

Parents praised him for bringing their children education and technology. "You should run for Senate," they said; he could claim the Rongelap seat that still existed, although no one lived there.

He wasn't interested. In Mejatto's elementary school, more tears came as he watched kids making stick charts from coconut strips and cowrie shells, peering into microscopes at water samples, solving STEM math problems on new laptops. In addition to school, ukulele lessons, and geology field trips on the reef, they'd woven a rope of palm branches hundreds of feet long, which they dragged into the lagoon to form a fish trap for a final banquet with the KASL engineers, and then lined up to hug Eric, Gregg, Dan, Veronica, and the young Pacific ALLIES teachers.

"We'll be back," Eric told them.

"We'll be here," Jelton said.

He couldn't imagine being anywhere else.

CHAPTER FIFTEEN

HOPE DIES LAST

i. Floor, Ceiling

In 1977, when John Platt was 14, he entered California State University, Long Beach, to study chemistry (it was the nearest college; he still lived with his parents). For graduate school, he chose the California Institute of Technology for its breakthroughs in getting machines to emulate how human brains memorize and learn. Amid neural networks and computer science, he also regularly found his way up 6,000-foot Palomar Mountain to work with Caltech astrogeologist Gene Shoemaker, photographing the cosmos through the observatory's fast-imaging telescope. By the time he'd finished his dissertation—on similarities between how artificial intelligence models minds and computer animation models bodies—he'd also discovered a pair of asteroids that he named for his mom and dad.

At Synaptics, where he directed research after Caltech, he helped put the touch in laptop touchpads and coaxed machines to recognize handwriting, including Chinese. In 1997, he joined friends doing computer graphics at Microsoft Research Redmond, where he invented an algorithm that vastly sped up the training of machine-learning models. In 2005, he won an Oscar for technical achievement by making computer-generated clothing in movies ripple realistically.

A big, affable, gentle-voiced polymath with short, wavy brown hair, Platt became deputy director of several elite Microsoft Research groups, including machine learning, computer graphics, computer vision, and cryptography. It was a fulfilling career, but in 2015, Google made him an irresistible offer: leading its artificial intelligence climate program.

At the time, both in scale and application to big problems, Google was more ambitious than Microsoft about AI. It was also the first big tech company to commit serious money and brains to averting climate calamity. During a trip to Morocco, seeing the advancing Sahara from atop a camel sealed John's decision. He had two teenagers. It terrified him to watch their future dangerously approach a cliff.

Eight years earlier, Google had announced achieving carbon neutrality by purchasing enough carbon credits to offset its operations, employee travel, construction, and server maintenance. It was an earnest step, although the value of most carbon offsets has since been challenged—such as for preserving forests on land trusts that never would have been cut anyway. Google's chief offset investment was capturing methane from landfills and pig farms to burn for fuel, with the rationale that the resulting CO_2 emissions are less potent than releasing straight methane.

Selling carbon credits involving industrial swine operations, however, becomes a climatic Ponzi scheme, as demand for such offsets encourages more of them, producing more emissions and requiring more synthetically fertilized grain to fatten more hogs. But other Google sustainability efforts were laudable, like using machine learning to optimize cooling in data centers—and, soon after John Platt arrived, matching its annual electricity use with renewable energy purchases.

He took a year to learn the complicated interactions of geophysics, energy, and the economics of climate change, to quantify what we're facing. Clearly, no more methane should be emitted, period—or anything else. It was also clear that the global economy had too much invested in existing energy infrastructure to abandon it. To even hold to just a 2°C increase, never mind 1.5°C, he told the podcast *Eye on A.I.*, "we essentially have to stop building fossil fuel infrastructure now, like today. We have no more time to waffle or think about it."

But that was a fantasy. China alone was building a new coal-fired plant each week. The answer to resolving this disconnect seemed clear: find a clean source of energy, available 24/7. But merely clean wasn't good enough. It had to be cheap.

"It has to cost less to build and run than the old system's fuel cost, so that anyone would be nuts not to chuck their obsolete fossil fuel plant and build the new thing instead."

Unfortunately, no one had yet discovered what John came to call "a strong energy miracle"—versus weak energy miracles like wind, solar, and batteries, whose costs, while becoming competitive, still weren't low enough to convince energy czars to strand their CO_2-belching assets. Meanwhile, global energy demand kept growing by 2 percent annually, meaning that by the end of the century, it would increase sixfold. Something had to happen, or we'd be cooked.

Of all possibilities for a strong miracle, fusion seemed the best bet, so he began advising a California company, TAE Technologies, that Google had invested in. Unlike the tokamak doughnut at Commonwealth Fusion Systems, TAE accelerates high-energy particle beams into a 30-meter tube of hydrogen-boron plasma.* It's potentially the safest fusion reaction, as it emits no neutrons, but it does require much higher temperatures: over a billion degrees Celsius.

* Google also invests in Commonwealth Fusion Systems; the now-retired Google VP who hired John Platt—Alan Eustace, world record holder for the highest skydive—is on its board.

Because their chamber was too hot for sensors, John was using machine learning to help them infer how the unruly plasma they'd been trying to tame since 1998 was behaving.

Even if their fusion worked tomorrow—"Or Commonwealth's, or anyone's; I just wanted *someone* to win"—he figured it would be 2040 before it made a significant dent in the climate crisis. Likewise, it would be the second half of the century before another strong miracle many dreamed of—capturing excess carbon dioxide directly from the air and locking it up—became widely enough employed, if ever. A Harvard engineer, David Keith, had patented a device he believed could do it for $100 per ton. That was a lot lower than Platt had guessed, but then he did the math and found that if this worked, lowering global temperatures by 1 degree would cost $200 trillion. Including the energy cost of mining and forging steel to manufacture enough carbon-capture machines to ship and deploy worldwide, then transporting and stowing all that CO_2 permanently in abandoned oil or gas reservoirs, or dissolving it in saline aquifers—the scale becomes hard to grasp.

Platt and colleagues tossed around another potential strong miracle: making the sea's surface more alkaline to absorb atmospheric CO_2 faster. Several papers have proposed several approaches, but big puzzles remain, including impacts on marine life, and how to do that without incurring big energy costs. Meanwhile, as our reality means we can't wait for fancy new technologies, these days he looks for weaker miracles to chip away at decarbonizing, as much and as fast as possible.

For example, artificial intelligence had begun improving supply chain efficiency by optimizing freight routes—and Google Maps directions—to maximize energy savings. Yet John regularly reminded his colleagues of an annoying economic precept, the Jevons Paradox, whose truth dates back to steam engines: Whenever technology increases energy efficiency, energy becomes cheaper. More people can then afford it, so demand doesn't decrease.

Via global satellite images, AI was also analyzing a hundred million rooftops to find the best angle for solar panels. One of its most promising applications involved jet contrails, the long, artificial cirrus clouds that jets leave in moisture-laden sections of the icy upper troposphere as condensation forms on their warm, sooty exhaust. Especially at night in the wake of red-eye flights, contrails trap enough heat to account for about 1 percent of global warming.

"From satellite images, AI now can predict where contrails will be." Six months of tests with American Airlines pilots using Google's AI-based predictions to avoid humid atmospheric layers reduced contrails by 54 percent.

"We just figure out if the plane should go right, left, or down a bit. I don't know if that's a scratch or a dent in the problem, but it reduces global warming by 1 percent without waiting for the energy transition."

One percent down, 99 to go.

When John accepted Google's offer, among his conditions was staying in Seattle for his family, but he frequently meets with team members at the Googleplex in Mountainview, California. On his way to grab dinner at Root Cafe, one of the many free restaurants for Google employees, the sand volleyball courts he hurries past on a hazy January night are full of barefoot software engineers, playing under cones of light. He takes a table overlooking the boulevard, watching a line of white minibuses that shuttle Google engineers to the homes they can afford in San Francisco, as he eats roti stuffed with quinoa and roasted vegetables, contemplating how AI might tackle the next 99 percent.

"There's an old physics axiom: You can't control what you can't measure." AI can scour satellite meteorological data and forecast floods in real time*—however, he laments, it can't stop 250 million people from being flooded each year. Nor can it yet fulfill, if ever, the Silicon Valley trope of a virtual Earth, which might warn us if we're approaching the point where we'll extinguish enough species to cause a cascade we can't escape ourselves.

"We don't have a good-enough model of all the ecosystems, because we don't have enough sensors on the ground or in the water"—only limited numbers of animals can be radio-collared, and microbes not at all. "Or we don't have enough computation to absorb all the information. We don't even know all the species on the planet, nor precisely where they actually are, let alone how they interact and how they support us."

John's family has a cabin in the Sierra Nevada. It worries him to watch alpine environments being pressured from below, as organisms at their temperature thresholds climb to beat the heat. "The ones at the top might just be crowded out of existence. Between bark beetles and droughts, there are already a hundred million dead trees in California."

He's seen the IPCC model that says that 99 percent of corals will disappear at 2°C. "I wish I could save the reefs. A billion people's livelihoods rely directly or indirectly on coral." He inhales the remaining half of his roti, then gazes sadly at his empty plate. "The current Earth is really neat. It'd be a shame to wreck it. We don't need a complete global model to predict that we shouldn't yank the tiger's tail—just the precautionary principle. It's a big tiger. Don't yank its tail. I can't make it plainer than that."

---

* Accessible at Flood Hub, https://sites.research.google/floods/l/0/0/3.

A single species, *Homo sapiens*, whose evolution depends on an ecosystem too vast to model, is also too complex for AI to foresee when and where its organisms will next attack each other.

"There are very crude social models, but we don't have psychohistory, like in Isaac Asimov's *Foundation Trilogy*. We don't have AI sophisticated enough to predict, let alone control, what goes on in human society. Nowhere near."

What he does know is that IPCC models also show that at 2.5°C, crop yields decline. Also, within the 2020s, the planet's refugees will double. As they flee elsewhere, demagogues will rise to block them, because people scared of change pick strongmen who promise to stop it all from happening, never mind how impossible.

"This isn't going to sound nice," John says, hoisting his pack. "Although we're passing 1.5°, the truth is 2° isn't the ceiling; it's the floor. I think we'll avoid the complete and utter dystopia of 3°C, but between 2° and 3° is a lot of human suffering. And it's a knife fight for every tenth of a degree Celsius between 2° and 3°, which is terrible. I try to engage in this knife fight. It's a good fight to have. Those tenths of degrees matter, because of suffering and because of unknown tipping points."

Although January in northern California is usually sweater weather, the volleyballers are in T-shirts, and clouds of moths skitter in the lights. John has just been in Germany to give a contrails talk and meet with colleagues working on increasing oceanic CO_2 absorption. Since the global COVID pandemic, AI's sudden stunning leaps have left him more optimistic than in years. It won't supplant the scientific method or replace scientists anytime soon, but it's making them new, incredibly fast tools. Machine-learning models are finally beating long-range weather models. But between artificial intelligence's bottomless appetite for data storage and its trillions of computations—a generative AI query uses 10 times the energy of a normal Google search—Google's AI energy demands alone will soon equal that of the Netherlands, jeopardizing its plan to be net-zero by 2030.

Is all this spectacular technology just the candle flame of civilization flaring brighter right before it extinguishes?

"The sad thing is," John says, arriving at his rental car, "we've just had the hottest year in the past 120,000. Arctic temperatures have already increased 4°C. Adaptation will become ever more important, because there will be more and more damage."

Even with AI, climate science is up against physical inertia in the systems it hopes to reorient—and social inertia.

"So much involves how quickly we can change society. If we had fusion, how rapidly could we switch, even if with clever AI to help? There's innate inertia in society. You probably don't want to get rid of it entirely, because that could break things even more."

His eyes squeeze shut. "We can't despair—that would just be giving up. We have no alternative."

Eyes open. "Everybody knows it's going to be terrible. But what else can we do? Even if most things we try don't work, we have to keep going."

ii. Art of Stubbornness

The inertia in Jae-Eun Choi's life began the year she was born, 1953, when Korea formally split in two. No one expected that division to last, let alone for generations, although each side had differing opinions about how things would resolve.

From the deck of the Cheorwon Peace Observatory, the South Korean artist gazes north over a misted wilderness of glowing fall oaks and maples: the setting for her proposed magnum opus. Her plan, she's been told, sounds impossible. "But we must," she insists, persuasively enough to convince a team of internationally renowned colleagues to join her.

Artists and architects, they hope to achieve what politicians have failed to do: bring the two enemy nations of the Korean peninsula together in peaceful common cause.

And maybe, help defuse an escalating threat of global nuclear war.

As she watches, six rare white-naped cranes cross the mountainous backdrop that is North Korea. Soon they're joined by even more endangered red-crowned cranes—symbols of luck and longevity that appear in silk paintings throughout the Far East. Choi's site, in the heart of the Korean Demilitarized Zone, is where more than half the world's remaining red-crowned cranes winter. Over seven decades, this 160-mile-long, 2.5-mile-wide buffer separating two of the world's most hostile armies has reverted to nature, ironically becoming one of Asia's most important wildlife sanctuaries: an accidental refuge to scores of imperiled plants and animals, from orchids to leopard cats, Eurasian otters, and the Amur goral, a nearly vanished antelope-goat.

"From the blood of many young soldiers who sacrificed their lives, a garden blossomed," says Jae-Eun Choi. Previous efforts to declare the DMZ an international peace park or a UNESCO World Heritage Site have failed to interest North Korea. But this time, she and her collaborators hope, may be different.

Jae-Eun, her short black hair parted on the left, still trim as a gymnast at 70, has always pushed the edge of what's possible. Spurning a career in fashion design after college, she went to study flower arranging in Japan with ikebana master and Oscar-nominated* film director Hiroshi Teshigahara. But instead of cutting flowers, she found that she wanted to root them. With the blessing of sculptor Isamu Noguchi, who designed Tokyo's indoor Sogetsu Plaza, she filled it with 13 tons of soil covered with grass seed, which sprouted during her exhibition into an undulating green carpet.

Freed from ikebana's intimate scale, she began turning entire buildings into vases: a tree bursting from a corroding steel tower to symbolize the end of military dictatorship, a church crowned with a bamboo forest. From 60,000 bottles she recycled herself from bars and restaurants, she built a 90-foot-high art expo pavilion. For the 46th Venice Biennale, she constructed Japan's pavilion from hundreds of stacked slabs of recycled plastic. Outside UNESCO's Paris headquarters, she lit an installation of a tea ceremony with a giant glowing, blood-orange moon. For three decades, she's buried sheets of her handmade rice paper on four continents, exhuming them to exhibit years later, transformed by soil, minerals, microbes, and time.

Her new project may require similar patience. In 1989, she'd wept at Brandenburg Gate as the Berlin Wall fell, wondering if it could ever happen in her own land. After participating in a 2014 group show about the DMZ, she called a friend in Tokyo with an idea.

The friend was Shigeru Ban, winner of that year's Pritzker Prize—architecture's Nobel—for building refugee and disaster relief camps from inexpensive, sustainable materials such as cardboard tubing and bamboo. At the 2016 Venice Architecture Biennale, Ban and Choi presented a scale model of a 13-kilometer, garden-lined bamboo walkway meandering between North and South Korea, elevated to protect visitors from ubiquitous DMZ land mines. Along its length would be towers for viewing nature and, every kilometer,

* *Woman in the Dunes*, 1965.

open-air Jung Ja meditation pavilions designed by different architects and artists, including several reserved for North Koreans. It would lead to a seed bank for preserving indigenous DMZ plants and a "knowledge bank" library of DMZ ecology.

Since actually building it anytime soon seemed unlikely, they titled it *Dreaming of Earth.*

The designs include Choi's own tower: an airy bamboo hourglass with an internal winding staircase. Icelandic Danish artist Olafur Eliasson, who'd dragged icebergs to Paris and Copenhagen climate talks and famously built a waterfall on the Brooklyn Bridge, collaborated with German architect Sebastian Behmann on a pavilion consisting of tall, overlapping bamboo circles webbed with shimmering, condensation-trapping mesh that symbolically appear to unite as visitors approach. Japanese artist Tadashi Kawamata, who lately had adorned several Paris buildings with oversized wooden swallow's nests, proposed to hang one from a DMZ cliff.

Korean minimalist Lee Ufan, fresh from exhibiting in New York's Guggenheim and London's Tate, devised a split arch where North and South Koreans would hover in facing tea rooms. His compatriot Lee Bul, whose shows usually involve polymers, mirrors, and metal, answered Choi's challenge to use native DMZ materials with reed cocoons suspended from the walkway. Mumbai architect Bijoy Jain, winner of the 2009 Global Award for Sustainable Architecture, submitted a cluster of 3D mandalas. The dean of Korean architects, Seung H-Sang, conceived a vertical monastery for birds as well as humans.

A perfect place for the seed and knowledge banks, decided fellow Seoul architect Minsuk Cho, 2014 winner of the Venice Biennale's Golden Lion for best pavilion, was a nearby invasion tunnel dug by North Korea, discovered when South Korean soldiers noticed smoke emerging from the ground. To give access to researchers from both countries, he envisioned a 10-meter-deep inverted spiral entering the tunnel exactly where it intersects the border. "Like a reverse Tower of Babel."

A hot spring that surfaces near that spot is one reason why so many rare birds winter in the DMZ. But the first assignment for Jaeseung Jeong, the scientist overseeing plans for the seed and knowledge banks for the DMZ's 2,700 species, will be land mine removal.

The US Army alone has planted more than a million in the DMZ, mostly plastic M14s that evade metal detectors. "It's the only plastic in the DMZ," says Jae-Eun. "This is both one of the purest and most dangerous places on Earth."

They investigated drone-based plastic-detection technology and using ballistic gel to smother land mine explosions. Removing land mines, she concluded, will cost more than the entire installation. "But cost isn't so serious. What is serious is world peace. Compared to that, cost is nothing."

Jeong, a jovial physicist in his 30s who models brain dynamics at KAIST, the Korea Advanced Institute of Science and Technology, agrees. He proposes that along with ecological data, the granite-lined tunnel should be a vault for analog versions of the world's most valuable knowledge, from Bibles to dictionaries to quantum mechanics textbooks, to help future generations rebuild civilization in the event of natural disasters or nuclear war.

"There may be some survivors," he says. "We hope they're not stupid enough to self-destruct. Like us."

In October 2017, Jae-Eun Choi formally presented *Dreaming of Earth* at the Seoul Museum of History, a venue that assured that Pyongyang would be listening. The hope was that Kim Jong-un might be enticed by the substantial international goodwill North Korea would accrue from a symbolic joint project with South Korea.

"He's not stupid," said Choi. "Even if he doesn't care about art or ecology, if he thinks it's good for him, why not?"

But even if Kim were to agree, with mine removal, construction logistics in a wilderness preserve, and eventual parleys with North Korean artists, she expected the process would take from 5 to 10 years. "Process is my kind of artwork. This is the interesting part."

After Shigeru Ban's tests revealed that bamboo grows poorly in the DMZ, for instance, they decided instead to build the walkway from timbers, charred to repel rot and insects. "We'll use wood from the countries who've most affected Korean history: the US, Japan, Russia, and China," proposed Jae-Eun. "Step by step, we'll solve these details." One day, she was convinced, it will be built.

That day has yet to come. An American president's courtship of Kim Jong-un proved humiliating. As the pandemic all but severed communications between the two Koreas, and the US bolstered Kwajalein Island's surveillance of intensified North Korean missile launches, Jae-Eun's scientific team discovered that the DMZ's ecology was even more vulnerable than previously believed. During 2020–21, new satellite images that her scientific collaborators consulted for a DMZ ecological status map revealed 683 barren patches totaling 72 square miles. The sole explanation was clandestine incursions by North Korean troops to cut firewood and timber.

Shocked by this unexpected loss of DMZ forest, Jae-Eun Choi pivoted. Rather than wait for politicians to evolve, she proposed to begin reuniting the two Koreas now—with trees. In a scheme she called Nature Rules, her team would plant specimens of the missing species along the southern fences for wind and birds to broadcast their seeds. To avoid land mines and further DMZ trespassing, they would use drones to drop "seed bombs"—balls of clay impregnated with seeds from native species that her colleagues had identified, including 106 DMZ plants designated as endangered.

The idea came from Japanese farmer-philosopher Masanobu Fukuoka, who used airplanes to fight desertification in Africa and South Asia. Drones that literally fly under the radar would allow Nature Rules to pinpoint where each species needs replenishing. Jae-Eun hoped to engage the entire world in the DMZ's recovery. Via social media, anyone on Earth would be able to donate a seed bomb.

Ultimately, nature does rule. Sooner or later—albeit sometimes much later—it bounces back from whatever calamitous mass extinction it endures, from asteroid strikes to ice ages to volcanic cataclysms. The war-torn DMZ, the Korean peninsula's bottleneck but also its richest vein of life, epitomizes how nature can recover with surprising swiftness from scars that humans wreak. To protect this global symbol of hope amid hell will ultimately require easing the tensions that squeeze it like a vice—and, after that, controlling the impulses of the most ravenous species in biological history to plunder its resources, and then pave and fill it with commercial clutter. While we still await leaders wise enough to make those things happen, Jae-Eun Choi has her audacious plan to act immediately.

With luck and relentless persistence, her greater dream will also someday take root, along with her seedlings. Artists and architects from both Koreas, joined by international colleagues, will build a bridge from tension to beauty in the garden that blossomed between them. *Dreaming of Earth*'s graceful towers and floating meditation pavilions, its repositories of seeds and ecological knowledge, will be so eloquent that the whole world will sigh to see that such a gentle, peaceful encounter between adversaries, despite their differences, is possible.

"We need it. They need it," says Jae-Eun. "We don't need war. With terrorism, hurricanes, and extinctions, the world is already dangerous enough. In such times, we must get together. Right now, this must be our art."

iii. Strategies

In the face of seemingly impossible odds, why don't people quit trying? Because it's the right, noble thing to do? Or the reflexive thing to do—like walking dead, they keep going through the motions? Because they're still being paid? Or, for artists like Jae-Eun Choi, because of an uncontrollable urge to express?

Or because they still believe, hoping against hope, that it could matter?

It can be any or all of those, swirling through the heads and hearts of people alert to the real stakes involved here: The world we have, becoming the world we once had.

From charred northern boreal forests to Brazil's tattered lungs and Basra's oily flares, we're close enough to smell it. Why this is happening involves another reason why some others don't quit: because what they do makes them excessively rich.

Excessive, because turning everything to gold is wondrous—until it isn't, as an aghast Midas discovered when his golden touch unexpectedly transmuted both his food and his daughter.

Were they to quit now and never make another cent, they'd die rich and so would their heirs. What keeps them drilling for more to burn—and what keeps us paying them trillions in subsidies to do it—is an ancient story of the most addictive drug of all: power. They have it. They don't want to lose it.

They will. It's already cheaper to produce electrons with wind and sun than with oil, gas, or coal. Way cheaper, after factoring in the damage from burning fossils: biblical storms, drought, and deluges; ruined crops and food price shocks; perverse air and insidious plastic; anoxic deltas, dead reefs, global heat stroke . . .

There are even more alternatives in the works. Something will stick—with luck, in time for us.

For decades, MIT has sought to exploit an elegantly simple concept: Wherever you are, drill down far enough and you'll tap the heat of Earth's mantle. Pour in water, and it returns as steam that can drive a turbine.

Unfortunately, conventional rotary drills can't go deep enough; the deepest hole ever drilled, in Russia, took nearly 20 years to reach 12 kilometers (7.6 miles), not even to the edge of constant 300°C to 500°C temperatures—but to where conventional equipment begins to melt.

Thus, "hot dry rock" never got off—or rather, out of—the ground. But in 2008, just down the hall from Dennis Whyte's fusion lab at MIT, it occurred to plasma physicist Paul Woskov to take the ultra-high-intensity, laserlike gyrotron used in tokamaks to heat plasma and point it at the ground. Its super-concentrated electron beam, Woskov realized, wouldn't melt.

He'd first had the idea in the 1980s, but back then there was no urgency and therefore no money to try. But now, when he tested it on chunks of basalt, another advantage became obvious. As the high-powered radio-frequency beam vaporized the rock, the bore hole's sides hardened into glass, eliminating the need for miles of steel well-casing. The relatively short time it took the beam to turn a rock into a doughnut also demonstrated it could reach depths of 20 kilometers in only a couple of months.

Unlike fusion, which requires specialized operators, says Woskov's former student Carlos Araque, CEO of Quaise Energy, another MIT spinoff that's begun boring test holes, petroleum roughnecks will need minimal training to man this new technology. "Drilling is drilling. The energy transition needs to happen so fast, you almost have to bet on an existing workforce. We can't afford waiting a generation to build one."

To that point, Dennis Whyte agrees. "Remember, we need to replace 82 percent of all human energy use. There are only two inexhaustible energy sources on Earth: fusion and the energy of Earth itself."

Like his fusion team's ARC design, Quaise's bore-hole generator would fit the footprint of a conventional fossil fuel plant, so it could connect to existing transmission lines. Their deep geothermal plans hold deep promise, but neither they nor fusion will be supplanting former coal-fired plants before the 2030s, if by then. Right now, solar energy, the least expensive technology of all, needs no centralized plant: our linked rooftops can become our utilities. If today's utility companies don't want to relinquish their control of our power and their profits, let them lease us and maintain our photovoltaic panels instead of suicidally burning more exhumed carbon.

Prior to the 2015 COP21 talks leading to the Paris Agreement to try to limit global temperature increase to well below 2°C, Gustavo Petro, then the mayor of Bogotá, hosted a Latin American summit to discuss how best to portray the threats that climate change posed to their region. Three decades earlier, Petro had belonged to the M-19, among the first guerrilla groups in Colombia's long civil war to lay down arms and join the political process. After earning degrees in economics and public administration, he became a congressman and a senator, championing poor people's rights to land, health care, and a clean environment. Before being elected mayor of Colombia's capital in 2011, he ran for president to challenge the incumbent's family ties to right-wing paramilitary groups and drug money.

His 2015 Bogotá summit ended with a panel on how to finance climate protections the global south would need. His own city, dramatically backdropped by the Andes, was built atop rivers that flowed down from their skirts, nearly all now buried under neighborhoods that increased rainfall could wash away. In coastal cities like Rio de Janeiro and Buenos Aires, millions might become displaced. Petro, moderating the panel, offered a simple way to pay: "A one-thousandth percent tax on all global transactions would finance everything."

A panelist from the French Development Agency objected. "Although my heart is green, we're looking for bankable climate solutions."

"What do you consider 'bankable'?"

"Anything the private sector finds profitable."

Petro, short, olive-skinned, open-collared, and rumple-haired, regarded the tall white European banker in the tailored suit. "Cartagena is already flooding," he replied. "If seas continue to rise, it will flood permanently. Where is the profit in moving a million people, 70 percent who are impoverished, to higher ground? Banks will happily finance moving Cartagena's oil refinery, which adds to the problem, but lives don't matter to markets. Solving this by indebting us more to lending banks, whose wealth comes from petroleum investments, is obscene."

Before the speechless development banker could reply, Petro pressed on. "At COP meetings, the people trying to save life keep colliding with people trying to make money. Unless we break with oil, COPs won't change anything."

In 2022, Petro was elected president of Colombia—the country's first chief executive from outside the two-party plutocracy that had traded the presidency for two centuries. That ruling class was not pleased when, as promised, he halted exploration by Colombia's state-owned coal and petroleum industries—South America's first- and second-biggest, respectively. He also ended the gasoline subsidy and announced an agenda to phase out fossil-fuel extraction entirely.

That same year, he addressed the United Nations: "My country has rainforests, the Amazon and Chocó. From these jungles, oxygen flows and CO_2 is absorbed. Among the CO_2-absorbing plants is coca, sacred to the Incas. Paradoxically, you destroy the forest you're trying to save by dumping glyphosate to kill it, yet you ask us for more coal, more oil to satisfy another addiction. To minimize deaths by overdose, the ruling powers deem cocaine a poison, but they protect poisonous coal and oil so they can extinguish all of humanity. The world powers have lost their reason."

A year later, 2023, at COP28—held in an OPEC country, chaired by a sultan who heads the state oil company—Petro added: "Since 2010, global CO_2 emissions have risen 12 percent. Rich states won't devalue their fossil capital by decarbonizing. The American dream, European comfort, and China's outreach depend on burning carbon. Transferring wealth from the north so that peoples of the south—who barely emit CO_2—might adapt to a deadly runaway climate is an alien idea to markets. So water scarcity is inciting an exodus from the tropics. The movement of entire peoples northward is underway. Today they number tens of millions. Tomorrow it will be hundreds of millions. What will happen to international law? What will happen to humanity?"

The following month, on a panel with Bill Gates and the prime minister of the Netherlands at the World Economic Forum in Davos, Switzerland, Petro offered a plan to save the remaining Amazon and reforest its denuded portions.

"It will take $2.5 billion a year, not the $10 to $15 million in alms for the poor that trickle in from developed countries. We propose financing it ourselves. How? By exchanging debt for climate action. Currently, Colombia must pay a premium on its foreign debt, because like all countries of the Amazon rainforest, it is considered risky. But the risky ones today aren't Colombia, Brazil, Peru, and Ecuador. You northern countries represent more risks to human life than we do. If our risk premium becomes zero, the money we free up would go to climate action. It's not a handout. It's a powerful climate financing mechanism."

Thus far, Colombia's Gustavo Petro is the only head of a major coal-, gas-, and oil-producing state bold enough to begin phasing out their production. His debt-for-nature plan was similar to one that then-Senator Joe Biden had supported years earlier, which the two had discussed in the White House before Davos. Yet wealthy powers in Colombia were suspected of trying to discredit Petro and destabilize his administration by inciting new waves of violence, especially against tourists, resulting again in US State Department warnings of the risk of being kidnapped in Medellín.

President Biden's own climate problems were partly self-inflicted, as he tried to have it both ways. He engineered the most sweeping environmental bill in US history, full of credits, tax breaks, and incentives for green energy and the transition to renewable electricity, and he killed the notorious Keystone XL tar sands pipeline. Yet he wouldn't stop Line 3, nor, to satisfy one West Virginia senator, the Mountain Valley natural gas pipeline, nor construction of liquefied natural gas export terminals—any of which would undermine the country's 2030 climate goals.

The laws of climate physics not being conducive to compromise, Biden's political bargaining skills turned on him, especially in a 21st-century America fragmented by a politically charged disinformation industry and social media hijacked by conspiracy fomenters. With the president too shielded by enabling advisors, it took people whose ideas weren't formed in the last millennium to finally get to him.

In 2009, 16-year-old Michael Greenberg read climate scientist James Hansen's *Storms of My Grandchildren: The Truth About the*

Coming Climate Catastrophe and Our Last Chance to Save Humanity and got scared. As he lived near Washington, DC, in Rockville, Maryland, he tried to get his high school environmental club a meeting with the assistant secretary of state charged with reviewing the Keystone XL pipeline. Her staff replied that assistant cabinet secretaries don't have time to meet with high schoolers.

A week later he called back, now representing 20 high schools. He got a meeting.

A few years later, studying economics at Columbia University, he organized XL Dissent, at which 400 college students were arrested at the White House in 2014, leading to President Obama rejecting Keystone.* After graduation, in 2021 he was working at Mighty Earth, an environmental NGO known for battling palm-oil-plantation deforestation, when he learned about Ojibwe women lying in front of bulldozers to block Line 3. Greenberg took time off to go to see for himself.

It was midwinter in northern Minnesota, but he trudged into camps where water keepers from various bands were trying to defend their treaty-granted lands and waters from the pipeline company. "You're also defending the climate," he said to the Indigenous leaders. "Suppose we got environmental organizations to support you?"

With their assent, he contacted the Minnesota chapter of 350.org. Named for the number of parts per million of atmospheric CO_2 climate experts consider to be safe,† 350.org had morphed from its founding by students in author Bill McKibben's climate seminar at Vermont's Middlebury College into a network reaching 188 countries. MN350's pipeline organizer remembered Greenberg from the White House protest and arranged a meeting in Minneapolis be-

* Trump later reinstated it. Biden then quashed it.
† 350 ppm was last seen in 1988. By 2021, we'd passed 420 ppm.

tween environmental NGOs and the Ojibwe group Honor the Earth. Due to COVID, they talked outside around a fire, eating burritos. Three months and the first round of COVID vaccinations later, more than 2,000 people gathered on treaty lands at the banks of the Mississippi to try to stop Enbridge's Line 3.

Their action received national media coverage, but it neither stopped the pipeline nor created a lasting strategic framework. Greenberg decided that the climate campaign needed an organization that could have breakthrough moments but also last for the long haul. In March 2023, he assembled a few veteran protester friends and coined a name: Climate Defiance. It only took 10 to 20 people, Greenberg had calculated, to upend things.

With their country now pumping more crude oil than any nation, ever, their first targets were Democrats who dithered while the planet roasted. Presidential advisors and congressional backers of fossil-fueled projects suddenly found themselves chased from podiums by Climate Defiance activists who leapt from audiences, shouting belligerent questions and unfurling banners accusing them of torching their own children's future. Within a few months, 50 million on social media had seen them chase the deputy interior secretary who'd signed off on ConocoPhillips's Alaskan Willow Project down New York streets; shut down a luncheon honoring ExxonMobil CEO Darren Woods; force coal-baron Senator Joe Manchin to flee through a diner's kitchen; crash Fed chair Jerome Powell's speech at the annual Jackson Hole symposium; interrupt the launch of Senator Amy Klobuchar's book *The Joy of Politics*; and hound Transportation Secretary Pete Buttigieg off a stage for backing two new Texas crude oil terminals.

When Deputy Interior Secretary Tommy Beaudreau resigned after the fifth time they publicly branded him a climate criminal for approving Willow, Greenberg wrote to Climate Defiance's growing membership, "Today I feel hope." His hopes grew when, instead of arresting him again, the president's administration reached out, inviting him to meet with the nation's number two energy official. He accepted warily, lest their scrappy movement become "yet another boring nonprofit" that softens its tactics to avoid jeopardizing newly won access.

"The world is literally burning, and we don't have time to trade business cards and recite scripted talking points," he declared—yet he also accepted the next invitation: to the White House itself, with President Biden's senior clean energy innovation and implementation advisor, John Podesta.

In that meeting, he repeatedly demanded that the administration quash permits for Calcasieu Pass 2, a $10 billion liquefied-natural-gas export terminal at Lake Charles, Louisiana. CP2 was the biggest of several proposed LNG export terminals, whose cumulative emissions, the Sierra Club had calculated—from energy needed to compress fracked gas for shipping, methane leaks, and end-use CO_2—would equal the excretions of 532 coal-fired plants. Poor Black residents of Louisiana's Cancer Alley had been fighting it alone for years, until environmentalists realized that CP2 was a climate megabomb 20 times bigger than Willow, venting as much carbon annually as all of Costa Rica, and made it an international cause célèbre.

At every opportunity, Climate Defiance had been harassing Energy Secretary Jennifer Granholm and her staff about it. "Climate Defiance," Podesta told Greenberg when they happened to meet on the Washington Metro, "can be a real pain in the ass."

"Thank you," Greenberg replied.

On January 25, 2024, 18 Climate Defiance volunteers, some arriving by train from New York and DC, crowded into the basement rec room of a sympathizer in Wellesley, Massachusetts, mostly teens and twentysomethings. Their target that night was speaking at a Rotary dinner at Wellesley Country Club: Brian Moynihan, CEO of

Bank of America, the world's fourth-largest fossil-fuel financier, having spent nearly $300 billion since the 2015 Paris Agreement, including $999 million for Enbridge's pipelines and $1 billion for CP2.

The $100-per-plate admission, normally an effective filter for who attended, was no problem. Just seven months since its founding, Climate Defiance had raised a half million in donations. More challenging was the staid country club's dress code. "Business attire," they were warned, which elicited scrambling to borrow jackets and pants that weren't jeans. "And no earrings, necklaces, scarfs, ties—things security can grab." With a red Sharpie, they scrawled a jail-support phone number on their forearms. Each carried just $40 in cash, identification, prescription medications they'd need if arrested, and phones with facial recognition disabled for livestreaming their action.

Greenberg, sallow, wiry, and bushy haired, in white-soled black sneakers, sat to one side as Climate Defiance's full-time Boston organizer, Martin G, laptop open on their forearm, covered logistics and assigned the volunteers their designated police liaisons, de-escalators, "care bears" to safeguard everyone's welfare, and videographers.

"We act nonviolently. We hold our ground using our weight, collective presence, and linked arms. We don't fight back. We leave together if security gets really physical or when police give the last credible warning."

To avoid arriving at the sprawling neocolonial clubhouse en masse, they had staggered their departures. In the venue's second-floor ballroom, they followed the throng to the buffet of roast chicken and salmon, pasta primavera, Caesar salad, and brioche, then distributed themselves among tables at either side of the stage. After an hour, Moynihan, wearing a gray tailored suit that matched his hair, and his interlocutor, a reporter from *Bloomberg News*, ascended the short dais, settled into chairs, and faced 200 well-fed Rotarians and guests seated beneath crystal chandeliers.

Since Moynahan's home is in Wellesley, the first question was a warm-up, asking him to contrast local and international banking. As he started to answer, from either side Climate Defiance protesters silently filed behind him and unrolled the banners they'd smuggled under their coats:

BUSINESS AS USUAL IS A CLIMATE DISASTER

BANKING ON CLIMATE DESTRUCTION

BANK OF ATROCITIES

[with Bank of America's flag logo dripping blood]

A protester in a shirtwaist and blazer named Izzi faced Moynihan. "We have just one question: Will you commit to stop funding fossil fuels?" she loudly repeated, as his flummoxed three-man security detail, caught unaware, tried to move her away and pull down the banners, instead pulling down another female protester. For a full minute, Moynihan sat tight-lipped, literally twiddling his thumbs—later, a streaming hit—until they hustled him out. A bodyguard tried to block the protesters from following, but they wriggled around him and ran down the carpeted staircase to where Moynihan took refuge in an office.

For 20 minutes, they chanted outside its mullioned glass door: "Off fossil fuels, Brian, off fossil fuels! Not another billionaire, we want clean air!"—until Wellesley police arrived. After warning them to leave or be arrested, they departed peacefully.

Upstairs, the guests had filled dessert plates with pastries. Several had booed the protesters, but others confessed admiration for their planning and discipline. Moynihan resumed. He'd just returned from Davos, where he'd met with the presidents of Ukraine and Israel, and he listed current risks to the global economy. Besides wars, he noted grain and fertilizer prices, terrorism, US-China military standoffs, and inflation.

During the Q&A, a man in a tweed jacket asked about a risk he hadn't mentioned: "Given the unstable climate and anxious insurance companies, are your stockholders as nervous about stranded investments in things like CP2 and pipelines as those young protesters are about their future?"

Lips tightening anew, Moynihan skirted the question by touting corporate net-zero commitments and investments in green technology. At Davos, he said, Occidental Petroleum's CEO told him about billions they were pouring into carbon capture and storage—the technology that the petroleum industry hopes will let it keep selling combustible fuel forever, because its wastes would miraculously be vacuumed from the sky, buried, and presumably never leak.

He added that his company and others were burning SAF, sustainable aviation fuel, in their corporate jets. SAF, an attempt to recycle wastes like cooking oil from Burger King and Wendy's, can replace up to 50 percent of conventional aviation kerosene, reducing individual flight emissions significantly—but not necessarily emissions to the atmosphere. Feed-lot cattle, the source of those burgers, do nothing to help and much to harm the climate. SAF can use plant-based sources—especially yellow grease from french fries and onion-ring oil—but both lack the esters and fatty acids that beef tallow and pork fat provide; they also leave contrails, and cultivating vegetable oil for fuel would displace even more food and forests.

Converting our waste to fuel: It's all being tried. It all requires energy. The laws of thermodynamics are implacable.

Back in the basement rec room, the protesters passed around pizza (and, it turned out, COVID) and debriefed. The two women roughed up in the scuffling weren't injured, but were still distressed from being grabbed. As one named Lauren later blasted on social media:

> I didn't sign up for this to feel comfortable. . . . I chose to do this with Climate Defiance because humanity is dying out. The climate crisis is impending, we are totally fucked if we don't dramatically change course, and no one is doing remotely enough.
>
> I'm 18 years old and I want to live a long life. I want to have a family and I want to be optimistic about my future. I don't want to experience droughts, floods, or food shortages. I don't want to see the beauty of this planet die.

Shortly after midnight that evening, Michael Greenberg was still awake when the White House issued a proclamation from President Biden pausing all pending approvals of liquefied natural gas exports. The economics of CP2 and 16 other LNG plants, as calculated five years earlier, needed to be reassessed, given accelerating greenhouse effects. "This pause on new LNG approvals sees the climate crisis for what it is: the existential threat of our time." The president concluded:

> We will heed the calls of young people and frontline communities who are using their voices to demand action from those with the power to act. And as America has always done, we will turn crisis into opportunity—creating clean energy jobs, improving quality of life, and building a more hopeful future for our children.

"This is monumental," Greenberg wrote to his troops. "Mind-bogglingly big. Tears are in my eyes. We are doing it. We are winning. I'm inspired as hell to keep making trouble so we can stop ALL fossil fuel infrastructure."

He signed it: "With profound, profound hope."

In Europe, desperate for a distracted world's attention, chanting young Extinction Rebellion members glue their feet to tennis stadiums during Grand Slam events; others glue their hands to museum walls, or symbolically deface priceless paintings by spattering paint or tomato soup on the protective glass—"No Art on a Dead Planet" read one sign. Many were still teenagers in 2019 when the United Nations warned that the world must brake emissions increases by 2025, then actually cut them in half by 2030. Yet by mid-decade global emissions were still rising—more slowly, due to less coal being burned overall. But using fracked natural gas, liquefied under energy-sucking pressure to replace coal, is no cure, especially with more methane leakages constantly being discovered.

Instead of pulling back from the 1.5°C brink, temperature and wind speed records across the globe were being smashed annually, in nonlinear leaps that existing models couldn't predict—while Donald Trump vowed if reelected to unfreeze LNG terminal permits on day one and to auction new drilling leases from Alaska to the Gulf of Mexico.

Michael Greenberg's generation knows we will never have this chance again. "You need to stubbornly believe there might be a path forward, that your actions can have real consequences," he says. "Not having hope is a self-fulfilling prophecy."

iv. Hope at Last

Do other species hope? While our ancestors squatted around fires gnawing fresh-killed meat from bones, were our future canine companions circling in the shadows, hoping that these adept primates would share scraps from their hunt with them? Do similar parts of doggie brains light up as ours do when our hopes soar? Can neurologists who study this tell us anything useful—or is that as futile as dissecting songbirds to learn why they sing, because hope lies where our emotions touch what we call our souls, the existence of which science has yet to detect, much less define. Even though everyone senses they have one.

We humans are part of an evolutionary continuum on this green orb, so there's no telling where soul entered the picture. Follow your genealogy back to where it blurs into prehistory; consider if your hunting, gathering, grunting forebears had nascent souls—and you intuit that of course they did. That's how they became the mythmakers whose legacies still shape what we believe. But they, too, had forebears, back to the shrewlike, furry survivors of the Chicxulub asteroid strike that gave birth to all placental mammals—meaning both you and your neighbor's dog. Try convincing her that her adorable spaniel doesn't have a soul, bursting with hope that she'll take him for a walk.

From drumming chimpanzees to flickering fireflies, we know that other species tell stories. They can describe their immediate surroundings and issue precise warnings about who's lurking; they can mourn, woo, tell of their longing, have extensive memories, and teach their young. Our attempts to communicate with them—dolphins, octopi, equines—have been nearly always on our terms. We have barely tried to grasp theirs, much less peer into their souls, but from ants to crows to elephants, we have endless examples that they socialize, think, and feel—just differently than we do.

But maybe not so differently. We breathe the same air, eat from the same sources, share the same turf, need the same water. Once, we humans were just another animal, living off the savannas and shores. Until Molly Jahn's microbial miracle leaps from the lab onto our plates, we still do—and we believe otherwise at our peril. But—a big *but*—we're now so distracted by our planet's uncorked climate blowing away our rooftops, firebombing our forests, mauling our coasts, and scorching our crops that, too often, we forget that one of the biggest phase shifts in Earth's history, a global holocaust against life itself, is also underway.

Like two vipers coiled in a deadly double-helix, the unbridled climate and mass extinction are "one indivisible crisis—a global health emergency," as the world's 300 top medical journals wrote en masse to the UN in 2023. Meaning:

> A third of humanity's food supply doomed,
> should pollinators succumb.
>
> Protein famines after 2°C, when virtually all
> corals die along with fish and shellfish who
> depend on their reefs.
>
> Tropical plants for our medicines, withered.
>
> Monocultured crops, their genetic stocks extinct
> except for a few frozen relics in seed banks,
> defenseless from pandemics.

As all that happens, tempers shorten and violence heightens—any police blotter can correlate heat with mayhem. Wars break out over dwindling water. Refugees flee parched lands, stoking fears of hungry invading hordes, emboldening dictators and vigilantes to stop them. Like thawing glaciers, laws melt.

As pernicious new human pathogens race around a teeming, warming planet, governments like the US and the Dutch print extra money to sustain millions of quarantined citizens, quenching panic but inflating their economies. Even medical miracles that quash viruses can't cure grocery sticker shock, so angry, anxious people turn to autocrats who promise to revive an affordable, mythical past when no ravenous foreigners threatened to steal their jobs. The easiest way for the autocrats to fuel their reign is to drill more, so temperatures mount even faster. Fires spread, winds roar, rivers surge, coasts shrink, species fade, and panic returns.

Eventually, people overthrow self-serving, inept demagogues, but it's a changed world. Amid the madness, the Center for Biological Diversity's ecologists and lawyers keep on, submitting new candidates for protection: pygmy rabbits, Southern Plains bumblebees, two eastern salamander species, white-margined penstemon wildflowers, Railroad Valley toads. It's not just about them. Every weed, snail, or slug they shelter also means saving land and waters where they dwell.

The US still isn't a signatory to the UN's 30x30 goal of conserving at least 30 percent of lands and waters by 2030—the biggest conservation plan ever contemplated: more than doubling global protection of terrestrial, inland water, and coastal regions, and quadrupling marine protection. But in 2021, to circumvent entrenched fossil-fueled politics blocking participation, the US simply began its own 30x30 plan, titled America the Beautiful. Although so far just a few countries and US states—Gabon, Costa Rica, Vermont—are anywhere near full compliance, achieving 30x30 worldwide is now on a better track than meeting the UN's goal of holding the temperature increase to 1.5°C by that same year.

The 30 billion tons of CO_2 we still emit annually has us on a trajectory for a 4.3°C increase by the end of the century, unless business as usual radically changes.

It must. It comes down to two simple things humanity can do to rein in both runaway temperatures and the loss of at least a third of all species—likely more, as each loss cascades into many others:

Stop running modern life on ancient carbon,
and let forests return.

In 2023, more than a hundred scientists from six continents, writing in the journal *Nature*, calculated that letting all the planet's forests recover could absorb roughly a third of what we've added to the atmosphere. The forests, in turn, would soften the climate's extreme gyrations and reel species back from the brink of extinction. But that won't work until we also dramatically cut back on burning fuel.

The alternative is we lose control of the situation. Some might argue—they do—that even though humans killed off mammoths, mastodons, and nearly 70 entire genera of megafauna at the end of the Pleistocene, we've done pretty well since then—so does it really matter that more than two-thirds of all birds now are chickens we raise? Life seems to go on, right?

It's a question that's unanswerable until it's too late, when we find out the hard way. "No jobs on a dead planet," notes Center for Biological Diversity cofounder Peter Galvin, reminding us that it's best to err on the side of caution. It's never smart to see how many foundation bricks you can remove before the building falls on you.

Either way, the biodefenders' work will have been worth it. If humans aren't among the species that survive into this planet's next, hotter phase, some percentage of the flora and fauna they've managed to load into the ark will replenish the Earth, just as in five previous extinctions—like the birds who, against all hope, survived Chicxulub.

Or, if we're lucky enough to be among them, they'll be our companions. And our food.

May we thrive together.

EPILOGUE

On February 1, 2023, Jassim Al-Asadi headed to Baghdad to meet with the minister of water resources. Having accused the government of failing to protect Iraq's marshes from climate change, Jassim had launched a national campaign to save them, angering both farmers who coveted their water and oil interests who wanted to drill there. Five kilometers from the capital, his Toyota was blocked at a checkpoint by two carloads of armed men in plain clothes, who handcuffed and abducted him.

From Amman, Jordan, where he was visiting family, Azzam Alwash pressed Iraq's government to find his colleague—but it was Jassim's Asadi tribe that, through ancient channels, learned which militias were holding him. While his fate was negotiated, he was beaten daily with truncheons, tortured with electroshocks, double head-cuffed until his eardrums bled, had his face shoved underwater until he half-drowned, and was denied his prostate cancer medication—but he refused their demands to halt his environmental work. After two weeks, he was released just west of the marshes at Al-Fuhud, where thousands of cheering compatriots fired guns in the air and waved tribal flags, then formed a convoy to escort his car back to Chibayish.

"Now I owe the marsh people even more," Jassim told Azzam.

Warned to stay out of the country, Azzam appointed Jassim director of Nature Iraq. Work commenced on the "living machine" garden park designed by war-zone artist Meridel Rubenstein to treat Chibayish's sewage and return clean water to the marshes. But between relentless upstream dam construction in Turkey and Iran and more prolonged droughts than anyone alive had ever seen, the marshes shrank to canals barely wide enough for two mashoufs to pass each other. Fishermen were ruined; buffalo breeders kept relocating as water grew saltier and reed beds shriveled, or sold their herds and migrated.

"Buffalos ate your garden," Jassim reported to Meridel, who was fundraising in Europe. "But the roots are fine. It's now fenced and regrowing."

"I rather like that we fed starving buffalo," she replied. Soon, the water discharging from their first three acres of garden was so clear that at the ribbon-cutting, Jassim reported, the minister of water resources committed to building eight more.

"God knows what the next years will bring us," said Azzam from exile, "but Jassim's spirit remains unbeaten."

Sixty kilometers northwest of Ur, Abraham's birthplace, is Uruk, capital of the Sumerian empire, where Gilgamesh reigned. Founded more than 7,000 years ago, it was the world's first great city. At its zenith, around 3100 BCE, it had nearly 90,000 inhabitants. Uruk was linked by canals to the nearby Euphrates, where fishermen reaped and tradesmen ventured to the sea. So much commerce flowed to Uruk that to account for it, writing was invented here. A pictograph of a rolling sledge shows that the wheel was, too.

Jassim comes often, although it's now just a desiccated archaeological site. The tallest ziggurat in Mesopotamia, which once towered over city and river, is gone. The largest mound left here was once the temple of Inanna, the Sumerian goddess of both love and war, called Ishtar when Assyrians later ruled here. Atop it, Jassim Al-Asadi mulls Uruk's parched remains.

Faint quadrants etched in sand to the horizon show where grand brick houses once lined avenues and canals, where markets and neighborhoods spread. For 5,000 years, Uruk flourished, far longer than any other city. It might still—except by around AD 300, the Euphrates had deposited so much silt that it shifted west, leaving Uruk high and dry. Today, only mussel-shell fragments hint at what once flowed here.

Gazing on the ruins of history's most enduring civilization, pondering the fragility of empire and grandeur lost so quickly when water vanished, should humble us—but it's too distant, too preindustrial, too impossible to imagine our own capitals reduced to dust.

Not for Jassim, bearing daily witness to the fate of rivers that once bore the lifeblood of civilization's birth.

"The Euphrates and Tigris," he tells Azzam, "are now lower than any time in history."

Northwest of Uruk, along the shrinking Euphrates lie towns that once housed academies founded by the religious leaders of thousands of Jews captured by King Nebuchadnezzar, who sacked Jerusalem and destroyed Solomon's Temple. Exiled back to the Babylon that their patriarch Abraham had left behind for the Promised Land, with no more temple to maintain their oral tradition, uprooted Torah scholars compiled the Babylonian Talmud, today still the greatest compendium of Jewish law and rabbinical commentary.

Both language and historical allusions suggest that most of the Book of Joshua, the sixth book of the Old Testament following Moses' death, was also written here. Joshua tells of Israel's conquest and settlement of Canaan. With its triumph eased by miracles—the Jordan River parts; Jericho's walls collapse; giant hailstones clobber their foes; the sun pauses until Israel subdues an enemy—Joshua is generally considered more myth than history by biblical scholars. Either way, to anyone but Judeo-Christians, it's shocking. Thirty-three cities aren't just conquered, but annihilated. Except for the family of a harlot who shelters two Israelite spies and a tribe that Joshua keeps for slaves, the Hebrews massacre every man, woman, and child. This slaughter of innocents is justified because, the Bible says, God wants only Israel to possess the land. For failing to know His supremacy, all others must die.

The idea of divinely sanctioned savagery is chilling, except to the victors. But throughout history, we humans have invoked the will of God—or Allah, or a thousand other names—wherever we've laid claim, and by whatever bloody means. Just upriver from the ancient Jewish enclaves, in the city of Najaf is the tomb of Imam Ali, the third holiest site in Shia Islam after Mecca, Muhammad's birthplace, and Medina, where he made Islam a great power, ultimately by the sword.

Shi'ite Muslims believe that Ali, the Prophet's first cousin and son-in-law, was his chosen heir, and that he is buried alongside Adam and Noah. The vast Imam Ali shrine, with gilded dome, marble columns, tiled walls, bejeweled interior, and chandeliers dripping with crystals, attracts millions of the faithful yearly. (Just five miles from his tomb, another shrine bathed in gold and flanked by minarets encrusted with colored glazed tiles, with courtyards of reflective white marble so pilgrims don't scald their feet in the summer, marks where Ali was assassinated—and also where it's believed Nūh lived and built his ark. Still another opulent shrine in nearby Karbala, this one to Ali's martyred, decapitated son Imam Hussain, is the destination of the annual Arba'een pilgrimage, the world's biggest, which begins at the top of the Persian-Arabian Gulf and, by the time it arrives, is 22 million strong.)

Every day, cars arrive at Imam Ali's shrine with wooden coffins strapped to their roofs. Hoisted onto shoulders, they're borne inside for a few moments of prayer as close as possible to his tomb, surrounded by dazzling mirror mosaics, then taken outside to be buried as near to him as possible, considered the safest place a soul can rest. Over centuries, the surrounding Wadi al-Salam cemetery has grown into the world's largest: more than 6 million sand-colored tombs, stretching past the horizon. Many extra hectares were added during the two American wars, then many more during the war with ISIS—1,600 car-bombings in Najaf alone—and even more during COVID.

Two blocks from the cemetery's entrance is a nondescript yellow concrete building, its hot facade hatched by afternoon shadows from a cluster of cellular towers. It houses the offices of Ayatollah Sheikh Mohammad Hussein al-Ansari, the authority in this holy city on Imam Mahdi: the so-called 12th Imam who went into hiding after his father's poisoning in AD 874, and who, Shi'ites believe, still lives and will reemerge with the Prophet Jesus in the End Times, to defeat the Antichrist and rid the world of evil.*

Gray-bearded Ayatollah al-Ansari's broken nose, which cants sharply left beneath his white turban, and a Y-shaped gash in his forehead attest to evil he has known during Saddam's times and the bedlam years that followed. Conflation of spiritual passion with assassinations, vengeance, and the thrill of headless martyrs is routine to him in this land where death pervades. Nor should blood that humans spill in the name of God surprise any religious cleric: Hinduism's seminal text, the Bhagavad Gita, is the deity Krishna's battlefield blessing to a warrior; Jews and Hamas out-terrorize each other in Gaza; Christians fetishize a grisly, martyred Prince of Peace, impaled with nails and agonizing on a wooden cross. Even Buddhists wage genocide on Muslim Rohingyas.

Our histories are litanies of barbarity against each other. But now,

* Like Jews, Sunnis are still waiting for the Messiah to arrive.

Ayatollah al-Ansari tells his flock during Friday prayers, all our world's hostile tribes are battling a common enemy: ourselves.

Summers here are now so hot that people claim the nails in their houses are melting. Sleeping on the roof to cool off has grown impossible—even at night, the wind feels like a flame. In Al-Ansari's library, wrapped in bookshelves and carpeted with prayer rugs, a fly-zapper punctuates noon prayers like gunshots. Leather-bound hadith volumes and copies of his own books, including an Islamic perspective on human cloning, feel like heated bricks.

Heavily, he sinks back into the floral-patterned cushions of his brown upholstered bench. "Allah created a perfect world, which we have defiled," he says. "We tip Earth's balance. We lose fishes and animals. We change the very climate." From memory he recites from Ash Shams, Surah 91:6–10 in the Holy Qu'ran:

> By the sky and He who constructed it
> And the Earth and He who spread it
> And the soul and He who fashioned it
> And inspired it to tell wicked from righteousness,
> He has succeeded who purifies it,
> And he has failed who corrupts it.

"God gave humans a mind to tell good from bad, and free will to choose between them. We go through life with these two choices, sometimes making the right one, sometimes not."

But the choice facing us now, the ayatollah counsels, is humanity's ultimate decision.

"We have two options. The first is that this corruption will grow until it finishes humankind." This, he says, is illustrated later in the same surah. Sinners are forewarned not to harm an animal, but still do, so God brings destruction down upon them.

"Either that, or man uses his mind to understand that he must return to the Creator this perfect world as it was made for him. Like the Prophet Nūh, righteous humans will repair the Earth and restore nature's palace."

But is there still hope that can happen? What if, despite all the efforts by good people to save Earth's sublime sanctuaries like this ancient human cradle, climate and extinction have already gone too far?

He sips from a glass of lemon tea and readjusts his gray tunic. "That's why we believe in a savior. It will be time for our 12th Imam to reappear with Jesus. But nobody knows when they will, except Allah. We have some signs. It was said that Romans will enter Basra before He appears, and the Americans invaded. We don't know from these signs if the savior is close or far. But we mustn't sit and wait while all is destroyed. This should inspire us to try to fix the problems, to unify for the planet."

He turns his palm upward, brown prayer beads laced through his fingers. "We all live in the same ship. For it to be a home, we must cooperate. Learn to live without extravagance—that is *haram*: forbidden. We are asking Allah to guide to us."

Jassim Al-Asadi's wife, Souad, wrapped in her black abaya, is a devout Muslim. Jassim doesn't know what he believes, except that people don't need God to do the right thing. He and Souad respect each other's opinion, because the outcome is the same.

"I have many questions for which I cannot find a logical answer," Jassim says. "But goodness does not need religion. Goodness is the characteristic of humans reconciled with themselves and the world. Climate change, religious and political conflict, senseless wars: we need to choose good over bad."

The kidnapping and torture have chiseled his lines a little deeper, but his eyes are bright as always as he inhales the marsh, walking from Nature Iraq to where Abu Haider awaits in his mashouf. He climbs in, and they head to see how the new wetland garden is growing. There's been some rain; along the channel, fresh reed shoots are greening. When, after many dry months, Jassim hears the laughter of a Basra reed warbler, his furrowed face breaks into a rapturous smile.

They're still here.

ACKNOWLEDGMENTS

Early in my research, a famous environmentalist snapped at me, "Hope is for weaklings. What we need now is courage." I'd wondered myself if hope weren't just a form of denial that deludes us into mass-scale cognitive dissonance: believing civilization can go (and grow) on as it is, despite mounting evidence to the contrary.

So foremost, I thank my editor, John Parsley—now the publisher of Dutton at Penguin Random House—for urging me, over four years of biweekly phone discussions, to keep probing our species' realistic hopes in this make-or-break century. Meeting people who fight for our future against all odds, I've learned that hope is a prerequisite for their courage. Courage isn't fearlessness: it's setting fear aside to do what's needed. Everyone I've portrayed here is scared of how things could end up unless somebody acts—so they do. The kind of hope they embody is an action verb, not a passive state. They don't wait for miracles; they keep trying to make them.

Among the most disquieting symbols I've ever encountered of how we've fouled our own nest is the ravaging of the cradle that first birthed civilization and three major world religions—notably by my own country's disastrous invasion. Amid Iraq's lethal political instability, escalating temperatures, and overstressed rivers, Azzam Alwash, Jassim Al-Asadi, and Ameer Naji don't quit. I thank them for our time together and their example—and deep respect to artist Meridel Rubenstein, who persevered for 12 years until her waste treatment garden finally took root. Thanks also to Abu Haider and family, Fadhel Dwaidj, Amir Doshi, *New York Times* Baghdad bureau chief Alissa J. Rubin, Emily Garthwaite, Rashad Salim, and Zahra Souhail. Special gratitude to Najaf Cultural and Arts Palace director Rashid Jabar Al-hasani, who took me to the great shrines of his city and Karbala, and to the Wadi Al-Salam cemetery. And, for his frankness and kindness, *shukran jazeelan* to Ayatollah Sheikh Mohammad Hussein al-Ansari.

My thanks to Molly Jahn and Bill Rutherford, both as wickedly entertaining as they are wildly brilliant. Thanks also to Collin Timm and his team at Johns Hopkins Applied Physics Laboratory; APL Biological and Chemical Sciences program manager Sara Herman; DARPA's Stacey Wierzba and Eric Butterbaugh; and Cornucopia contractors Elisabeth Perea at SRI, Jay Schwalbe at Nitricity, and Ting Lu at the University of Illinois. Among other DARPA projects I explored, I thank UC–San Diego biochemist Seth Cohen at AWE (Atmospheric Water Extraction); Jennifer Sleeman at APL's Climate Tipping-point Modeling program; and two REEFense contractors I visited: APL's Jennie Boothby and the University of Hawaiʻi-Mānoa's director of Ocean Science and Technology, Ben Jones. Directly or indirectly, a billion people rely on the livelihood of coral reefs. REEFense intends to hold on to them.

The Center for Biological Diversity, I learned, drives many people crazy—which I find absolutely fitting in a world where insanity runs amok. Peter Galvin, Todd Schulke, Kierán Suckling, Russ McSpadden, Lydia Millet, Tierra Curry, Randy Serraglio (ret.), and Center attorneys Kassie Siegel, Brendan Cummings, Jean Su, Lori Ann Burd, and Brian Segee: thank you all for your generous time. Life is literally richer for your efforts. The same is true for Sky Island Alliance's Louise Misztal; Tucson biologist Chris Bugbee, who made public his video witness of the jaguar El Jefe; and the Northern Jaguar Project's Miguel Gómez, Carmina Gutiérrez, Turtle Southern, and Diana Hadley: thanks for your aid and your efforts.

Dennis Whyte, Brandon Sorbom, and Bob Mumgaard of MIT's Plasma Science and Fusion Center and Commonwealth Fusion Systems graciously and patiently explained to me what's needed to make and harvest our own starlight on Earth, and shared their dreams about what it could mean if we can. I thank them and MIT doctoral candidate Julia Witham, PSFC's Julianna Mullen, and CFS's Kristen Cullen for their help and clarity.

Along the Mesoamerican Reef, I'm grateful to have accompanied the Natural Capital Project's Katie Arkema, Stacie Wolny, and Jess Silver; the World Wildlife Fund's Ryan Bartlett, Abby Hehmeyer, Nadia Brood, and Luís Chévez; NASA's Manishka De Mel and Meridel Phillips; coastal engineers Christian Mario Appendini and Alec Torres-Freyermuth; reserve managers Jaicy Maldonado and Gustavo Cabrera; and all the other Mexican, Guatemalan, Belizean, and Honduran park and wildlife managers I was privileged to meet.

A few weeks later, Manishka and Meridel hosted me at NASA's Goddard Institute of Space Studies in New York, where I benefited from the wisdom of climate director Cynthia Rosenzweig, economist Malgosia Madajewicz, crop modeler Alex Ruane, and climatologist Kate Marvel. Later, Kate would collaborate with Project Drawdown, whose director, Jon Foley, helped guide my thinking about our realistic hopes for reducing excess carbon.

I am ever grateful to my research assistant, Claudine LoMonaco, for crucially connecting me with Jaap de Heer, Henk Ovink, and Deltares engineers Frans Klijn, Jaap Kwadijk, and Roelof Stuurman, all who took time to show me how the Netherlands has stayed the tide for the past eight centuries. My tours of the country's ingenious waterworks included meeting Rijkswaterstaat's Richard Jorissen and Eric van der Weegen, whose candor contributed enormously to my understanding, as did visiting the Zeeland studio of photographer Rem van den Bosch, whose work both exalts and alerts his country. Thanks, too, to journalists Rosanne Kropman and Veerle Schyns, who guided me to Limburg to meet dike count Patrick van der Broeck and visit the beautiful town of Valkenburg, mopping up after an unprecedented flood.

It was a singular honor in the Netherlands to meet Urgenda founder Marjan Minnesma, and then Milieudefensie director Donald Pols. Should humanity win the battle to hold climate change within manageable limits, they will be among the heroes we'll thank for contriving how to finally hold governments and companies accountable.

The Netherlands was my first foreign trip following an almost two-year travel hiatus. I'd originally planned to go in early 2020, and then accompany Jaap de Heer to see Bangladesh's Delta Plan 2100, but a global pandemic intervened. During that first COVID year, I made one trip, to the "mask-free zone" north of Utica, where, thanks to Andy Mower of Performance Premixes, I rode corn-choppers with Joe Van Lieshout of Brabant Farm in Verona, New York, and Jake and Randal Conway of Conway Dairy Farm in Turin. Many thanks to them and to Dan Mower.

Previously, I'd visited the Berkeley, California, company making the microbial seed treatment they were testing. Meeting Pivot Bio's scientists, mission-driven to eliminate synthetic nitrogen, was uplifting. Thank you Karsten Temme, Alvin Tamsir, Keira Havens, Neal Shah, Allison Nicole Johnson, Kent Wood, Shayin Gottlieb, David Brown, Sarah Bloch, Rosemary Clark, and Tracy Willits. Thanks, too, to Brandon Hunnicutt, Emerson Nafziger, and Trenton Roberts for their added perspective.

Journalists are often asked what's the most dangerous country we've worked in. For this book, it turned out to be the US. After

COVID vaccines eased domestic travel restrictions, I went to northern Minnesota to witness the mass protests of Enbridge Line 3. At one point, despite wearing press tags and verbally identifying myself to law enforcement as a journalist (I was also writing an essay for the *Los Angeles Times*), I was snared by overzealous sheriffs along with 185 activists who had commandeered a pump station. In the Hubbard County jail in Park Rapids, I was strip-searched, given orange prison clothing, never allowed a telephone call, and held in solitary confinement in a windowless cell.

Before my mobile phone was confiscated, however, I'd messaged another journalist, resulting in a barrage of calls protesting the violation of my press freedom rights. After several hours, I was released, but, like the protesters, charged with gross misdemeanor trespassing on critical public service facilities, punishable by a year in prison and a $3,000 fine. I'm grateful to the Committee to Protect Journalists, PEN America, Authors Guild, and colleagues Sandy Tolan, Mort Rosenblum, Amy Wilentz, Jacqueline Sharkey, John Parsley, and Stuart Krichevsky who wrote outraged letters on my behalf; to Louise and Pallas Erdrich, who had wine waiting after I was released; and to attorney Steven Meshbesher, who eventually got my charges dismissed.

Recognizing that I'm not the first journalist to be detained while covering the petroleum cabal and won't be the last, with the assistance of the Cyrus R. Vance Center for International Justice, Homelands Productions, a nonprofit journalism cooperative I cofounded, established the Public Interest Reporting Legal Defense Fund, which provides a mechanism for raising tax-deductible donations to help defray journalists' legal expenses incurred from being arrested while doing their jobs, or who face harassment lawsuits intended to squelch their reporting. (If that's you: www.homelands.org/defense-fund.)

I'm indebted to Charles and Julie Davidson, Judith Spiegel, Cindy Kalland, John Courtright, Antonia Malchik, Bill Posnick, Pat Lanier, Nubar Alexanian, Rebecca Koch, Jacqueline Sharkey, John Parsley, and Mary Caulkins of the Tooth Fairy Fund, whose contributions helped me square my own legal bills and get back to reporting. I hope this book is worthy of your generosity.

On the White Earth Reservation and in Ojibwe Treaty Lands, deepest thanks to Winona LaDuke, Nancy Beaulieu, Debra Topping, Gina Peltier, Dawn Goodwin, Simone Senogles, Sherry Couture, Mary Lyons, Robert Shimek, Frank Bibeau, and Big Wind for enlightening me about their heritage and ongoing resistance. My conversation with Couchiching First Nation tribal attorney Tara Houska unfortunately had barely begun when a low-flying Department of Homeland Security helicopter scattered us. I also thank Nizhóní Begay of the Water Protector Legal Collective, Akilah Sanders, Martin Keller, and Kevin Whelan for their assistance, and wish blessings onto Katy Grisamore and Wayne Eimers of Hubbard County Jail Ministries, who kindly drove me back into the fray.

While in Bangladesh, I was invited by Syed Naved Husain, CEO of BEXIMCO, the country's largest conglomerate and private-sector employer, to see its plan for helping our species live far more sensibly on this planet, involving something nearly all humans do: wear clothes. Bangladesh is the world's second-biggest exporter of clothing, and textiles are BEXIMCO's biggest product—annually, it makes 60 million yards of denim alone. Husain wanted to show me how each year they're making more remarkably soft fabric from recycled plastic bottles and packaging waste. In partnership with pioneering Spanish recycling firm Recover, they're also using scraps gleaned from cutting-room floors of textile manufacturers worldwide—and they're recycling clothing itself. Every year, he explained, 73 percent of the 48 million tons of disposed clothing worldwide ends up in landfills or incinerated. The rest mostly becomes rags; only 1 percent is recycled. "It's a textile waste crisis."

If every clothing manufacturer followed their recycling plan, his team calculated that 73 percent would drop to 23 percent, reducing production emissions accordingly. "A sizable step toward the UN's 1.5°C pathway to 2030."

Increasingly, he added, they're also substituting hemp fiber for water-intensive cotton. Touring BEXIMCO's immense plant with its sustainability director and several department heads reminded me that any path to a livable future must involve commerce, because it's so pervasive—we all buy things—and so human: before language, trade was probably the first form of peaceful communication between strangers. But is believing that consuming green products can save us just wishful hoping?

Recycling plastic into clothing fibers only postpones its eventual arrival in our air, water, and bodies. A textile industry's prosperity still depends on a fashion industry that encourages us to discard perfectly wearable clothes and buy new ones. I was impressed that BEXIMCO was building the country's biggest solar field to power its machines—but I looked down at it from the helicopter that daily flies Husain to work, because Dhaka is so choked that his 10-minute commute would otherwise take 3 hours by car.

Yet it's hard not to be hopeful in Bangladesh: facing inundation by both water and its own population, it's among the most dynamic places I've ever reported from: figuratively or literally, everybody is pedaling their rickshaws, finding a way. My thanks to Sharif Jamil and Sultana Kamal; Anu Muhammad and the late Saleemul Huq; to Jaap de Heer's colleagues Giasauddin Choudhury and Joost Meijer; BEXIMCO's Syed Naved Husain, Saquib Shakoor, and Mohidus Samad Khan; SOLShare's Sebastian Groh and cofounder Daniel Ciganovic, Salma Islam, Hannes Kirchhoff, Anisha Hassan, Tasnim Tabassum, Mashiat Fariha Alam, and Abu Hasan. In the Sundarbans, I thank Hasan Mehedi, Tonny Nowshon, Capt. Kader Mulla, Bablu Hajra, forester Howlader Azad Kabir, and the villagers of Banishanta and Kutubdia. Special gratitude goes to Mohammed Shah Nawaz Chowdhury for showing me so much of his country and for facilitating my entry to the UNHCR's Rohingya refugee camps, and to his University of Chittagong students and faculty colleagues Shahadat Hossain and Sayedur R. Chowdhury.

In the Comcáac village of Punta Chueca, Sonora, Mexico, *¡yooz ma samsisíin xo!* to Erika Barnett, Alberto Mellado, Laura Molina, Betina and Lupita Romero Morales, Filomena Barnett, Chapo Barnett, and Ernesto Molina. Thanks, too, to Lorayne Meltzer and Héctor Pérez of Prescott College's Kino Bay Center, linguist Cathy Moser Marlett, master baker Don Guerra, and the late Choctaw filmmaker Phil Lucas. In Andalusia's El Puerto de Santa María, *mil gracias a* Juan Martín Bermúdez, Ricardo Ariza, Juan Ariza, Isabel Ramírez, Chef Ángel León, and Aponiente's incomparable crew. Gary Paul Nabhan and Laurie Monti deserve not just my thanks, but everyone's: they epitomize how to savor and steward life.

I thank Buckminster Fuller Institute director Amanda Ravenhill for introducing me to Bren Smith, who navigated me through his Thimble Island Ocean Farm and, in thoughtful conversations and correspondence, the eco-socioeconomics of raising aquatic vegetation. Thanks also to GreenWave's Michelle Stephens, Jill Pegnataro, Kat McSweeney, Toby Sheppard Bloch, Lindsay Olsen, and Ron Gautreau. For further help in understanding how seaweed can improve our relationship with this blue planet, I thank Atlantic Seafood's Brianna Warner, lobsterman Keith Miller, AKUA's Courtney Boyd Myers, Barnacle Foods' Matt Kern, Macro Oceans' Matthew Perkins, Stony Brook University's Shellfish Restoration & Aquaculture director Michael Doall, the World Bank's Karlotta Rieve, Montauk Seaweed Supply's Sean Barrett, Kodiak Island kelp farmers Al

Pryor and Nick Mangini, Loliware's Sea Fawn Briganti, The Crop Project's Casey Emmett, AgriSea's Tane and Clare Bradley, and the Lloyd's Register Foundation's Vincent Doumeizel.

I've never forgotten riding through Medellín in 1988 with Universidad de Antioquia faculty chair Beatriz Ortiz de Turizo, whose five predecessors had been assassinated. While pausing at stoplights, we'd duck our heads, lest armed sicarios on motor scooters drive by. More than 30 years later, I was pleased to find her alive and heartened by her city's renaissance. I thank her, my adept fixer Diana Patiño, journalist Adriaan Alsema, ecologist Carlos Delgado, ornithologist Iván Lau, sociologist Margarite Alzate, Comuna 13's Angela González, and Área Metropolitana's Eugenio Prieto, Daniela García, and Carlos David Hoyos for orienting me to a new Medellín.

Colombian novelist Yolanda Reyes introduced me to Medellín librarians Gloria Rodríguez and Luís Bernardo Yepes, whose brave efforts sowed peace in the once-murderous comunas. Geographer Felipe Osorio Vieira and urban ecologist María Angélica Mejía connected me to Parque Explora's Carolina Sanín and architects Oscar Santana and Juan Sebastián Bustamante, which led me to biologist Camilo Jaramillo, arborist Mauricio Jaramillo, artists Fredy Serna and Camila Flores, Ruta N's Sergio Naranjo, and C40's Lina López. For valuable context, I also thank journalist Constanza Vieira Quijano and, at Bogotá's Instituto Humboldt, ecologists Hernando García, Germán Corzo, and Brigitte Baptiste.

For his time and candidness, special thanks to Colombian President Gustavo Petro.

In Ithaca, New York, I thank Jon Miller, Rebecca Nelson, Peter Bardaglio, Courtney Ann Roby, and especially Luis Aguirre-Torres. For showing me how future cities might rise with seas rather than be engulfed by them, much thanks to MIT's Nick Makris, Bjarke

Ingels Group's Daniel Sundlin, Biorock creator Tom Goreau, and for many thoughtful conversations, especially to Oceanix's Itai Madamombe and Marc Collins Chen.

My last day on Ebeye, I stood near the pier at sunset, watching kids playing on three different courts, wondering if the only good thing my country ever did for the Marshall Islands was to introduce basketball. Eric Rasmussen and Gregg Nakano are trying hard, in the face of history and encroaching waves, to right that record. I first met Eric at the 2002 retreat that he and energy visionary Amory Lovins held to design sustainable refugee camps, just before Eric would lead US military humanitarian services during the Iraq War. It was compelling to accompany him and Gregg on their latest efforts to restore order amid disaster, this time preemptively in the Marshalls Islands. I thank both for hours of soul-searching discussion and for their efforts.

In Majuro, my thanks to editor Giff Johnson of the *Marshall Islands Journal*, the late Irene Taafaki of the College of the Marshall Islands, master outrigger builder Alson Kelen, Senator David Paul, then–US ambassador Roxanne Cabral, Ben Chufaro, Nika Wase, Riyad M. Mucadam, and, at the environmental NGO Jo-Jikum, Loredel Faye Areieta, Konea Ishimura, Jobod Silk, and their colleagues. On Ebeye, my understanding owes greatly to Kwajalein Atoll Disaster Response Manager Abon Arelong and Sonia Tagoilelagi of the International Organization for Migration. My great thanks and admiration to Jelton Anjain, and I wish good luck to his students Lowa Kabua and Kobie Loeak. Thanks, too, to John Silk, Anjo Kabua, Kenneth Kady, Father Lando Cuasito, Lae Atoll canoes builders Season Langda and Lawday Elisha—and, of course, to *Iroojlaplap* Mike Kabua. To poet Kathy Jetñil-Kijiner, eloquent daughter of her beautiful, beleaguered atoll nation and a conscience for our times: my deepest respect and thanks.

I'm grateful to and forever inspired by artist Jae-Eun Choi, who walked me to North Korea through a military tunnel beneath the DMZ. Thanks also to her Dreaming of Earth collaborators Seung H-Sang, Lee Bul, Minsuk Cho, Jaeseung Jeong, Bijoy Jain, Rumi Okazaki, and Sebastian Behmann.

Claudine LoMonaco arranged my visit to the Google campus with artificial intelligence savant John Platt, whom I thank for explaining his vision of what AI might achieve and his honest concerns about its limitations. Another imaginative scientist I was fortunate to meet is MIT physicist Paul Woskov, who realized that gyrotrons powerful enough to wrangle plasma can be pointed at clean energy lurking right beneath our feet. I thank him, his protégé CEO Carlos Araque of Quaise Energy, and Matthew Houde who toured me through a Quaise test facility at Oak Ridge National Laboratory.

It's become clichéd to say that young people embody hope, but clichés get repeated so often because they're true. Nowhere was that clearer to me than in a country club in Wellesley, Massachusetts, watching a bunch of determined American teenagers and twenty-somethings stand up to their country's petroleum industry. Whether Climate Defiance's tactics inspire or offend doesn't matter to them, because their future is on the line. Being hauled to jail after disrupting an annual congressional baseball game sponsored by Chevron hasn't stopped more from joining. Days after the 2024 election, eleven new chapters formed nationwide. My thanks to Michael Greenberg, Martin Gioannetti, and their colleagues for trusting and inviting me along. Everyone should thank them for screaming truth right into power's face, because we're all in this one together: no one will escape the heat.

My initial research included several conversations with David Pollack and Lise Van Susteren, cofounders of the Climate Psychiatry Alliance. In 2019, believing that climate change would soon become a principal underlying cause for anxiety was still an outlier theory in their field. Largely due to their efforts, recognizing eco-anxiety is now mainstream. CPA's materials on helping climate-stricken patients cope, including by encouraging them to take positive action, are widely consulted by psychiatrists and psychologists alike.

During 2020, Dr. Pollack, a public health psychiatrist and author of protocols for handling mass trauma in the wake of disaster, found his own Oregon home surrounded by fires. Luckily, his family didn't have to evacuate—but their near-miss underscores that everyone is now a close call away from calamity. My great thanks to him and Dr. Van Susteren—with luck, some of her Washington, DC, clientele have leverage to make a difference.

Others whose input and insights helped shape this book include Dan Kammen and Diana Liverman, both on the 2007 IPPC panel that won the Nobel Peace Prize; 350.org and Third Act cofounder Bill McKibben; eco-entrepreneur Paul Hawken; Natural Capital Project founder Gretchen Daily; Whole Earth visionary Stewart Brand and his wife Ryan Phelan, director of Revive & Restore, whose extinction-defying efforts will have something to teach us, regardless of their outcome; Silicon Valley sustainability crusader John Picard; Switch It Green's Anna Chirico; Henry L. Stimson Center global security expert David Bray; Rebuild by Design's Amy Chester; Leibniz Institute astrophysicist Sydney Barnes; University of Arizona biogeographer Greg Barron-Gafford; Canary Islands botanists Jaime Gil González and Marta Peña Hernández; Dr. Ye Tao, who gave up his Harvard nanotech lab to found MEER: Mirrors for Earth's Energy Rebalancing; Rabbi Baruch Clein; and Cal State–Chico social worker Darci Yartz, who took me to the scorched remains of her hauntingly named hometown, Paradise, and amid her own grief counseled its survivors.

Much credit for their skilled help with this book goes to cartographer David Lindroth, bibliographer Jenny Lee Wildermuth, Penguin Random House legal counsel Patricia Clark, book designer Lorie Pagnozzi, copy editor Maureen Klier, and Dutton associate editor David Howe.

For morale support I critically needed while writing, thanks to my deep soul friends Francie Rich, Stephen Philbrick, Connie Talbot, Jon Miller, Ruxandra Guidi, Bear Guerra, Sandy Tolan, Mort Rosenblum, Jacqueline Sharkey, Bill Posnick, Debra Gwartney, Jay Dusard, Jim and Deb Hills, Sylvia Plachy, Dick Kamp, Feng Mohan, Alan Stesin, John Courtright, Jeannine Relly, Alison Hawthorne Deming, Nubar Alexanian, Rebecca Koch, Priscilla Pierce Goldstein, Roz Driscoll, Alton Wasson—and, alas posthumously, Jeff Jacobson, Ronn Spencer, Anthony Weller, and Barry Lopez.

The contract for this book was negotiated by my long-time agent Nick Ellison, a former boxer who studied at the Sorbonne, which kind of says it all about him. In 2021, Nick succumbed to debilitating illness; his friendship and advocacy remain in my heart. To my great fortune, agents David Patterson and Stuart Krichevsky of the Stuart Krichevsky Literary Agency took me in, providing all the warmth, encouragement, and savvy wisdom an author could hope for. I am profoundly grateful to them and their colleagues Aemilia Phillips, Hannah Schwartz, and Chandler Wickers for welcoming me so well.

Long before I was finished, I'd already composed the dedication to *Hope Dies Last* to my two greatest muses—my wife, Beckie Kravetz, and my older sister, Rochelle Hoffman. I'm glad that I wrote it back then, because months later when Rochelle, my sole sibling, was diagnosed with an aggressive inoperable cancer, I could show her its timestamp to prove I wasn't honoring her because she was dying, but for all she did while living: constantly hand me books, teach me poetry, and champion my writing life, no matter how precarious. She loved me as unflaggingly as she loved literature, and she died relieved to know that I'd finally completed a draft of this book. "Damn,

I'll miss you," I told her. "Good," she replied, "I want you to." Which I do, daily. My thanks to our family who held us both: Brian and Pahoua Yang Hoffman; Olivia Thanadabout; and Peter, Zeynep Turan, and Onat Hoffman . . .

. . . and, for her unfailing everything, to Beckie: an artist who fiercely defends beauty as a source of hope, because we can always create more of it. Whenever I'd fret while struggling for words that all seemed hopeless, her sheer existence would remind me that it's not.

I pray that I've done justice to her and to the people I've portrayed. As long we let them keep on, there is hope.

SELECT BIBLIOGRAPHY

CHAPTER ONE: CRADLE TO GRAVE

Books

Alwash, Suzanne. *Eden Again: Hope in the Marshes of Iraq.* Fullerton, CA: Tablet House, 2013.

Finkel, Irving. *The Ark Before Noah.* New York: Anchor Books, 2014.

Helle, Sophus. *Gilgamesh: A New Translation of the Ancient Epic.* New Haven, CT: Yale University Press, 2021. Kindle edition.

Lonergan, Steve, and Jassim Al-Asadi. *The Ghosts of Iraq's Marshes: A History of Conflict, Tragedy, and Restoration.* Cairo: American University Cairo Press, 2024.

Pournelle, Jennifer. "Physical Geography." In *The Sumerian World,* edited by Harriet Crawford, chapter 1. Oxfordshire: Routledge, 2012.

Ramli, Muhsin al-. *Daughter of the Tigris.* Translated by Luke Leafgren. London: Quercus, 2019. Kindle edition.

———. *The President's Gardens.* Translated by Luke Leafgren. London: Quercus, 2016. Kindle edition.

Thesiger, Wilfred. *The Marsh Arabs.* London: Penguin Books, 1977.

Articles

Altaweel, Mark, Anke Marsh, Jaafar Jotheri, and Carrie Hritz. "New Insights on the Role of Environmental Dynamics Shaping Southern Mesopotamia: From the Pre-Ubaid to the Early Islamic Period." *Iraq* 81 (2019), page 1 of 24. https://doi.org/10.1017/irq.2019.2.

Alwash, Azzam. "Iraq's Climate Crisis Requires Bold Cooperation." The Century Foundation, December 14, 2020. https://tcf.org/content/report/iraqs-climate-crisis-requires-bold-cooperation.

American University of Iraq, Sulaimani. "Azzam Alwash." https://auis.edu.krd/?q=board/azzam-alwash.

Anderson, Tim. "Wet Bulb Temperature Is the Scariest Part of Climate Change You've Never Heard Of." Medium, July 2, 2021. https://medium.com/the-infinite-universe/wet-bulb-temperature-is-the-scariest-part-of-climate-change-youve-never-heard-of-8d85bef1ca98.

Andreoni, Manuela. "Climate Change Is Causing Severe Drought in a Volatile Mideast Zone, Study Finds." *New York Times,* November 8, 2023. https://www.nytimes.com/2023/11/08/climate/climate-change-drought-fertile-crescent.html?smid=nytcore-ios-share&referringSource=articleShare.

Ansari, Nadhir al-, Sven Knutsson, and Ammar A. Ali. "Restoring the Garden of Eden, Iraq." *Journal of Earth Sciences and Geotechnical Engineering* 2, no. 1 (2012): 53–88.

Asafu-Adjaye, John, Linus Blomqvist, Stewart Brand, Barry Brook, et al. "An Ecomodernist Manifesto," 2015. https://www.ecomodernism.org.

Billing, Lynzy. "After the Wars in Iraq, 'Everything Living Is Dying.'" *Inside Climate News*, December 29, 2021. https://insideclimatenews.org/news/29122021/iraq-ecocide.

Brasington, Leigh. "The Two Biblical Stories of Creation." Revised December 15, 2017. https://www.leighb.com/genesis.htm.

Bryner, Jeanna. "Lost Civilization May Have Existed beneath the Persian Gulf." Live Science, December 9, 2010. https://www.livescience.com/10340-lost-civilization-existed-beneath-persian-gulf.html.

Carrington, Damian. "Extreme Heatwaves Could Push Gulf Climate beyond Human Endurance, Study Shows." *The Guardian*, October 26, 2015. https://www.theguardian.com/environment/2015/oct/26/extreme-heatwaves-could-push-gulf-climate-beyond-human-endurance-study-shows.

Circle of Blue. "Iraq's First National Park: A Story of Destruction and Restoration in the Mesopotamian Marshlands." September 4, 2013. https://www.circleofblue.org/2013/world/iraqs-first-national-park-a-story-of-destruction-and-restoration-in-the-mesopotamian-marshlands.

Dance, Scott. "The Heat Index Reached 152 Degrees in the Middle East—Nearly at the Limit for Human Survival." *Washington Post*, updated July 18, 2023. https://www.washingtonpost.com/weather/2023/07/18/extreme-heat-record-limits-human-survival/#.

Day, John W., Jr., Joel D. Gunn, William J. Folan, Alejandro Yáñez-Arancibia, et al. "The Influence of Enhanced Post-Glacial Coastal Margin Productivity on the Emergence of Complex Societies." *Journal of Island and Coastal Archaeology* 7, no. 1 (2012): 23–52. https://doi.org/10.1080/15564894.2011.650346.

Dilleen, Connor. "Turkey's Dam-Building Program Could Generate Fresh Conflict in the Middle East." *The Strategist*, November 5, 2019. https://www.aspistrategist.org.au/turkeys-dam-building-program-could-generate-fresh-conflict-in-the-middle-east.

Dziadosz, Alexander. "The End of Eden." *Harper's Magazine*, August 2018.

Feinman, Peter. "Where Is Eden? An Analysis of Some of the Mesopotamian Motifs in Primal J." Academia.edu, 2013. https://www.academia.edu/39535508/Where_Is_Eden_An_Analysis_of_Some_of_the_Mesopotamian_Motifs_in_Primeval_J.

Ganopolski, A., R. Winklemann, and H. J. Schnellnhuber. "Human-Made Climate Change Suppresses the Next Ice Age." Potsdam Institute for Climate Impact Research, January 13, 2016. https://www.pik-potsdam.de/en/news/latest-news/human-made-climate-change-suppresses-the-next-ice-age.

Gasche, Hermann, ed. "The Persian Gulf Shorelines and the Karkheh, Karun, and Jarrahi Rivers: A Geo-Archaeological Approach; A Joint Belgo-Iranian Project First Progress Report—Part 3." *Akkadica* 128 (2007): 1–62.

George, Susannah, and Sam McNeil. "Iraq's Vast Marshes, Reborn after Saddam, Are in Peril Again." AP News, October 26, 2017. https://apnews.com/article/a8c150aa24ff4a58a95a8d761d3a9d37.

Goldman Environmental Prize. "Azzam Alwash." 2013. https://www.goldmanprize.org/recipient/azzam-alwash/#recipient-bio.

Got Questions Ministries. "Why Are There Two Different Creation Accounts in Genesis Chapters 1–2?" https://www.gotquestions.org/two-Creation-accounts.html.

Hamblin, Dora Jane. "Has the Garden of Eden Been Located at Last?" *Smithsonian* 18, no. 2 (May 1987). Accessed at https://www.ldolphin.org/eden.

Hockenos, Paul. "Turkey's Dam-Building Spree Continues, at Steep Ecological Cost." *Yale Environment 360*, October 3, 2019. https://e360.yale.edu/features/turkeys-dam-building-spree-continues-at-steep-ecological-cost.

Humat Dijlah. https://humatdijlah.org/en/main.

International Media Center. "Arbaeen Festival Launched from the Farthest Point in Southern Iraq." September 7, 2021. https://imhussain.com/english/newsandreports/3371.

Iraqi Presidency. "Mesopotamia Revitalization Project: A Climate Change Initiative to Transform Iraq and the Middle East." Presidential documents, October 17, 2021. https://presidency.iq/en/Details.aspx?id=3437.

Kaufman, Alexander C. "The Savior of Iraq's Garden of Eden Says He Knows How to Stop the Next Big War." *HuffPost*, March 19, 2018. https://www.huffpost.com/entry/azzam-alwash-iraq-war_n_5aa840e1e4b018e2f1c27649.

Kullab, Samya. "In Iraq's Iconic Marshlands, a Quest for Endangered Otters." AP News, May 28, 2021. https://apnews.com/article/europe-middle-east-iraq-environment-and-nature-c5e6c8ae4a4ea24f11fdd54417d986e5#.

Livingston, Ian. "These Places Baked the Most During Earth's Hottest Month on Record." *Washington Post*, August 2, 2023. https://www.washingtonpost.com/weather/2023/08/02/july-hottest-month-global-temperatures/#.

Loveluck, Louisa, and Salim Mustafa. "From Cradle to Grave: Where Civilization Emerged between the Tigris and Euphrates, Climate Change Is Poisoning the Land and Emptying the Villages." *Washington Post*, October 21, 2023. https://www.washingtonpost.com/world/interactive/2021/iraq-climate-change-tigris-euphrates.

Ludwig, Mike. "20 Years after US Invasion, Iraq Faces Cascading Climate and Water Crises." *Truthout*, March 20, 2023. https://truthout.org/articles/20-years-after-us-invasion-iraq-faces-cascading-climate-and-water-crises.

McCarron, Leon. "How Climate Change Is Devastating Ancient Mesopotamian Marshlands." *Noēma*, October 19, 2021. https://www.noemamag.com/the-last-of-the-marsh-arabs.

Mühl, Simone. "The Intentional Destruction of Cultural Heritage in Iraq." Research Assessment and Safeguarding of the Heritage of Iraq in Danger. https://www.ohchr.org/sites/default/files/Documents/Issues/CulturalRights/DestructionHeritage/NGOS/RASHID.pdf.

NASA. "Panama: Isthmus That Changed the World." March 23, 2008. https://www.nasa.gov/image-article/panama-isthmus-that-changed-world.

Nature Iraq. http://www.natureiraq.org.

———. "Eden in Iraq: The Wastewater Garden Project; Ecological and Cultural Restoration in the Mesopotamian Marshes." http://www.natureiraq.org/wastewater-garden-project.html.

Nissen, Hans J., Elizabeth Lutzeier, and Kenneth J. Northcott. "Review: The Early Settlement of Southern Mesopotamia; A Review of Recent Historical, Geological, and

Archaeological Research." *Journal of the American Oriental Society* 112, no. 1 (1992). https://www.jstor.org/stable/604585.

O'Dea, Aaron, Harilaos A. Lessios, Anthony G. Coates, et al. "Formation of the Isthmus of Panama." *Science Advances* 2, no. 8 (August 17, 2016). https://www.science.org/doi/10.1126/sciadv.1600883.

Pearce, Fred. "Fertile Crescent Will Disappear This Century." *New Scientist*, July 27, 2009. https://www.newscientist.com/article/dn17517-fertile-crescent-will-disappear-this-century.

Polly, David, Samir Patel, Richard Chang, Adia Jackson, Pia Sorenson, Ying-Ying Wu, and David Smith. "The Pliocene Epoch." UC Museum of Paleontology, Berkeley. Last updated June 10, 2011. https://ucmp.berkeley.edu/tertiary/pliocene.php.

Potts, D. T. "Shatt Al-Arab. Encyclopedia Iranica." Academia.edu, 2004. https://www.academia.edu/37316820/Potts_2004_Shatt_al_Arab_Encyclopaedia_Iranica.

———. "Uruk and the Origins of the Sacred Economy." *Engelsberg Ideas*, November 15, 2021. https://engelsbergideas.com/essays/uruk-and-the-origins-of-the-sacred-economy.

Rose, Jeffrey. "Lost Civilization under Persian Gulf?" *Science Daily*, December 8, 2010. https://www.sciencedaily.com/releases/2010/12/101208151609.htm.

———. "New Light on Human Prehistory in the Arabo-Persian Gulf Oasis." *Current Anthropology* 51, no. 6 (December 2010). https://www.jstor.org/stable/10.1086/657397.

Salim, Mudhafar A., Omar Fadhil Al-Sheikhly, Korsh Ararat Majeed, and R. F. Porter. "An Annotated Checklist of the Birds of Iraq." *Sandgrouse* 34 (2012): 4–24.

Save the Tigris. "Iraq and Iran's Hawizeh Marshes: Threats and Opportunities." https://savethetigris.org/iraq-and-irans-hawizeh-marshes-threats-and-opportunities.

Sheikhly, Omar F. al-, and Iyad A. Nader. "The Status of Iraq Smooth-Coated Otter and Eurasian Otter in Iraq." *IUCN Otter Specialist Group Bulletin* 30, no. 1 (January 3, 2013). https://www.iucnosgbull.org/Volume30/AlSheikhly_Nadar_2013.pdf.

Sullivan, Paul. "It Won't Be Oil That Decides Iraq's Future." *National News*, July 21, 2022. https://www.thenationalnews.com/opinion/comment/2022/07/21/it-wont-be-oil-that-decides-iraqs-future.

Universita di Bologna. "Climate Crisis in Mesopotamia Prompted the First Stable Forms of State." *Newswise*, April 27, 2021. https://www.newswise.com/articles/climate-crises-in-mesopotamia-prompted-the-first-stable-forms-of-state.

Water Technology. "Atatürk Dam, Euphrates River, Anatolia." https://www.water-technology.net/projects/ataturk-dam-anatolia-turkey.

World Water Atlas. "Saddam's Regime Dried Up the Famous Arab Marshes." https://www.worldwateratlas.org/narratives/marsh/marshes-in-the-river-tigres-and-euphrates-dried-up/#arab-marsh-disappearance.

CHAPTER TWO: AGAINST ALL HOPE

Books

Dixson-Decleve, Sandrine, Owen Gaffney, Jayati Ghosh, Jørgen Randers, Johan Rockstrom, and Per Espen Stoknes. *Earth for All: A Survival Guide for Humanity.* Gabriola Island, BC: New Society, 2022, by the Club of Rome. Kindle edition.

Hawken, Paul, ed. *Drawdown: The Most Comprehensive Plan Ever Proposed to Reverse Global Warming.* New York: Penguin Books, 2017.

The Holy Bible: New Revised Standard Version. Oxford: Oxford University Press, 1998.

Johnson, Ayana Elizabeth, and Katharine Wilkinson, eds. *All We Can Save: Truth, Courage, and Solutions for the Climate Crisis*. New York: Random House, 2020. Kindle edition.

Wallace-Wells, David. *The Uninhabitable Earth*. New York: Tim Duggan Books, 2019.

Weisman, Alan. *Countdown: Our Last, Best Hope for a Future on Earth?* New York: Little, Brown, Inc., 2013.

Articles

American Geophysical Union. "Earth's Cryosphere Shrinking by 87,000 Square Kilometers per Year." *Newswise*, July 2, 2021. https://news.agu.org/press-release/earths-cryosphere-shrinking-by-87000-square-kilometers-per-year.

———. "The Greenland Ice Sheet Is Close to a Melting Point of No Return, Says New Study." Phys.org. https://phys.org/news/2023-03-greenland-ice-sheet.html.

Black, Riley. "What Happened in the Seconds, Hours, Weeks after the Dino-Killing Asteroid Hit Earth?" *Smithsonian*, August 9, 2016. https://www.smithsonianmag.com/science-nature/what-happened-seconds-hours-weeks-after-dino-killing-asteroid-hit-earth-180960032.

Boerner, Leigh Krietsch. "Industrial Ammonia Production Emits More CO_2 Than Any Other Chemical-Making Reaction. Chemists Want to Change That." *Chemical and Engineering News*, June 15, 2019. https://cen.acs.org/environment/green-chemistry/Industrial-ammonia-production-emits-CO2/97/i24.

Bradshaw, Corey J. A., Paul R. Ehrlich, Andrew Beattie, Gerardo Ceballos, et al. "Underestimating the Challenges of Avoiding a Ghastly Future." *Frontiers in Conservation Science* 1 (January 2021): article 615419. https://doi.org/10.3389/fcosc.2020.615419.

Brigham-Grette, Julie, and Steve Petsch. "The Arctic Hasn't Been This Warm for 3 Million Years—and That Foreshadows Big Changes for the Rest of the Planet." *The Conversation*, September 30, 2020. https://theconversation.com/the-arctic-hasnt-been-this-warm-for-3-million-years-and-that-foreshadows-big-changes-for-the-rest-of-the-planet-144544.

Byrd, Deborah. "Methane Release Likely Caused Mystery Crater on Yamal Peninsula." *EarthSky*, July 31, 2014. https://earthsky.org/earth/second-mysterious-crater-reported-from-yamal.

Cafaro, Philip, Pernilla Hansson, and Frank Götmark. "Overpopulation Is a Major Cause of Biodiversity Loss and Smaller Human Populations Are Necessary to Preserve What Is Left." *Biological Conservation* 272 (2022): 109646.

Center for Climate Repair at Cambridge. https://web.archive.org/web/20220628222434/https:/www.climaterepair.cam.ac.uk/marine-cloud-brightening-mcb.

Ceres. https://www.ceres.org.

Chao, Julie. "More Potent Than Carbon Dioxide, Nitrous Oxide Levels in California May Be Nearly Three Times Higher Than Previously Thought." *News from Berkeley Lab*, December 4, 2012. https://newscenter.lbl.gov/2012/12/04/nitrous-oxide-levels-in-california.

Coren, Michael J. "The Plant Protein That Could Push Meat Off Your Plate." *Washington Post*, June 27, 2023. https://www.washingtonpost.com/climate-environment/2023/06/27/new-plant-based-meat-developing-rubisco-duckweed.

DeConto, Robert M., David Pollard, Richard B. Alley, Isabella Velicogna, et al. "The Paris Climate Agreement and Future Sea-Level Rise from Antarctica." *Nature* 593 (May 6, 2021): 83–99. https://doi.org/10.1038/s41586-021-03427-0.

Dessandier, P.-A., J. Knies, A. Plaza-Faverola, C. Labrousse, M. Renoult, and G. Panieri. "Ice-Sheet Melt Drove Methane Emissions in the Arctic during the Last Two Interglacials." *Geology* 49, no. 7 (July 1, 2021). https://doi.org/10.1130/G48580.1.

Dvorak, M. T., K. C. Armour, D. M. W. Frierson, C. Proistosescu, et al. "Estimating the Timing of Geophysical Commitment to 1.5 and 2.0°C of Global Warming." *Nature Climate Change* 12 (2022): 547–552. https://doi.org/10.1038/s41558-022-01372-y.

Froitzheim, Nikolaus, Jaroslaw Majka, and Dmitry Zastrozhnov. "Methane Release from Carbonate Rock Formations in the Siberian Permafrost Area during and after the 2020 Heat Wave." *Proceedings of the National Academy of Sciences* 118, no. 32 (August 2, 2021): e2107632118. https://doi.org/10.1073/pnas.2107632118.

German, Senta. "Ziggurat of Ur." SmartHistory. https://smarthistory.org/ziggurat-of-ur.

Grunwald, Michael. "Why Vertical Farming Just Doesn't Work." *Canary Media*, June 28, 2023. https://www.canarymedia.com/articles/food-and-farms/why-vertical-farming-just-doesnt-work.

Haley, Jeff T., and Matthew J. Nicklas. "Damping Storms, Reducing Warming, and Capturing Carbon with Floating, Alkalizing, Reflective Glass Tiles." *London Journals Press* 21, no. 6, comp. 1.0 (2021). https://journalspress.com/LJRS_Volume21/Damping-Storms-Reducing-Warming-And-Capturing-Carbon-with-Floating-Alkalizing-Reflective-Glass-Tiles.pdf.

Hansen, James. "Climate Change in a Nutshell: The Gathering Storm." Expert opinion in *Juliana v. the United States*, December 18, 2018. https://www.columbia.edu/~jeh1/mailings/2018/20181206_Nutshell.pdf.

Hansen, James E., Makiko Sato, Leon Simons, Larissa S. Nazarenko, et al. "Global Warming in the Pipeline." *Oxford Open Climate Change* 3, no. 1 (November 2, 2023): kgad008. https://doi.org/10.1093/oxfclm/kgad008.

Harvey, Chelsea. "Rate of Arctic Warming Faster Than Previously Thought." *E&E News*, July 11, 2022. https://www.eenews.net/articles/rate-of-arctic-warming-faster-than-previously-thought.

Heller, Marty, and Iana Salim. *Beyond Burger 3.0: Life Cycle Assessment.* Blonck Consultants. https://investors.beyondmeat.com/static-files/758cf494-d46d-441c-8e96-86ddb57fbed4.

Holthaus, Eric. "Siberia's Permafrost Is Exploding. Is Alaska's Next?" *Slate*, April 2, 2015. https://slate.com/technology/2015/04/exploding-methane-holes-in-siberia-linked-to-climate-change-is-alaska-next.html#.

Ice 911 Research. "Restoring Arctic Sea Ice with Silica Glass Sand." https://www.youtube.com/watch?v=mao2uRDSFzs.

Jackson, R. B., et al. "Human Activities Now Fuel Two-Thirds of Global Methane Emissions." *Environmental Research Letters*, no. 19 (September 10, 2024): 101002.

Jarvis, Brooke. "The Insect Apocalypse Is Here." *New York Times*, November 27, 2018. https://www.nytimes.com/2018/11/27/magazine/insect-apocalypse.html.

Jet Propulsion Laboratory. "Greenland, Antarctica Melting Six Times Faster Than the 1990s." NASA, March 16, 2020. https://www.nasa.gov/centers-and-facilities/jpl/greenland-antarctica-melting-six-times-faster-than-in-the-1990s.

Kaplan, Sarah. "A Forgotten Cold War Experiment Has Revealed Its Icy Secret. It's Bad News for the Planet." *Washington Post*, March 15, 2021. https://www.washingtonpost.com/climate-environment/2021/03/15/greenland-ice-sheet-more-vulnerable.

Kemp, Luke, Chi Xu, Joanna Depledge, and Timothy M. Lenton. "Climate Endgame: Exploring Catastrophic Climate Change Scenarios." *Proceedings of the National Academy of Sciences* 119, no. 34 (August 1, 2022): e2108146119. https://doi.org/10.1073/pnas.2108146119.

Kramer, Andrew E. "Land in Russia's Arctic Blows 'like a Bottle of Champagne.'" *New York Times*, September 20, 2021. https://www.nytimes.com/2020/09/05/world/europe/russia-arctic-eruptions.html.

Lark, Tyler J., Nathan P. Hendricks, Aaron Smith, and Holly K. Gibbs. "Environmental Outcomes of the US Renewable Fuel Standard." *Proceedings of the National Academy of Sciences* 119, no. 9 (February 14, 2022): e2101084119. https://doi.org/10.1073/pnas.2101084119.

Madhusoodanan, Jyoti. "Top US Scientist on Melting Glaciers: 'I've Gone from Being an Ecologist to a Coroner.'" *The Guardian*, July 21, 2021. https://www.theguardian.com/environment/2021/jul/21/climate-crisis-glacier-diana-six-ecologist.

Mann, Adam. "Life after the Asteroid Apocalypse." *Proceedings of the National Academy of Sciences* 115, no. 23 (May 30, 2018): 5820–5823. https://doi.org/10.1073/pnas.1807339115.

McCoy, Terrence. "Scientists May Have Cracked the Giant Siberian Crater Mystery—and the News Isn't Good." *Washington Post*, August 5, 2014. https://www.washingtonpost.com/news/morning-mix/wp/2014/08/05/scientists-may-have-cracked-the-giant-siberian-crater-mystery-and-the-news-isnt-good.

McKibben, Bill. "Truly 'Uncharted Territory.'" *Crucial Years*, April 2, 2023. https://billmckibben.substack.com/p/truly-uncharted-territory/comments.

MEER. https://www.meer.org.

Mooney, Chris. "Greenland's Ice Losses Have Septupled and Are Now in Line with Its Highest Sea-Level Scenario, Scientists Say." *Washington Post*, December 10, 2019. https://www.washingtonpost.com/climate-environment/2019/12/10/greenland-ice-losses-have-septupled-are-pace-sea-level-worst-case-scenario-scientists-say.

———. "Two Major Antarctic Glaciers Are Tearing Loose from Their Restraints, Scientists Say." *Washington Post*, September 14, 2020. https://www.washingtonpost.com/climate-environment/2020/09/14/glaciers-breaking-antarctica-pine-island-thwaites.

Mosbergen, Dominique. "Insects Are Dying en Masse, Risking 'Catastrophic' Collapse of Earth's Ecosystems." *HuffPost*, February 11, 2019. https://www.huffpost.com/entry/insect-population-decline-extinction_n_5c611921e4b0f9e1b17f097d.

Moseman, Andrew. "Why Do We Compare Methane to Carbon Dioxide over a 100-Year Timeframe? Are We Underrating the Importance of Methane Emissions?" *Climate Portal*, January 4, 2024. MIT. https://climate.mit.edu/ask-mit/why-do-we-compare-methane-carbon-dioxide-over-100-year-timeframe-are-we-underrating.

Mufuson, Steven. "Scientists Expected Thawing Wetlands in Siberia's Permafrost. What They Found Is 'Much More Dangerous.'" *Washington Post*, August 2, 2021. https://www.washingtonpost.com/climate-environment/2021/08/02/climate-change-heat-wave-unleashes-methane-from-prehistoric-siberian-rock.

Mufson, Steven, and Sarah Kaplan. "Climate Warming Methane Emissions Rising Faster Than Ever." *Washington Post*, October 26, 2022. https://www.washingtonpost.com/climate-environment/2022/10/26/united-nations-climate-pledges-report.

Next Animation. "A New 3D Model Bolsters the Case That Climate Change Is Causing Mysterious Siberia Craters." *Arctic Business Journal*, February 23, 2021. https://www.arctictoday.com/a-new-3d-model-bolsters-the-case-that-climate-change-is-causing-mysterious-siberia-craters/?wallit_nosession=1.

Penn, Justin L., Curtis Deutsch, Jonathan L. Payne, and Erik A. Sperling. "Temperature-Dependent Hypoxia Explains Biogeography and Severity of End-Permian Marine Mass Extinction." *Science* 362, no. 6419 (December 7, 2018). https://doi.org/10.1126/science.aat1327.

Pinho, Bárbara. "Is Modern Food Lower in Nutrients?" *Chemistry World*, December 5, 2023. https://www.chemistryworld.com/features/is-modern-food-lower-in-nutrients/4018578.article.

Plumer, Brad. "How Much of the World's Cropland Is Actually Used to Grow Food?" *Vox*, December 16, 2014. https://www.vox.com/2014/8/21/6053187/cropland-map-food-fuel-animal-feed.

Quadir, Shamim. "One Hamburger Takes 2,400 Litres of 'Hidden' Water to Make." City University of London, November 6, 2020. https://www.city.ac.uk/news-and-events/news/2019/10/one-hamburger-takes-2400-litres-of-hidden-water-to-make.

Ravindran, Jeevan. "The Ozone Hole over the South Pole Is Now Bigger Than Antarctica." CNN, September 16, 2021. https://www.cnn.com/2021/09/16/world/climate-ozone-antarctica-hole-scn-scli-intl/index.html.

Rees, William E. "Yes, the Climate Crisis May Wipe Out Six Billion People." *The Tyee*, September 18, 2019. https://thetyee.ca/Analysis/2019/09/18/Climate-Crisis-Wipe-Out.

Ridgwell, Henry. "This African City May Be the First Ever with 100 Million People Living in It." *Voice of America*, March 29, 2018. https://www.globalcitizen.org/es/content/100-million-people-city-lagos-nigeria-africa.

Ripple, William J., Christopher Wolf, Jillian W. Gregg, Kelly Levin, Johan Rockström, et al. "World Scientists' Warning of a Climate Emergency 2022." *BioScience* 72 (October 2022): 1149–1155. https://doi.org/10.1093/biosci/biac083.

Roberts, David. "Dr. Ye Tao on a Grand Scheme to Cool the Earth." *Volts*, June 8, 2022. Podcast. https://www.volts.wtf/p/volts-podcast-dr-ye-tao-on-a-grand.

Rosen, Yereth. "Arctic Council Affirms It: The Arctic Is Warming at Three Times the Global Rate." *Arctic Business Journal*, May 25, 2021. https://www.arctictoday.com/arctic-council-affirms-it-the-arctic-is-warming-at-three-times-the-global-rate.

Seibert, M. K., and W. E. Rees. "Through the Eye of a Needle: An Eco-Heterodox Perspective on the Renewable Energy Transition." *Energies* 14 (2021): 4508. https://doi.org/10.3390/en14154508.

Siegert, Martin, Mike J. Bentley, Angus Atkinson, Thomas J. Bracegirdle, Peter Convey, Bethan Davies, Rod Downie, et al. "Antarctic Extreme Events." *Frontiers* 11 (2023). https://doi.org/10.3389/fenvs.2023.1229283.

Steffen, Will, Johan Rockström, Katherine Richardson, Timothy M. Lenton, et al. "Trajectories of the Earth System in the Anthropocene." *Proceedings of the National Academy of Sciences* 115, no. 33 (August 14, 2018): 8252–8259. www.pnas.org/cgi/doi/10.1073/pnas.1810141115.

Ucak, Ayhan. "Adam Smith: The Inspirer of Modern Growth Theories." *Procedia—Social and Behavioral Sciences* 195 (July 3, 2015): 663–672. https://doi.org/10.1016/j.sbspro.2015.06.258.

University of Colorado at Boulder. "Extreme Melt on Antarctica's George VI Ice Shelf." *EurekaAlert*, February 25, 2021. https://www.eurekalert.org/news-releases/703647.

Weisman, Alan. "Why the Earth Is Farting." CNN, August 12, 2014. https://www.cnn.com/2014/08/12/opinion/weisman-craters-methaneindex.html.

Xu, Chi, Timothy A. Kohler, Timothy M. Lenton, Jens-Christian Svenning, et al. "Future of the Human Climate Niche." *Proceedings of the National Academy of Sciences* 117, no. 21 (May 26, 2020): 11350–11355. https://www.pnas.org/cgi/doi/10.1073/pnas.1910114117.

You, Xiaoying. "Could Giant Underwater Curtains Slow Ice-Sheet Melting?" *Nature*, January 17, 2024. https://www.nature.com/articles/d41586-024-00119-3.

Zimmer, Katarina. "The Daring Plan to Save the Arctic Ice with Glass." BBC, September 23, 2020. https://www.bbc.com/future/article/20200923-could-geoengineering-save-the-arctic-sea-ice#.

CHAPTER THREE: FOOD FROM THIN AIR

Books

Brosig, Max, Parker Frawley, Andrew Hill, Molly Jahn, et al. *Implications of Climate Change for the U.S. Army.* Carlisle, PA: US Army War College, 2019.

Articles

Air Protein. https://airprotein.com.

Andersen, Suzanne Z., Viktor Čolić, Sungeun Yang, Jay A. Schwalbe, Adam C. Nielander, Joshua M. McEnaney, Kasper Enemark-Rasmussen, et al. "A Rigorous Electrochemical Ammonia Synthesis Protocol with Quantitative Isotope Measurements." *Nature* 570 (September 26, 2019): 504–508. https://doi.org/10.1038/s41586-019-1260-x.

Baskerville Willy. "Molly's Gonna Save the World." Song written and sung by Bill Rutherford. https://jahnresearchgroup.net/what-we-do/190-2.

Bryce, Emma. "A New Discovery Could Unleash the Full Potential of Switchgrass for Making Biofuel." *Anthropocene*, March 10, 2023. https://www.anthropocenemagazine.org/2023/03/a-new-discovery-could-unleash-the-potential-of-switchgrass-for-making-biofuel.

D'Agati, Caroline. "5 DARPA Inventions That Changed the World." ClearanceJobs, July 13, 2019. https://news.clearancejobs.com/2019/07/13/5-darpa-inventions-that-changed-the-world.

DARPA. "A Cornucopia of Microbial Foods." December 2, 2021. https://www.darpa.mil/news-events/2021-12-02.

———. "DARPA Selects Teams to Develop Novel Hybrid Reef-Mimicking Structures." June 15, 2022. https://www.darpa.mil/news-events/2022-06-15.

———. "Teams Begin Work to Develop Tasty Food from Air, Water and Electricity." February 3, 2023. https://www.darpa.mil/news-events/2023-02-03.

Derouin, Sarah. "Did Bacterial Enzymes Cap the Oxygen in Early Earth's Atmosphere?" *EOS*, September 25, 2019. https://eos.org/articles/did-bacterial-enzymes-cap-the-oxygen-in-early-earths-atmosphere.

Ditlevsen, Peter, and Susanne Ditlevsen. "Warning of a Forthcoming Collapse of the Atlantic Meridional Overturning Circulation." *Nature Communications* 14 (July 25, 2023): article 4254. https://doi.org/10.17894/ucph.b8f99b67-d4e6-4a2e-b518-00bddeed323b.

Douglas, Leah. "U.S. Corn-Based Ethanol Worse for Planet Than Gasoline, Study Finds." Reuters, February 14, 2022. https://www.reuters.com/business/environment/us-corn-based-ethanol-worse-climate-than-gasoline-study-finds-2022-02-14.

Eckelkamp, Margy. "Nitrogen Made Different Than Ever Before." *The Scoop*, November 4, 2022. https://www.thedailyscoop.com/authors/margy-eckelkamp?page=5.

The Economist. "Cloning DARPA; Inventing the Future." June 5, 2021, p. 67.

Farming Monthly National. "Nitrogen Ideas Lab." August 22, 2013.

Fassler, Joe. "The Revolution That Died on Its Way to Dinner." *New York Times*, February 9, 2024. https://www.nytimes.com/2024/02/09/opinion/eat-just-upside-foods-cultivated-meat.html.

Food Agility. "Cybersecurity in Food Supply Chains: Keynote Address by Dr Molly Jahn." https://youtu.be/MhGLyMJCaKo?si=rWBKcpO7RWjgzeE0.

Food and Agriculture Organizations of the United Nations, International Fund for Agricultural Development, UNICEF, World Food Programme, and the World Health Organization. *The State of Food Security Systems and Nutrition in the World 2021: Transforming Food Systems for Food Security, Improved Nutrition, and Affordable Healthy Diets for All*. 2020. https://openknowledge.fao.org/server/api/core/bitstreams/191ee56e-2d4f-46cb-9710-0ca167ca314d/content/cb4474en.html.

Foroohar, Rana. "The US Military Has a Plan to Make Food from Thin Air. No Really." *Financial Times*, December 11, 2021. https://www.ft.com/content/d06bfccb-e43b-4840-b4b4-ce95c2c1ecb4.

Friedlander, Blaine. "Cornell Delicata Squash, Disease-Resistant Version of Heirloom Winter Variety, Named 2002 All-America Selection." *Cornell Chronicle*, October 22, 2001. https://news.cornell.edu/stories/2001/10/cornell-delicata-squash-named-2002-all-america-selection.

Garland, Chad. "Pentagon Looks to Microbes to Feed Troops with Air, Water, and Maybe Even Trash Turned to Protein." *Stars and Stripes*, December 15, 2011. https://www.stripes.com/theaters/us/2021-12-15/darpa-cornucopia-air-water-microbe-foods-military-deployments-3978766.html.

Hann, Elizabeth C., Sean Overa, Marcus Harland-Dunaway, Andrés F. Narvaez, Dang N. Le, Martha L. Orozco-Cárdenes, Feng Jiao, and Robert E. Jinkerson. "A Hybrid Inorganic-Biological Artificial Photosynthesis System for Energy-Efficient Food Production." *Nature Food* 3 (2022): 461–471. https://doi.org/10.1038/s43016-022-00530-x.

Heartland Stories Radio. "Episode 82: Molly Jahn." June 7, 2022. Health Research Alliance. https://hh-ra.org/hs-radio-episode/episode-83-molly-jahn.

Henning, Mark J., Henry M. Munger, and Molly M. Jahn. "'Hannah's Choice F': A New Muskmelon Hybrid with Resistance to Powdery Mildew, Fusarium Race 2, and Potyviruses." *HortScience* 40, no. 2 (April 2005): 492–493. https://doi.org/10.21273/HORTSCI.40.2.492.

HigherGov. "Broad Agency Announcement: Cornucopia." DARPA Defense Sciences Office, HR001122S0012. December 10, 2021. https://www.highergov.com/grant-opportunity/cornucopia-336937.

Imperial College of London. "Professor Bill Rutherford." https://www.imperial.ac.uk/people/a.rutherford.

Jahn, Molly. CV, September 10, 2020. https://jahnresearchgroup.net/wp-content/uploads/2023/01/MollyCV010223-1.pdf.

———. "How 'Multiple Breadbasket Failure' Became a Policy Issue." *Issues in Science and Technology* 37, no. 2 (Winter 2021): 80–86. https://issues.org/global-food-security-molly-jahn-darpa.

Janetos, Anthony, Christopher Justice, Molly Jahn, Michael Obersteiner, et al. *The Risks of Multiple Breadbasket Failures in the 21st Century: A Science Research Agenda.* Frederick S. Pardee Center for the Study of the Longer-Range Future at Boston University Research Report. March 2017. https://www.bu.edu/pardee/the-risks-of-multiple-breadbasket-failures-in-the-21st-century-a-science-research-agenda.

Johns Hopkins Ralph O'Connor Sustainable Energy Institute. "Startup Co-led by ROSEI Researcher Raises $3M to Scale Carbon-Removal Tech." August 21, 2023. https://energyinstitute.jhu.edu/startup-co-led-by-rosei-researcher-raises-3m-to-scale-carbon-removal-tech.

Johns Hopkins University Applied Physics Laboratory. "Climate Security." https://www.jhuapl.edu/work/projects-and-missions/climate-security.

———. "Creating Food from Thin Air: The Future of Nutrition." June 1, 2023. https://www.jhuapl.edu/news/news-releases/230601-apl-using-microbes-to-make-food-from-thin-air.

Kitchlew, Iffah. "Is Super-Polluting Pentagon's Climate Plan Just 'Military-Grade Greenwash'?" *The Guardian,* March 10, 2022. https://www.theguardian.com/us-news/2022/mar/10/pentagon-us-military-emissions-climate-crisis.

Kiverdi. https://www.kiverdi.com.

Kohler, Nan. "Marquis Wheat." *Grist and Toll.* https://www.gristandtoll.com/marquis-wheat.

Lloyd's. *Emerging Risk Report—2015.* Innovation Series. https://assets.lloyds.com/assets/pdf-risk-reports-praedicat-final/1/pdf-risk-reports-Praedicat-FINAL.pdf.

Meridian Biotech. https://www.meridian-bio.com.

Morrison, Malcolm J. "Sir Charles Edward Saunders, Dominion Cerealist." *Genome,* May 20, 2008. https://doi.org/10.1139/G08-028.

Nitricity. "Renewable Nitrogen Fertilizer Pioneer Nitricity Raises $20 Million in Series A Funding." October 18, 2022. https://www.nitricity.co/renewable-nitrogen-fertilizer-pioneer.

Oliver, Thomas, Tom D. Kim, Joko P. Trinugroho, Violeta Cordón-Preciado, Nitara Wijayatilake, Aaryan Bhatia1, A. William Rutherford, and Tanai Cardona. "The Evolution and Evolvability of Photosystem II." *Annual Review of Plant Biology* 74 (May 2023). https://doi.org/10.1146/annurev-arplant-070522-062509.

ORCID. "Molly Jahn." https://orcid.org/0000-0003-0181-0374.

Reed, John S., and Lisa Dyson. "Use of Oxyhydrogen Microorganisms for Non-photosynthetic Carbon Capture and Conversion of Inorganic and/or C1 Carbon

Sources into Useful Organic Compounds." Google Patents. 2013. https://patents.google.com/patent/US20130149755A1/en?oq=US20130149755A1.

Ritchie, Hannah. "Can We Reduce Fertilizer Use without Sacrificing Food Production?" OurWorldInData.org. September 9, 2021. https://ourworldindata.org/reducing-fertilizer-use.

Rutherford, A. W., H. Raine, and A. Fantuzzi. "Reality Checks for Biofuels as Replacements for Fossil Fuels: Sustainable Avion Fuels as Case Study." International Conference on Clean Energy for Carbon Neutrality, 2023. https://www.cityu.edu.hk/hkice/iccecn2023/RUTHERFORD.html.

Schramski, John R., David K. Gattie, and James H. Brown. "Human Domination of the Biosphere: Rapid Discharge of the Earth-Space Battery Foretells the Future of Humankind." *Proceedings of the National Academy of Sciences* 112, no. 31 (August 4, 2015): 9511–9517. https://doi.org/10.1073/pnas.1508353112.

Sciubba, Enrico. "A Thermodynamic Measure of Sustainability." *Frontiers in Sustainability* 2 (December 17, 2021): article 739395. https://doi.org/10.3389/frsus.2021.739395.

Sleeman, Jennifer, David Chung, Anand Gnanadesikan, Jay Brett, et al. "A Generative Adversarial Network for Climate Tipping Point Discovery (TIP-GAN)." arXiv:2302.10274 v1 [cs.LG] (February 16, 2023).

The Sociable. "DARPA Looks to Make Microbial Food from Electricity, Air and Water to Ease Cargo Logistics." December 7, 2021. https://sociable.co/military-technology/darpa-microbial-food-electricity-air-water-cargo-logistics.

SRI International. https://www.sri.com.

Swift, Duncan, and Molly Jahn (advisor). *Evolving Risks in Global Food Supply.* London: Lloyd's and Moore Stephens, 2019. https://www.moore.co.uk/MediaLibsAndFiles/media/MooreStephensUK/Documents/Evolving-Risks-in-Global-Food-Supply-Report_2.pdf.

University of Illinois Urbana-Champaign. "Professor Ting Lu Jointly Presented with 1 Million Euro Future Insight Prize for Converting Waste into Food." July 13, 2021. https://bioengineering.illinois.edu/news/ting-lu-future-insight-prize-2021.

US National Science Foundation. "US and UK Scientists Collaborate to Design Crops of the Future." News release 12-147, August 21, 2013. https://www.nsf.gov/news/news_images.jsp?cntn_id=128878&org=NSF.

Venhuizen, Harm. "8 Weird DARPA Projects That Make Science Fiction Seem like Real Life." *Army Times*, September 4, 2020. https://www.armytimes.com/off-duty/military-culture/2020/09/04/8-weird-darpa-projects-make-science-fiction-seem-like-real-life.

Viola, Stefania, William Reoseby, Stefano Santabarbara, Dennis Nürnberg, Ricardo Assunção, Holger Dau, Julien Sellés, et al. "Impact of Energy Limitations on Function and Resilience in Long-Wavelength Photosystem II." *eLife*, July 19, 2022. https://doi.org/10.7554/eLife.79890.

Voices from DARPA. "Future of Food: Meals from Microbes." Podcast episode 56. https://youtu.be/9om-s4ISFMY?si=4aZ0as0VoICj3w9c.

Wackett, Lawrence P. "Microbial Meat Substitutes." *Microbial Biotechnology* 13, no. 4 (July 2020): 1284–1285. https://doi.org/10.1111/1751-7915.13610.

Zarook, Ruqaiyah. "Why the Pentagon Is the World's Biggest Single Greenhouse Gas Emitter." *Mother Jones*, October 7, 2022. https://www.motherjones.com/environment/2022/10/pentagon-climate-change-neta-crawford-book.

Zócalo. "Agricultural Scientist Molly Jahn." March 16, 2021. https://www.zocalopublicsquare.org/2021/03/16/agricultural-scientist-molly-jahn-darpa-delicata-squash/personalities/in-the-green-room.

CHAPTER FOUR: ARK BUILDERS

Books

Brown, David E., and Carlos A. López González. *Borderland Jaguars*. Salt Lake City: University of Utah Press, 2001.

Glenn, Warner. *Eyes of Fire: Encounter with a Borderlands Jaguar*. El Paso, TX: Printing Corner Press, 1996.

Su, Jean, and Maya Golden-Krasner. *The Climate President's Emergency Powers: A Legal Guide to Bold Climate Action from President Biden*. Tucson: Center For Biological Diversity, 2022.

Articles

Abbott, David. "Hudbay Ramps Up Excavation for Copper World Complex as Local Resistance Continues and Expands." *AZ Mirror*, January 6, 2023. https://azmirror.com/2023/01/06/hudbay-ramps-up-excavation-for-copper-world-complex-as-local-resistance-continues-and-expands.

Avila-Villegas, Sergio, and Jessica Lamberton-Moreno. "Wildlife Survey and Monitoring in the Sky Island Region with an Emphasis on Neotropical Felids." *USDA Forest Service Proceedings*, RMRS-P-67 (2013): 441–467.

Bale, Rachel. "FRONTLINE/CIR Exposed How Public Lands Are Still Ruled by 1872 Mining Law." *Reveal*, February 2, 2015. https://revealnews.org/article/frontline-exposed-how-public-lands-are-still-ruled-by-1872-mining-law.

Barchfield, Vanessa. "How the Center for Biological Diversity Grew in Its Fight against the 'Status Quo.'" AZPM, November 12, 2019. https://news.azpm.org/p/newsfeature/2019/11/12/161496-how-the-center-for-biological-diversity-grew-in-its-fight-against-the-status-quo.

Bar-On, Yinon M., Rob Phillips, and Ron Milo. "The Biomass Distribution on Earth." *Proceedings of the National Academy of Sciences* 15, no. 25 (June 19, 2018): 6506–6511. https://doi.org/10.1073/pnas.1711842115.

Berwin, Bob. "'Rewilding' Parts of the Planet Could Have Big Climate Benefits." *Inside Climate News*, March 27, 2023. https://insideclimatenews.org/news/27032023/rewilding-animals-carbon-storage.

Cagle, Alison. "Tribes Halt Major Copper Mine on Ancestral Lands in Arizona." Earthjustice. November 7, 2022. https://earthjustice.org/article/rosemont-mine-arizona-tribes-tohono-oodham.

Carol, Tim, Zeke Rowe, Joel Berger, Philippa Wholey, and Andrew Dobson. "An Inconvenient Misconception: Climate Change Is Not the Principal Driver of Biodiversity Loss." *Conservation Letters* 15, no. 3 (2022). https://doi.org/10.1111/conl.12868.

Ceballos, Gerardo, Paul R. Ehrlich, and Peter H. Raven. "Vertebrates on the Brink as Indicators of Biological Annihilation and the Sixth Mass Extinction." *Proceedings of the National Academy of Science* 117, no. 24 (March 22, 2020): 13596–13602. https://doi.org/10.1073/pnas.1922686117.

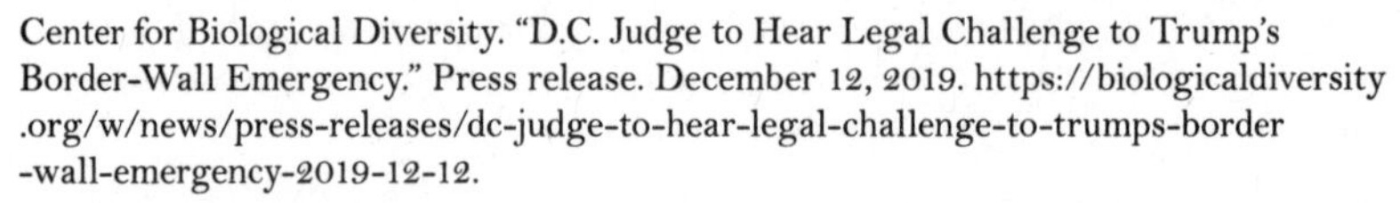

Center for Biological Diversity. "D.C. Judge to Hear Legal Challenge to Trump's Border-Wall Emergency." Press release. December 12, 2019. https://biologicaldiversity.org/w/news/press-releases/dc-judge-to-hear-legal-challenge-to-trumps-border-wall-emergency-2019-12-12.

———. "Judge Overturns Wildlife Agency's Approval of Rosemont Copper Mine in Arizona." Press release. February 10, 2020. https://biologicaldiversity.org/w/news/press-releases/judge-overturns-wildlife-agencys-approval-of-rosemont-copper-mine-in-arizona-2020-02-10/?_gl=1*1gxnxsg*_gcl_au*MTIxODE2NzA3LjE3MTU5NTQ5OTU.

———. "Lawsuit Launched to Protect Home of America's Only Known Jaguar from Copper Mine." Press release. June 29, 2016. https://www.biologicaldiversity.org/news/press_releases/2016/jaguar-06-29-2016.html.

———. "New Video Shows America's Only Known Wild Jaguar." Press release. February 3, 2016. https://www.biologicaldiversity.org/news/press_releases/2016/jaguar-02-03-2016.html.

———. "Saving the Coleman's Coralroot." https://www.biologicaldiversity.org/species/plants/Colemans_coralroot/index.html.

———. "Two Southern Arizona Plants Proposed for Endangered Species Act Protection." Press release. December 15, 2019. https://biologicaldiversity.org/w/news/press-releases/two-southern-arizona-plants-proposed-endangered-species-act-protection-2019-12-05.

Center for Biological Diversity; Defenders of Wildlife; and Animal Legal Defense Fund, Plaintiffs, v. Donald J. Trump et al. United States District Court for the District of Columbia. Case 1:19-cv-00408-TNM Document 31. Filed May 31, 2019.

Chung, Andrew. "U.S. Top Court Snubs Environmental Challenge to Trump's Border Wall." Reuters, December 3, 2018. https://www.reuters.com/article/idUSKBN1O21XL.

Climate One. "Kassie Siegel." https://www.climateone.org/people/kassie-siegel.

Corbett, Jessica. "'Nature Is Unraveling': New WWF Report Reveals 'Alarming' 68% Plummet in Wildlife Populations Worldwide since 1970." *Common Dreams*, September 10, 2020. https://www.commondreams.org/news/2020/09/10/nature-unraveling-new-wwf-report-reveals-alarming-68-plummet-wildlife-populations.

Curry, Tierra. "America's Rivers Need Help. I Should Know—I Swam in 108 of Them This Summer." *Washington Post*, September 28, 2021. https://www.washingtonpost.com/opinions/2021/09/28/americas-rivers-need-help-i-should-know-i-swam-108-them-this-summer.

Davis, Tony. "Arizona Jaguar 'Biologically Insignificant' Wildlife Manager Says." *Arizona Daily Star*, June 30, 2023. https://tucson.com/news/local/arizona-jaguar-biologically-insignificant-wildlife-manager-says/article_a39244d5-81a9-5b4c-9447-5b9d31b77884.html.

———. "Biologists Who Warned of Harm to Jaguar Were Overruled." *Arizona Daily Star*, August 16, 2014. https://tucson.com/news/science/environment/biologists-who-warned-of-harm-to-jaguar-were-overruled/article_ec21be6d-57ef-533f-b1ba-b3f35ffc3346.html.

———. "Federal Judge Again Rules against Rosemont Mine." *Arizona Daily Star*, February 11, 2020. https://tucson.com/news/local/federal-judge-again-rules-against-rosemont-mine/article_1fed0837-0af1-5fc2-a8c5-6a63fcb5179d.html.

———. "Forest Service Gives Crucial OK to Rosemont Mine." *Arizona Daily Star*, June 7, 2017. https://tucson.com/news/local/forest-service-gives-crucial-ok-to-rosemont-mine/article_5838ee16-4ba9-11e7-b45b-3b71ecf899f3.html.

———. "Jaguar Video Sparked Controversy That Rages Six Months Later." *Arizona Daily Star*, August 15, 2016. https://tucson.com/news/local/jaguar-video-sparked-controversy-that-rages-six-months-later/article_ec7d2a63-d75d-535c-9f69-a6f62afce178.html.

———. "Rosemont Mine Would Cause Water Quality Violations, Corps District Says." *Arizona Daily Star*, December 13, 2019. https://tucson.com/news/local/rosemont-mine-would-cause-water-quality-violations-corps-district-says/article_ca6fd7d9-8da0-56ca-8ac7-e4341d5ad09f.html.

———. "Unit of Federal Agency Recommends Denying Key Rosemont Mine Permit." *Arizona Daily Star*, July 28, 2016. https://tucson.com/news/science/environment/unit-of-federal-agency-recommends-denying-key-rosemont-mine-permit/article_2595e9a7-7898-5021-828e-00b4f7119c33.html.

Ehrlich, Paul R. "Ghastly Future: A Survival Revolution Response." Millennium Alliance for Humanity and the Biosphere, February 10, 2021. https://mahb.stanford.edu/blog/ghastly-future-a-survival-revolution-in-response.

Einhorn, Catrin. "A Quarter of Freshwater Fish Are at Risk of Extinction, New Assessment Finds." *New York Times*, December 11, 2023. https://www.nytimes.com/2023/12/11/climate/climate-change-threatened-species-red-list.html.

Environmental News Service. "Endangered U.S. Jaguars to Get Critical Habitat, Recovery Plan." January 12, 2010. http://www.ens-newswire.com/ens/jan2010/2010-01-12-01.html.

Erickson, Jim. "Study Suggests Ground-Dwelling Mammals Survived Mass Extinction 66 Million Years Ago." *Michigan News*, October 11, 2021. University of Michigan. https://news.umich.edu/study-suggests-ground-dwelling-mammals-survived-mass-extinction-66-million-years-ago.

Fears, Daryl. "The Endangered Species Act May Be Heading for the Threatened List. This Hearing Confirmed It." *Washington Post*, February 15, 2017. https://www.washingtonpost.com/news/energy-environment/wp/2017/02/15/the-endangered-species-act-may-be-heading-for-the-threatened-list-this-hearing-confirmed-it.

Fischer, Claire. "Waving Hello to the Wall: The Supreme Court's Denial of a Constitutional Challenge to Environmental Law Waivers at the U.S.-Mexico Border." *Georgetown Environmental Law Review*, February 15, 2019. https://www.law.georgetown.edu/environmental-law-review/blog/waiving-hello-to-the-wall-the-supreme-courts-denial-of-a-constitutional-challenge-to-environmental-law-waivers-at-the-u-s-mexico-border.

Forest History Society. "1991: Judge Dwyer Decisions." https://foresthistory.org/research-explore/us-forest-service-history/policy-and-law/wildlife-management/spotted-owl-timeline/1991-judge-dwyer-decisions.

Gaskill, Melissa. "The Environmental Impact of the U.S.-Mexico Border Wall." *Newsweek*, February 14, 2016. https://www.newsweek.com/environmental-impact-us-mexico-border-wall-426310.

Griffin, Andrea S., Alex Callen, Kaya Klop-Toker, Robert J. Scanlon, and Matt W. Hayward. "Compassionate Conservation Clashes with Conservation Biology: Should Empathy, Compassion, and Deontological Moral Principals Drive Conservation

Practice?" *Frontiers in Psychology* 11 (May 26, 2020). https://doi.org/10.3389/fpsyg.2020.01139.

Hoch, Heather. "Night of the Jaguar at Barrio Brewing Co. to Celebrate Everyone's Favorite Feline." *Tucson Weekly*, February 8, 2016. https://www.tucsonweekly.com/TheRange/archives/2016/02/08/night-of-the-jaguar-at-barrio-brewing-co-to-celebrate-everyones-favorite-feline.

Johnson, Warren E., Eduardo Eizirik, Jill Pecon-Slattery, William J. Murphy, et al. "The Late Miocene Radiation of Modern Felidae: A Genetic Assessment." *Science* 311, no. 5757 (January 6, 2006): 73–77. https://doi.org/10.1126/science.1122277.

Jones, Benji. "Why the US Won't Join the Single Most Important Treaty to Protect Nature." *Vox*, May 20, 2021. https://www.vox.com/22434172/us-cbd-treaty-biological-diversity-nature-conservation.

Jung, Yoohyun. "Tucson Students Name America's Only Known Wild Jaguar." Center for Biological Diversity, via the *Arizona Daily Star*, November 2, 2015. https://www.biologicaldiversity.org/news/center/articles/2015/arizona-daily-star-11-02-2015.html.

Kamp, Dick. "Presence of Jaguar Has Foes of Mining Worried." *Nogales International*, March 19, 2017. https://www.nogalesinternational.com/news/presence-of-jaguar-has-foes-of-mining-worried/article_23d48bca-9ccc-5007-8cbe-5b2b7e587c62.html.

Kimball, Renee. "Going for the Throat." *Albuquerque Journal*, September 7, 1993, p. 1.

Kurth, Torsten, Gerd Wübbels, Adrien Portafaix, Alexander Meyer zum Felde, and Sophie Zieckle. *The Biodiversity Crisis Is a Business Crisis*. Boston Consulting Group, March 2, 2021. https://web-assets.bcg.com/fb/5e/74af5531468e9c1d4dd5c9fc0bd7/bcg-the-biodiversity-crisis-is-a-business-crisis-mar-2021-rr.pdf.

Lallensack, Rachael. "North America Has Lost Nearly 3 Billion Birds since 1970." *Smithsonian*, September 19, 2019.

Main, Douglas. "Why a New Jaguar Sighting near the Arizona-Mexico Border Gives Experts Hope." *National Geographic*, March 23, 2021. https://www.nationalgeographic.com/animals/article/jaguar-near-arizona-border-wall-mexico.

MiningWatch Canada. "HudBay Minerals Environmental and Human Rights Track Record in Guatemala and Canada." July 2014. https://miningwatch.ca/sites/default/files/_mwc_brief_re_hudbay_july_10_2014_0.pdf.

Mirroff, Nick. "Trump's Border Wall, Vulnerable to Flash Floods, Needs Large Storm Gates Left Open for Months." *Washington Post*, January 30, 2020. https://www.washingtonpost.com/immigration/trumps-border-wall-vulnerable-to-flash-floods-needs-large-storm-gates-left-open-for-months/2020/01/30/be709346-3710-11ea-bb7b-265f4554af6d_story.html.

———. "Where Trump's Border Wall Rises, These Ranchers See Defeat." *Washington Post*, March 6, 2020. https://www.washingtonpost.com/immigration/2020/03/06/where-trumps-border-wall-rises-ranchers-see-scar-range.

Mulkern, Anne C., Allison Winter, and Robin Bravender. "Brazen Environmental Upstart Brings Legal Muscle, Nerve to Climate Debate." *New York Times*, March 30, 2010. https://archive.nytimes.com/www.nytimes.com/gwire/2010/03/30/30greenwire-brazen-environmental-upstart-brings-legal-musc-82242.html.

Nature Conservancy. "Why We're Committing to 30x30." https://www.nature.org/en-us/what-we-do/our-priorities/protect-water-and-land/land-and-water-stories/committing-to-30x30.

Northern Jaguar Project. https://www.northernjaguarproject.org.

Paddison, Laura. "Global Loss of Wildlife Is 'Significantly More Alarming' Than Previously Thought, According to New Study." CNN, May 22, 2023. https://www.cnn.com/2023/05/22/world/wildlife-crisis-biodiversity-scn-climate-intl/index.html.

Rees, William. "The Human Eco-predicament: Overshoot and the Population Conundrum." *Vienna Yearbook of Population Research* 21 (2023): 21–39. http://doi.org/10.1111/conl.12868.

Rosenberg, Kenneth, Adriaan M. Dokter, Peter J. Blancher, John R. Sauer, et al. "Decline of the North American Avifauna." *Science* 366, no. 6461 (October 2019): 120–124.

Rothberg, Daniel. "The Curious Case of a Rare Plant's Destruction Raises Further Questions about the Extinction Crisis, Climate Change and the Role of Humans." *Nevada Independent*, January 10, 2021. https://thenevadaindependent.com/article/the-curious-case-of-a-rare-plants-destruction-raises-further-questions-about-the-extinction-crisis-climate-change-and-the-role-of-humans.

Rozsa, Matthew. "Extremely Rare Wild Jaguar Spotted in Arizona." *Salon*, January 11, 2024. https://www.salon.com/2024/01/11/extremely-rare-wild-jaguar-spotted-in-arizona.

Saunders, Sarah P., Joanna Grand, Brooke L. Bateman, Mariah Meek, et al. "Integrating Climate-Change Refugia into 30 by 30 Conservation Planning in North America." *Frontiers in Ecology and the Environment*, January 2023: 1–8. https://doi.org/10.1002/fee.2592.

Shanahan, Mike. "Biologists Warn 'Extinction Denial' Is the Latest Anti-science Conspiracy Theory." *Mongabay*, September 14, 2020. https://news.mongabay.com/2020/09/biologists-warn-of-extinction-denial-as-latest-anti-science-conspiracy.

Shearer, Dan. "Rosemont Mine Project Dealt Major Setback." *Green Valley News*, May 14, 2022. https://www.gvnews.com/news/rosemont-mine-project-dealt-major-setback/article_618c25f6-d31b-11ec-a24f-0bfdf082bcda.html.

Sitlhou, Makepeace, Trilce Estrada Olvera, and Hakob Karapetyan. "The Fall of an Arizona Border Wall." *Yes!*, November 27, 2023. https://www.yesmagazine.org/video/wall-arizona-border-resistance.

Sky Island Alliance. https://skyislandalliance.org.

SlideShare.net. "Rosemont Copper—Mining Law of 1872." March 22, 2012. https://www.slideshare.net/Rosemont-Copper/rosemont-copper-mining-law-of-1872?from_search=0.

Talks at Google. "Polar Bears: Jenny Ross & Kassie Seigel." https://www.youtube.com/watch?v=QCRoWnpI110.

US Environmental Protection Agency. "Summary of the Endangered Species Act." https://www.epa.gov/laws-regulations/summary-endangered-species-act.

Wagner, David L., Eliza M. Gramesa, Matthew L. Forister, May R. Berenbaum, and David Stopak. "Insect Decline in the Anthropocene: Death by a Thousand Cuts." *Proceedings of the National Academy of Sciences* 118, no. 2 (2021): e2023989118. https://doi.org/10.1073/pnas.2023989118.

Watts, Jonathan. "Stop Biodiversity Loss or We Could Face Our Own Extinction." *The Guardian*, November 6, 2018. https://www.theguardian.com/environment/2018/nov/03/stop-biodiversity-loss-or-we-could-face-our-own-extinction-warns-un.

Weisman, Alan. "Cat Fight." *Pacific Standard*, August/September, 2017.

Whitcomb, Theo. "Copper Mine near Tucson Dealt a Blow." *High Country News*, June 9, 2022. https://www.hcn.org/articles/south-latest-copper-mine-near-tucson-dealt-a-blow.

Wiens, John J., and Joseph Zelinka. "How Many Species Will Earth Lose to Climate Change?" *Global Change Biology* 30, no. 1 (2024): e17125, 1–19. https://doi.org/10.1111/gcb.17125.

Wildlands Network. "Arizona Jaguar 'El Jefe' Reappears in Central Sonora." August 4, 2022. https://www.wildlandsnetwork.org/news/arizona-jaguar-central-sonora.

CHAPTER FIVE: STAR TIME

Books

Turrell, Arthur. *The Star Builders: Nuclear Fusion and the Race to Power the Planet*. New York: Scribner, 2021. Kindle edition.

Articles

Ball, Philip. "The Chase for Fusion Energy." *Nature*, November 17, 2021. https://www.nature.com/immersive/d41586-021-03401-w/index.html.

Baltz, E. A., E. Trask, M. Dkikovsky, H. Gota, R. Mendoza, J. C. Platt, and P. F. Riley. "Achievement of Sustained Net Plasma Heating in a Fusion Experiment with the Optometrist Algorithm." *Scientific Reports* 7, no. 1 (2017): 6425. https://scholar.google.com/citations?view_op=view_citation&hl=en&user=z0FP5kcAAAAJ&citation_for_view=z0FP5kcAAAAJ:b0M2c_1WBrUC.

BerkeleyNUC. "Superconducting Magnets and the Path to Fusion Energy: Brandon Sorbom." UC Berkeley Nuclear Engineering Weekly Colloquiums. November 5, 2021. https://www.youtube.com/watch?v=eMVdBQMb0PU.

Booksfact. "Age of Universe and Brahma as Calculated in Vedas." April 5, 2013. https://www.booksfact.com/vedas/age-of-universe-and-brahma-as-calculated-in-vedas.html.

Business Wire. "Helion Energy Achieves 100 Million Degrees Celsius Fusion Fuel Temperature and Confirms 16-Month Continuous Operation of Its Fusion Generator Prototype." June 22, 2021. https://www.businesswire.com/news/home/20210622005366/en/Helion-Energy-Achieves-100-Million-Degrees-Celsius-Fusion-Fuel-Temperature-and-Confirms-16-Month-Continuous-Operation-of-Its-Fusion-Generator-Prototype.

Chandler, David. "MIT and Newly Formed Company Launch Novel Approach to Fusion Power." *MIT News*, March 9, 2018. https://news.mit.edu/2018/mit-newly-formed-company-launch-novel-approach-fusion-power-0309.

———. "MIT-Designed Project Archives Major Advance Toward Fusion Energy." *MIT News*, September 8, 2021. https://news.mit.edu/2021/MIT-CFS-major-advance-toward-fusion-energy-0908.

Choi, Charles Q. "Nuclear Fusion Reactor Could Be Here as Soon as 2025." *Live Science*, October 1, 2020. https://www.livescience.com/nuclear-fusion-reactor-sparc-2025.html.

Commonwealth Fusion Systems. https://cfs.energy.

———. "Commonwealth Fusion Systems Creates Viable Path to Commercial Fusion Power with World's Strongest Magnet." September 8, 2021. https://cfs.energy/news-and-media/cfs-commercial-fusion-power-with-hts-magnet.

———. "Cost-Sharing Partnership with the Private Sector in Fusion Energy." May 15, 2020.

———. "New Scientific Papers Predict Historic Results for Commonwealth Fusion Systems Approach to Commercial Fusion Energy." September 29, 2020. https://cfs.energy/news-and-media/new-scientific-papers-predict-historic-results-for.

———. "The 20 T HTS Magnet in Context." September 8, 2021. https://www.youtube.com/watch?v=INErFI3uVRw.

Commonwealth Fusion Systems and MIT Plasma Science and Fusion Center. "A Record of Progress and Milestones of the HTS Magnet Demonstration." https://htsmagnet.cfs.energy.

Energy Impact Center. "Ep.123—Dennis Whyte, MIT." June 19, 2020. https://www.youtube.com/watch?v=m4JTpiqABp8.

Engine Ventures. https://engineventures.com/team/founder-advisors.

Foley, Jonathan. "Occam's Razor for the Planet." *Medium*, March 5, 2021. https://globalecoguy.org/occams-razor-for-the-planet-b3a720cc961c.

Fountain, Henry. "Can Fusion Solve the Climate Crisis?" *New York Times*, December 13, 2022. https://www.nytimes.com/2022/12/13/climate/fusion-climate-change.html.

Gainor, Danya, and Angela Dewan. "A Giant Donut-Shaped Machine Just Proved a Near-Limitless Clean Power Source Is Possible." CNN, February 10, 2022. https://www.cnn.com/2022/02/09/uk/nuclear-fusion-climate-energy-scn-intl/index.html.

Galchen, Rivka. "Can Nuclear Fusion Put the Brakes on Climate Change?" *New Yorker*, October 4, 2021. https://www.newyorker.com/magazine/2021/10/11/can-nuclear-fusion-put-the-brakes-on-climate-change.

Greenwald, Martin. "Status of the SPARC Physics Basis." *Journal of Plasma Physics* 86 (August 14, 2020): 1–3. https://doi.org/10.1017/S0022377820001063.

Hayman, C. A., P. J. Wegner, J. M. Auerbach, M. W. Bowers, S. N. Dixit, G. V. Erbert, G. M. Heestand, et al. "National Ignition Facility Laser Performance Status." *Optica Pro Publishing* 46, no. 16 (2007): 3276–3303. https://doi.org/10.1364/AO.46.003276.

Helman, Christopher. "Fueled by Billionaire Dollars, Nuclear Fusion Enters a New Age." *Forbes*, January 2, 2022. https://www.forbes.com/sites/christopherhelman/2022/01/02/fueled-by-billionaire-dollars-nuclear-fusion-enters-a-new-age/?sh=3313f53329f3.

Journal of Plasma Physics. "Status of the SPARC Physics Basis." https://www.cambridge.org/core/journals/journal-of-plasma-physics/collections/status-of-the-sparc-physics-basis.

Kramer, David. "New MIT Design Revives Interest in High-Field Approach to Fusion." *Physics Today* 68, no. 10 (2015): 23–25. https://doi.org/10.1063/PT.3.2941.

Lex Fridman Podcast. "Dennis Whyte: Nuclear Fusion and the Future of Energy." Episode 353. January 21, 2023. https://www.youtube.com/watch?v=aJoRMFWn2Jk.

Magee, R. M., K. Ogawa, T. Tajima, and I. Allfrey. "First Measurements of p11B Fusion in a Magnetically Confined Plasma." *Nature Communications* 14 (February 10, 2023): article 955. https://doi.org/10.1038/s41467-023-36655-1.

McDowell, Rachel. "Scientists Use Supercomputers to Study Reliable Fusion Reactor Design, Operation." *Newswise*, February 18, 2021, https://www.olcf.ornl.gov/2021/02/18/scientists-use-supercomputers-to-study-reliable-fusion-reactor-design-operation.

McKibben, Bill. "The Fusion Breakthrough Suggests That Maybe Someday We'll Have a Second Sun." *New Yorker*, December 12, 2022. https://www.newyorker.com/news/daily-comment/the-fusion-breakthrough-suggests-that-maybe-someday-well-have-a-second-sun.

MIT Corporate Relations. "Accelerating Fusion Energy's Development through Innovative Technology Transfer—Whyte and Sorbom." https://www.youtube.com/watch?v=HuE_yzZCrBY.

MIT Plasma Science and Fusion Center. "MIT Pathway to Fusion Energy (IAP 2017)—Zach Hartwig." https://www.youtube.com/watch?v=L0KuAx1COEk.

———. "MIT Students Contribute to Success of Historic Fusion Experiment." *MIT News*. September 6, 2022. https://news.mit.edu/2022/mit-students-contribute-success-historic-fusion-experiment-0906.

———. "New Record for Fusion." *MIT News*, October 14, 2016. https://news.mit.edu/2016/alcator-c-mod-tokamak-nuclear-fusion-world-record-1014.

MIT News. "3Q: Zach Hartwig on MIT's Big Push on Fusion." March 9, 2018. https://news.mit.edu/2018/3q-zach-hartwig-mit-big-push-fusion-0309.

Murphy, Meg. "High-Intensity Fusion." *MIT News*, October 14, 2016. https://news.mit.edu/2016/high-intensity-fusion-1014.

Nuclear News. "MIT Ramps 10-Ton Magnet Up to 20 Tesla in Proof of Concept for Commercial Fusion." Nuclear Newswire, September 10, 2021. https://www.ans.org/news/article-3240/mit-ramps-10ton-magnet-up-to-20-tesla-in-proof-of-concept-for-commercial-fusion.

O'Neill, Kathryn M. "3 Questions: Why Fusion Research Is Needed to Help the World Reduce Carbon Emissions." *MIT News*, June 16, 2017. https://newenergytimes.com/v2/sr/iter/mit/MIT-Fusion-NEWS-20170616-screenshot.pdf.

Osborne, Hannah. "China Is about to Fire Up Its 'Artificial Sun' in Quest for Fusion Energy." *Newsweek*, December 17, 2019. https://www.newsweek.com/china-about-fire-its-artificial-sun-quest-fusion-energy-1477705.

Plasma Science and Fusion Center, Massachusetts Institute of Technology. https://www.psfc.mit.edu/research.

———. "Dennis G. Whyte." https://www.psfc.mit.edu/whyte. Reuters. "Italy's Eni and CFS Speed Up Plans for Fusion Energy." March 9, 2023. https://www.reuters.com/business/energy/italys-eni-cfs-speed-up-plans-fusion-energy-2023-03-09.

Revkin, Andrew C. "Heck of a Class Project: An 'Affordable, Robust, Compact' Fusion Reactor Design, Buildable in a Decade." *New York Times*, August 11, 2015. https://archive.nytimes.com/dotearth.blogs.nytimes.com/2015/08/11/heck-of-a-class-project-an-affordable-robust-compact-fusion-reactor-design-buildable-in-a-decade.

SciLinks from NTSA. "The History of Superconductors." December 2019. http://www.superconductors.org/History.htm.

Sorbom, B. N., J. Ball, T. R. Palmer, F. J. Mangiarotti, et al. "ARC: A Compact, High-Field, Fusion Nuclear Science Facility and Demonstration Power Plant with Demountable Magnets." *Fusion Engineering and Design* 100 (November 2015): 378–405. https://www.sciencedirect.com/science/article/abs/pii/S0920379615302337.

St. John, Jeff. "The Problem with Making Green Hydrogen to Fuel Power Plants." *Canary Media*, October 10, 2023. https://www.canarymedia.com/articles/hydrogen/the-problem-with-making-green-hydrogen-to-fuel-power-plants.

Talk to a Geek. "Brandon Sorbom Explains MIT's ARC Fusion Reactor." Facebook video. February 16, 2018. https://www.facebook.com/TalkToGeek/videos/357491848069214.

Vaughn, Adam. "UK Nuclear Fusion Reactor Will Fire Up for the First Time in 23 Years." *New Scientist*, January 24, 2020. https://www.newscientist.com/article/2231341-uk-nuclear-fusion-reactor-will-fire-up-for-the-first-time-in-23-years.

Videmšek, Boštjan. "Bottling the Sun." CNN.com, May 30, 2022. https://edition.cnn.com/interactive/2022/05/world/iter-nuclear-fusion-climate-intl-cnnphotos.

Waldrop, M. Mitchell. "Can the Dream of Fusion Power Be Realized?" *Canary Media*, January 15, 2024. https://www.canarymedia.com/articles/nuclear/can-the-dream-of-fusion-power-be-realized.

Washington Post Editorial Board. "Fusion Power Is Tantalizing, but It Won't Save the Planet." *Washington Post*, December 14, 2022. https://www.washingtonpost.com/opinions/2022/12/14/fusion-power-climate-energy-renewables.

Wesoff, Eric. "Fusion Breakthrough Thrills Physicists, but Won't Power Your Home Soon." *Canary Media*, December 13, 2022. https://www.canarymedia.com/articles/nuclear/fusion-breakthrough-thrills-physicists-but-wont-power-your-home-soon.

Wolchover, Natalie. "Physicists Debate Hawking's Idea That the Universe Had No Beginning." *Quanta Magazine*, June 6, 2018. https://www.quantamagazine.org/physicists-debate-hawkings-idea-that-the-universe-had-no-beginning-20190606.

Zuber, Maria T. "A New Approach to Fusion Energy Starts Today." *Boston Globe*, March 9, 2018. https://www.bostonglobe.com/opinion/2018/03/09/new-approach-fusion-energy-starts-today/cc7kpF93xLaopO5xdobKIO/story.html.

CHAPTER SIX: ON THE REEF

Articles

Albeck-Ripka, Livia, and Emily Schmall. "A Giant Blob of Seaweed Is Heading to Florida." *New York Times*, March 16, 2023. https://www.nytimes.com/2023/03/14/us/seaweed-blob-florida-mexico.html.

Aleína, Fabio Cresto. "Climate Change and Smart Coasts in Mesoamerica." *EntreMundos*. https://www.entremundos.org/revista/environment/climate-change-and-smart-coasts-in-mesoamerica/?lang=en.

———. "A Guatemalan Jewel under Threat: Discovery and Exploration of the Corona Caimán Coral Reef." *EntreMundos*. https://www.entremundos.org/revista/environment/climate-change/a-guatemalan-jewel-under-threat-discovery-and-exploration-of-the-cayman-crown-coral-reef/?lang=en.

Alpízar, Francisco, Róger Madrigal, Irene Alvarado, Esteban Brenes Vega, Ashley Camhi, Jorge Higinio Maldonado, Jorge Marco, Alejandra Martínez, Eduardo Pacay, and Gregory Watson. *Mainstreaming of Natural Capital and Biodiversity into Planning and Decision-Making: Cases from Latin America and the Caribbean*. International Development Bank, September 2020. https://doi.org/10.18235/0002667.

Arkema, Katie. "As Hurricane Season Ends, Now Is the Time to Take Local Action to Rebuild and Recover." *Mongabay*, December 3, 2019. https://news.mongabay.com/2019/12/as-hurricane-season-ends-now-is-the-time-to-take-local-action-to-rebuild-and-recover-commentary.

———. "Consequences of Kelp Forest Structure and Dynamics for Epiphytes and Understory Communities." Ph.D. thesis, University of California, Santa Barbara, 2008. ProQuest 3330405. https://www.proquest.com/openview/d4366c9e46ab240b7206f7eeaac1ea4f/1?pq-origsite=gscholar&cbl=18750.

Arkema, Katie, Greg Guanne, Gregory Verutes, Spencer A. Wood, et al. "Coastal Habitats Shield People and Property from Sea-Level Rise and Storms." *Nature Climate Change* 3, no. 10 (October 2013): 913–918. https://doi.org/10.1038/nclimate1944.

Arkema, Katie, and Mary Ruckelshaus. "Transdisciplinary Research for Conservation and Sustainable Development Planning in the Caribbean." In *Conservation for the Anthropocene Ocean*, 333–357. Edited by Phillip S. Levin and Melissa R. Poe. Academic Press, 2017. https//doi.org/10.1016/B978-0-12-805375-1.00016-7.

Broad, William J., and Kenneth Chang. "Fossil Site Reveals Day That Meteor Hit Earth and, Maybe, Wiped Out Dinosaurs." *New York Times*, March 29, 2019. https://www.nytimes.com/2019/03/29/science/dinosaurs-extinction-asteroid.html.

Broom, Douglas. "Only 15% of the World's Coastlines Remain in Their Natural State." World Economic Forum. February 15, 2022. https://www.weforum.org/agenda/2022/02/ecologically-intact-coastlines-rare-study.

Butvill, David Brian. "Model Island." *BioGraphic*, May 16, 2017. https://www.biographic.com/model-island.

Carvalho, Mónica, Carlos Jaramillo, Felipe de la Parra, et al. "Extinction at the End-Cretaceous and the Origin of Modern Neotropical Rainforests." *Science* 372, no. 6537 (April 2, 2021): 63–68.

Choi, Charles Q. "Chicxulub Asteroid Impact: The Dino-Killer That Scientists Laughed At." Space.com, February 7, 2013. https://www.space.com/19681-dinosaur-killing-asteroid-chicxulub-crater.html.

Coastlines. "Waterlogged: From Oyster Beds near Boston to Kelp Forests of Santa Catalina Island, Research Scientist Katie Arkema Has Always Kept an Eye on the Oceans." U.C. Santa Barbara alumni magazine, Spring 2020. https://www.alumni.ucsb.edu/coastlines/spring-2020/water-logged.

Conrad, Hannah. "A Dune Deal: Using Nature-Inspired Designs to Protect Coastal Communities." Texas A&M Engineering, September 2, 2020. https://engineering.tamu.edu/news/2020/09/a-dune-deal-using-nature-inspired-designs-to-protect-coastal-communities.html.

De Mel, M., M. Phillips, R. Bartlett, M. A. Porta, et al. *Climate Risk Information for the Mesoamerican Reef Region*. New York: Center for Climate Systems Research at Columbia University, WWF-US, and WWF-Mesoamérica, 2022.

Depalma, Robert, Jan Smit, and David A. Burnham. "A Seismically Induced Onshore Surge Deposit at the KPg Boundary, North Dakota." *Proceedings of the National Academy of Sciences* 116, no. 17 (April 1, 2019): 8190–8199. https://doi.org/10.1073/pnas.1817407116.

Dickie, Gloria. "Marine Heatwaves Are Sweeping the Seafloor around North America." Reuters, March 17, 2023. https://www.reuters.com/business/environment/marine-heatwaves-are-sweeping-seafloor-around-north-america-2023-03-13.

During, Melanie A. D., Jan Smit, Dennis F. A. E. Voeten, Camille Berruyer, Paul Tafforeau, Sophie Sanchez, Koen H. W. Stein, et al. "The Mesozoic Terminated in Boreal Spring." *Nature* 603 (February 23, 2022): 91–94. https://www.nature.com/articles/s41586-022-04446-1.

Einhorn, Catrin, and Christopher Flavelle. "A Race against Time to Rescue a Reef from Climate Change." *New York Times*, December 5, 2020. https://www.nytimes.com/2020/12/05/climate/Mexico-reef-climate-change.html.

EurekaAlert. "Burned Organic Matter in Chicxulub Impact." September 28, 2020. https://www.eurekalert.org/news-releases/587898.

Gustin, Georgina. "Ravaged by Drought, a Honduran Village Faces a Choice: Pray for Rain or Migrate." *Inside Climate News*, July 8, 2019. https://insideclimatenews.org/news/08072019/climate-change-migration-honduras-drought-crop-failure-farming-deforestation-guatemala-trump.

International Analog Forestry Network. "Cuyamel-Omoa National Park." https://analogforestry.org/impacts-of-deforestation-of-wetlands-in-the-cuyamel-omoa-national-park.

International Climate Initiative. "Climate-Smarting Marine Protected Areas and Coastal Management in the Mesoamerican Reef Region." May 2024. https://www.international-climate-initiative.com/en/project/climate-smarting-marine-protected-areas-and-coastal-management-in-the-mesoamerican-reef-region-18-ii-152-lac-a-mesoamerican-reef.

International Coral Reef Initiative. "Don't Miss: Mesoamerican Reef 2022 Health Report—Poor Reef Health and Fish Decline." February 14, 2023. https://icriforum.org/mesoamerican-reef-health-report-2022.

———. "Mesoamerican Reef Health Report by the Healthy Reefs Initiative (HRI) Shows Decline for 2021." June 9, 2022. https://icriforum.org/event-types/other-relevant-events.

IUCN. "Omoa Conservation Corps Joins IUCN to Implement Regional Coastal Biodiversity Project." May 7, 2019. https://www.iucn.org/es/news/mexico-america-central-y-el-caribe/201905/cuerpos-de-conservacion-omoa-se-une-a-la-uicn-para-implementar-el-proyecto-regional-de-biodiversidad-costera.

Jacobs, Bonnie, and Ellen D. Currano. "The Impactful Origin of Neotropical Rainforests." *Science* 372, no. 6537 (April 2, 2021): https://doi.org/10.1126/science.abh2086.

Jones, Lewis A., Philip D. Mannion, Alexander Farnsworth, Paul J. Valdes, Sarah-Jane Kelland, and Peter A. Alison. "Coupling of Paleontological and Neontological Reef Coral Data Improves Forecasts of Biodiversity Responses under Global Climatic Change." *Royal Society Open Science*, April 1, 2019. https://doi.org/10.1098/rsos.182111.

Kitreoff, Natalie, and Daniele Volpe. "'We Are Doomed': Devastation from Storm Fuels Migration in Honduras." *New York Times*, November 28, 2021. https://www.nytimes.com/2021/04/06/world/americas/migration-honduras-central-america.html.

Knox, Pam. "2 Hurricanes Devastated Central America. Will the Ruin Spur a Migration Wave?" *New York Times*, December 4, 2020. https://www.nytimes.com/2020/12/04/world/americas/guatemala-hurricanes-mudslide-migration.html.

Koch, Wendy. "Dunes, Reefs Protect U.S. Coasts from Climate Change." *USA Today*, July 14, 2013. https://www.usatoday.com/story/news/nation/2013/07/14/dunes-reefs-protect-us-coastlines-from-climate-change-storms/2513299.

Kulp, Scott L., and Benjamin H. Strauss. "New Elevation Data Triple Estimates of Global Vulnerability to Sea-Level Rise and Coastal Flooding." *Nature Communications* 10 (2009): article 4844. https://doi.org/10.1038/s41467-019-12808-z.

Lakhani, Nina. "'It Won't Be Long': Why a Honduran Community Will Soon Be Under Water." *The Guardian*, July 31, 2019. https://www.theguardian.com/global-development/2019/jul/31/honduras-community-coastal-towns-rising-sea-le.

Lustgarten, Abrahm. "Palm Oil Was Supposed to Help Save the Planet. Instead It Unleashed a Catastrophe." *New York Times*, November 20, 2018. https://www.nytimes.com/2018/11/20/magazine/palm-oil-borneo-climate-catastrophe.html.

Marine Biological Laboratory. "Wetlands Will Keep Up with Sea Level Rise to Offset Climate Change." *Newswise*, December 20, 2019. https://www.newswise.com/articles/wetlands-will-keep-up-with-sea-level-rise-to-offset-climate-change.

McField, Melanie, Nadia Bood, Ana Fonseca, Alejandro Arrivillaga, Albert Franquesa Rinos, and Rosa María Loreto Viruel. "Status of the Mesoamerican Reef after the 2005 Coral Bleaching Event." Chapter 5 in *The Status of Caribbean Coral Reefs after Bleaching and Hurricanes in 2005*. Edited by Clive Wilkinson and David Souter. Global Coral Reef Monitoring Network and Reef and Rainforest Research Centre. https://www.coris.noaa.gov/activities/caribbean_rpt/SCRBH2005_05.pdf.

NASA Earth Observatory. "Relief Map, Yucatan Peninsula, Mexico." February 7, 2003. https://earthobservatory.nasa.gov/images/3267/relief-map-yucatan-peninsula-mexico.

National Centers for Environmental Information. "Global Ocean Absorbing More Carbon." March 15, 2019. https://www.ncei.noaa.gov/news/global-ocean-absorbing-more-carbon.

National Oceanic and Atmospheric Administration. "Global Atmospheric Carbon Dioxide Levels Continue to Rise." November 15, 2022. https://research.noaa.gov/2022/11/15/no-sign-of-significant-decrease-in-global-co2-emissions.

Natural Capital Project, Stanford University. https://naturalcapitalproject.stanford.edu.

———. "Climate-Smart Coastal Planning and Sustainable Development in Latin America and the Caribbean." Stanford University Earth Day Event. https://naturalcapitalproject.stanford.edu/events/natcap-conversation/climate-smart-coastal-planning-and-sustainable-development-latin-america.

———. "Ensuring Coastal Resilience for the Bahamas." September 24, 2019. https://naturalcapitalproject.stanford.edu/news/ensuring-coastal-resilience-bahamas.

Nature Conservancy. "Nature-Based Coastal Defenses in Southeast Florida." https://www.nature.org/media/florida/natural-defenses-in-southeast-florida.pdf.

Palencia, Gustavo. "Health of Vast Mesoamerican Reef Declines after Years of Improvement: Study Finds." Reuters, February 13, 2020. https://www.reuters.com/article/idUSKBN208056.

Palomo, Areli. "Climate Change Haunts a Ghostly Border in Honduras." North American Congress on Latin America, November 25, 2020. https://nacla.org/news/2020/11/25/climate-change-haunts-ghostly-border-honduras.

Papadovassilakis, Alex. "Cocaine Spike Puts Spotlight on Honduras Atlantic." *Insight Crime*, April 16, 2021. https://insightcrime.org/news/cocaine-spike-puts-spotlight-on-honduras-atlantic.

Phillips, Meridel Murphy, Manishka De Mel, Anastasia Romanou, David Rind, Alex C. Ruane, and Cynthia Rosenzweig. "Catastrophic Bleaching Risks to Mesoamerican Coral Reefs in Recent Climate Change Projections." *ESS Open Archive*, June 8, 2022. https://doi.org/10.1002/essoar.10511564.1.

Preston, Douglas. "The Day the Dinosaurs Died." *New Yorker*, March 29, 2019. https://www.newyorker.com/magazine/2019/04/08/the-day-the-dinosaurs-died.

Sack, Melinda. "The Environment and the Bottom Line." *Medium*, March 5, 2018. https://medium.com/stanford-magazine/the-value-of-conservation-stanford-biology-professor-gretchen-daily-and-the-natural-capital-project-d9d0ceef8e95.

Schoepf, Verena, Michael Stat, James L. Falter, and Malcolm T. McColloch. "Limits to the Thermal Tolerance of Corals Adapted to a Highly Fluctuating, Naturally Extreme Temperature Environment." *Scientific Reports* 5 (2015): article 17639. https://doi.org/10.1038/srep17639.

Semple, Kirk. "Climate Change and Political Chaos: A Deadly Mix in Honduras Dengue Epidemic." *New York Times*, December 29, 2019.

Sheridan, Mary Beth, and Alejandro Cegarra. "The Seaweed Invasion." *Washington Post*, August 15, 2019. https://www.washingtonpost.com/graphics/2019/world/amp-stories/seaweed-invasion.

Smart Coasts / Costa Listas Project. "Climate-Smarting Marine Protected Areas and Coastal Management in the Mesoamerican Reef Region." World Wildlife Foundation. https://iki-alliance.mx/wp-content/uploads/Factsheet-Smart-Coasts-ingles.pdf.

Solie, Stacey. "Belize Coastal Plan a Model for the World." Natural Capital Project, April 20, 2017. https://naturalcapitalproject.stanford.edu/news/belize-coastal-plan-model-world.

Speelman, E. N., M. M. L. Van Kempen, J. Barke, H. Brinkhuis, G. J. Reichart, A. J. P. Smolders, J. G. M. Roelofs, et al. "The Eocene Arctic *Azolla* Bloom: Environmental Conditions, Productivity and Carbon Drawdown." *Geobiology* 7, no. 2 (March 2009). https://doi.org/10.1111/j.1472-4669.2009.00195.x

Spring, Jake. "Explainer: Scientists Come Closer to Solving Caribbean Seaweed Mystery." Reuters, September 29, 2021. https://www.reuters.com/world/americas/scientists-come-closer-solving-caribbean-seaweed-mystery-2021-09-29.

Sustainable Travel International. "Fighting Coral Disease on the Mesoamerican Reef." October 1, 2019. https://sustainabletravel.org/fighting-coral-disease-mesoamerican-reef.

Urrutia-Fucugauchi, Jaime, and Ligia Pérez-Cruz. "Post-Impact Carbonate Deposition in the Chicxulub Impact Crater Region, Yucatan Platform, Mexico." *Current Science* 95, no. 2 (July 25, 2008): 248–252.

World Wildlife Foundation. "8 Things to Know about Palm Oil." https://www.wwf.org.uk/updates/8-things-know-about-palm-oil.

———. "Living Planet Report 2022." https://livingplanet.panda.org/en-US.

———. "Mesoamerica." https://www.wwfca.org.

———. "Mesoamerican Reef." https://www.worldwildlife.org/places/mesoamerican-reef.

———. "Smart Coasts." https://www.wwfca.org/en/smartcoastsmar.

CHAPTER SEVEN: GOING DUTCH

Books

Cox, Roger. *Revolution Justified.* Translated by Elizabeth H. D. Manton. Maastricht: Planet Prosperity Foundation, 2011.

Kennedy, James C. *A Concise History of the Netherlands.* Cambridge: Cambridge University Press, 2017. Kindle edition.

Ovink, Henk, and Jelte Boeijenga. *Too Big: Rebuild by Design; A Transformative Approach to Climate Change.* Rotterdam: NAI010 Publishers, 2018.

Rosenzweig, Cynthia, et al., eds. *Climate Change and Cities: Second Assessment Report of the Urban Climate Change Research Network.* Cambridge: Cambridge University Press, 2018.

Van den Bosch, Rem. *Zeeuws Meisje.* Borsele, Zeeland: Zeus Art Projects, 2022.

Vos, Peter, Michiel van der Meulen, Henk Weerts, and Jos Bazelmans. *Atlas of the Holocene: Netherlands.* Translated by Annette Visser. Amsterdam University Press, 2020.

Articles

Beley, Celeste. "Why Are Dutch People So Tall?" *Fisher Science Education.* https://www.fishersci.com/us/en/education-products/publications/headline-discoveries/2015/issue-3/why-are-dutch-people-so-tall.html.

Berg, Nate. "How a Design Competition Changed the US Approach to Disaster Response." *The Guardian,* January 18, 2017.

Berwyn, Bob. "New Federal Report Warns of Accelerating Impacts from Sea Level Rise." *Inside Climate News,* February 16, 2022. https://insideclimatenews.org/news/16022022/sea-level-rise-noaa-report.

Bioneers. "Henk Ovink: Water by Design." October 26, 2017. https://bioneers.org/henk-ovink-water-by-design-bioneers.

Boztas, Senay. "'Every Tree Counts': Dutch Come Up with Cunning Way to Create Forests for Free." *The Guardian,* November 26, 2021. https://www.theguardian.com/environment/2021/nov/26/every-tree-counts-amsterdam-forest-leads-way-with-sapling-donation-plan.

Carrington, Damian. "Shell Directors Personally Sued over 'Flawed' Climate Strategy." *The Guardian,* February 9, 2023. https://www.theguardian.com/environment/2023/feb/09/shell-directors-personally-sued-over-flawed-climate-strategy.

Changerism. "Dirty Pearls: Exposing Shell's Hidden Legacy of Climate Change Accountability, 1970–1990." 2020. https://changerism.com/portfolio/dirty-pearls-exposing-shells-hidden-legacy-of-climate-change-accountability-1970-1990.

City of New York. *PlaNYC: A Stronger, More Resilient New York.* City of New York, 2013. http://s-media.nyc.gov/agencies/sirr/SIRR_singles_Lo_res.pdf.

Cleary Gottlieb. "Dutch Court Orders Shell to Reduce Emissions in First Climate Change Ruling against Company." June 30, 2021. https://www.clearygottlieb.com/news-and-insights/publication-listing/dutch-court-orders-shell-to-reduce-emissions-in-first-climate-change-ruling-against-company.

Climate-ADAPT. "Sand Motor—Building with Nature Solution to Improve Coastal Protection along Delfland Coast (the Netherlands)." European Commission and the European Environment Agency. https://climate-adapt.eea.europa.eu/en/metadata/case-studies/sand-motor-2013-building-with-nature-solution-to-improve-coastal-protection-along-delfland-coast-the-netherlands.

Corbett, Jessica. "After Shell CEO Claims 'We Have No Choice' but to Invest in Fossil Fuels, McKibben Says, 'We Have No Choice But to Try and Stop Them.'" *Common Dreams,* October 15, 2019. https://www.commondreams.org/news/2019/10/15/after-shell-ceo-claims-we-have-no-choice-invest-fossil-fuels-mckibben-says-we-have.

Corder, Mike. "Water Wizards: Dutch Flood Expertise Is Big Export Business." Associated Press, November 12, 2017. https://apnews.com/article/1bc1f137cb134fdca67afdff08c242bc.

Deltares. "About Jaap Kwadijk." https://www.deltares.nl/en/expertise/our-people/jaap-kwadijk.

Dietze, Michael, and Ugur Ozturk. "A Flood of Disaster Response Challenges." *Science* 373, no. 6561 (September 16, 2021): 1317–1318. https://doi.org/10.1126/science.abm061.

Doxey, John. "Maeslantkering." *As They Said*, May 2018. Blog. https://astheysaid.com/innovators/2018/6/1/maeslantkering.

Dutch News. "As Sea Levels Rise, How Long until the Netherlands Is Under Water?" December 23, 2019. https://www.dutchnews.nl/2019/12/as-sea-levels-rise-how-long-until-the-netherlands-is-under-water.

Dutch Water Sector. "Export Growth Dutch Water Sector Back on Track." January 17, 2019. https://www.dutchwatersector.com/news/export-growth-dutch-water-sector-back-on-track.

Erdbrink, Thomas. "To Avoid River Flooding, Go with the Flow, the Dutch Say." *New York Times*, September 7, 2021. https://www.nytimes.com/2021/09/07/world/europe/dutch-rivers-flood-control.html.

Erickson, Peter, et al. "Expert Letter: The Likely Effect of Shell's Reduction Obligation on Oil and Gas Markets and Greenhouse Gas Emissions." Requested by Milieudefensie on the effect of the reduction obligation imposed on Shell by the District Court, 2023. https://en.milieudefensie.nl/news/first-expert-statement.

Euronews. "'Deep Concern': Shell Employees Urge CEO to Rethink Shift from Renewables in Rare Letter." September 28, 2023. https://www.euronews.com/green/2023/09/28/deep-concern-shell-employees-urge-ceo-to-rethink-shift-from-renewables-in-rare-letter.

———. "The Dutch Masters of Flood Safety." July 28, 2017. https://www.euronews.com/2017/07/28/the-dutch-masters-of-flood-safety.

Furioso, Dante. "Environmental Justice?" *Archinect*, July 13, 2021. https://archinect.com/features/article/150270301/trashing-the-community-backed-big-u-east-side-coastal-resilience-moves-forward-despite-local-opposition-will-nyc-miss-another-opportunity-to-lead-on-climate-and-environmental-justice.

———. "Trashing the Community-Backed BIG U: East Side Coastal Resilience Moves Forward Despite Local Opposition. Will NYC Miss Another Opportunity to Lead on Climate and Environmental Justice?" *Archinet*, July 13, 2021. https://archinect.com/features/article/150270301/trashing-the-community-backed-big-u-east-side-coastal-resilience-moves-forward-despite-local-opposition-will-nyc-miss-another-opportunity-to-lead-on-climate-and-environmental-justice.

Giambusso, David. "Design Contest Leaders Seek Disaster-Minded Urban Planners." *Politico*, February 22, 2016. https://www.politico.com/states/new-york/city-hall/story/2016/02/design-contest-leaders-seek-disaster-minded-urban-planners-031475.

Gibbens, Sarah. "Experts Fear Germany's Deadly Floods Are a Glimpse into Climate Future." *National Geographic*, July 16, 2021. https://www.nationalgeographic.com/environment/article/experts-fear-germany-deadly-floods-glimpse-into-climate-future.

Goldman Environmental Prize. "2022 Goldman Prize Winner: Marjan Minnesma." https://www.goldmanprize.org/recipient/marjan-minnesma.

Gornitz, Vivien, and Cynthia Rosenzweig. "Severe Storms and Sea Level Rise in New York City." *Water Resources Impact* 11, no. 1 (January 2009): 10–14.

Gouda, H. B. W. "Ramspol Inflatable Surge Barrier, Kampen." ZJA Architectural Studio Design and Engineering. https://www.zja.nl/en/Ramspol-inflatable-dam-Kampen.

Green Interview. "Marjan Minnesma: The Dutch Climate Case; Inspiring the World." November 2015. https://thegreeninterview.com/interview/minnesma-marjan.

Gregg, Aaron. "Shell to Move Headquarters to U.K., Revamp Share Structure and Drop 'Royal Dutch.'" *Washington Post*, November 15, 2021. https://www.washingtonpost.com/business/2021/11/15/royal-dutch-shell-netherlands-uk.

Groeskamp, Sjoerd, and Joakim Kjellsson. "NEED: The Northern European Enclosure Dam for If Climate Change Mitigation Fails." *Bulletin of the American Meteorological Society* 101, no. 7 (July 2020). https://doi.org/10.1175/BAMS-D-19-0145.1.

Gustin, Georgina. "Complex Models Now Gauge the Impact of Climate Change on Global Food Production. The Results Are 'Alarming.'" *Inside Climate News*, March 27, 2022.

Holland—Land of Water. "St. Felix's Flood of 1530." 2024. https://www.hollandlandofwater.com/sint-felix-vloed-van-het-jaar-1530.

Holmes, Damian. "Mound Plan | Overdiepse Polder The Netherlands | Bosch Slabbers Landscape + Urban Design." *World Landscape Architecture*, April 8, 2014. https://worldlandscapearchitect.com/mound-plan-overdiepse-polder-the-netherlands-bosch-slabbers-landscape-urban-design/?v=3a1ed7090bfa.

H+N+S Landscape Architects. "Room for the River IJssel Delta." https://www.hnsland.nl/en/projects/room-river-ijssel-delta.

Jacobs, Karrie. "The Top 10 Post-Sandy Ideas from Rebuild by Design." *Architect Magazine*, April 10, 2014. https://www.hud.gov/sandyrebuilding/rebuildbydesign.

Kimmelman, Michael. "The Dutch Have Solutions to Rising Seas. The World Is Watching." *New York Times*, June 15, 2017. https://www.nytimes.com/interactive/2017/06/15/world/europe/climate-change-rotterdam.html.

———. "What Does It Mean to Save a Neighborhood?" *New York Times*, June 15, 2023. https://www.nytimes.com/2021/12/02/us/hurricane-sandy-lower-manhattan-nyc.html.

Klijn, Frans, Nathalie Asselman, and Dennis Wagenaar. "Room for Rivers: Risk Reduction by Enhancing the Flood Conveyance Capacity of the Netherlands' Large Rivers." *Geosciences* 8, 224 (June 20, 2018). https://doi.org/10.3390/geosciences8060224.

Klinenberg, Eric. "The Seas Are Rising. Could Oysters Help?" *New Yorker*, August 2, 2021. https://www.newyorker.com/magazine/2021/08/09/the-seas-are-rising-could-oysters-protect-us.

Krajcik, Kevin. "She Led Scientists Advising New York on Climate Change. Did the City Listen?" Columbia Climate School, October 31, 2022. https://news.climate.columbia.edu/2022/10/29/she-led-scientists-advising-new-york-on-climate-change-did-the-city-listen.

Kuper, Simon. "Can the Dutch Save the World from the Danger of Rising Sea Levels?" *Financial Times Magazine*, January 20, 2020. https://www.ft.com/content/44c2d2ee-422c-11ea-bdb5-169ba7be433d.

Leaders. "Water as a Driver of Change: An Interview with Henk Ovink, Special Envoy for International Water Affairs for the Kingdom of the Netherlands, and Sherpa to the United Nations/World Bank High Level Panel on Water." https://www.leadersmag.com/issues/2020.4_Oct/Resilience/LEADERS-Henk-Ovink_Netherland-Water.html.

Limarev, Nina. "In Face of Rising Sea Levels the Netherlands 'Must Consider Controlled Withdrawal.'" *Vrij Nederland*, February 7, 2019. https://www.vn.nl/rising-sea-levels-netherlands.

Marvel, Kate. "We Need Courage, Not Hope, to Face Climate Change." *On Being*, March 1, 2018. https://onbeing.org/blog/kate-marvel-we-need-courage-not-hope-to-face-climate-change.

Meredith, Sam. "What the Dutch Can Teach the World about Flood Preparedness." CNBC, July 30, 2021. https://www.cnbc.com/2021/07/30/europe-floods-what-the-dutch-can-teach-the-world-about-preparedness.html.

Milieudefensie. "We Are Suing ING in a Ground Breaking New Climate Case." https://en.milieudefensie.nl/climate-case-ing/our-climate-case-against-ing.

Minnesma, Marjan. "Not Slashing Emissions? See You in Court." *Nature*, December 20, 2019. https://www.nature.com/articles/d41586-019-03841-5.

Muggah, Robert. "The World's Coastal Cities Are Going Under. Here's How Some Are Fighting Back." World Economic Forum. January 16, 2019. https://www.weforum.org/agenda/2019/01/the-world-s-coastal-cities-are-going-under-here-is-how-some-are-fighting-back.

National Delta Programme. https://english.deltaprogramma.nl.

Nazaruk, Zuza. "Parts of Rotterdam Lie 7 Metres below Sea Level. Now It's a Global Leader in How to Stay Afloat." *Euronews*. December 16, 2023. https://www.euronews.com/green/2023/12/16/parts-of-rotterdam-lie-7-metres-below-sea-level-now-its-a-global-leader-in-how-to-stay-afl.

Newswise. "Earliest Known Coastal Seawall Uncovered at Neolithic Settlement Tel Hreiz." December 11, 2019. https://www.newswise.com/articles/earliest-known-coastal-seawall-uncovered-at-neolithic-settlement-tel-hreiz.

New York City Mayor's Office of Climate and Environmental Justice. *Climate Resiliency Design Guidelines*. Version 4.1, May 2022. https://www.nyc.gov/assets/sustainability/downloads/pdf/publications/CRDG-4-1-May-2022.pdf.

———. "Coastal Surge Flooding." https://climate.cityofnewyork.us/challenges/coastal-surge-flooding.

NL Times. "Code Orange: Netherlands to Shut Storm Surge Barriers with High Water, 120 km/h Winds Expected." January 20, 2022. https://nltimes.nl/2022/01/30/code-orange-netherlands-shut-storm-surge-barriers-high-water120-kmh-winds-expected.

O'Leary, Naomi. "When Will the Netherlands Disappear?" *Politico*, December 16, 2019. https://www.politico.eu/article/when-will-the-netherlands-disappear-climate-change.

Pols, Donald. "This Is How We Took On One of the World's Biggest Polluters and Won." *The Independent*, June 2, 2021. https://www.independent.co.uk/climate-change/opinion/shell-climate-change-victory-pollution-b1855754.html#.

Port of Amsterdam. "Gasoline." https://www.portofamsterdam.com/en/business/cargo-flows/liquid-bulk/gasoline.

Rebuild by Design. rebuildbydesign.org.

Reiley, Laura. "Cutting-Edge Tech Made This Tiny Country a Major Exporter of Food." *Washington Post*, November 21, 2022. https://www.washingtonpost.com/business/interactive/2022/netherlands-agriculture-technology.

Renaldi, Adi. "Indonesia's Giant Capital City Is Sinking. Can the Government's Plan Save It?" *National Geographic*, July 29, 2022. https://www.nationalgeographic.com/environment/article/indonesias-giant-capital-city-is-sinking-can-the-governments-plan-save-it.

Reuters. "Dutch Government 'Unpleasantly Surprised' by Shell HQ Move to Britain." November 15, 2021. https://www.reuters.com/article/idUSKBN2I0OTB.

———. "Thousands of Dutch Urged to Leave Their Homes as Rivers Flood." July 15, 2021. https://www.reuters.com/world/europe/thousands-dutch-urged-leave-their-homes-rivers-flood-2021-07-15.

Revolution Justified. "The Author—Roger Cox." Book website. Accessed May 7, 2024. https://www.revolutionjustified.org/roger-cox-author-of-revolution-justified.

Rijkswaterstaat Ministry of Infrastructure and Water Management. "Eastern Scheldt Barrier." https://www.rijkswaterstaat.nl/en/projects/iconic-structures/eastern-scheldt-barrier.

———. "Room for the River." https://www.rijkswaterstaat.nl/en/projects/iconic-structures/room-for-the-river.

SCAPE Landscape Architecture. "Living Breakwaters." https://www.scapestudio.com/projects/living-breakwaters.

Shorto, Russell. "How to Think like the Dutch in a Post-Sandy World." *New York Times*, April 9, 2014. https://www.nytimes.com/2014/04/13/magazine/how-to-think-like-the-dutch-in-a-post-sandy-world.html.

Sverige Radio. "The Environmental Organization Urgenda Won a High-Profile Climate Case against the Dutch State." March 10, 2023. https://sverigesradio.se/artikel/miljo organisationen-urgenda-vann-uppmarksammat-klimatmal-mot-den-nederlandska-staten.

United States Department of Housing and Development. "Hurricane Sandy Rebuilding Task Force / Rebuild by Design." https://www.hud.gov/sandyrebuilding/rebuildbydesign.

Urgenda. https://www.urgenda.nl/en/themas/climate-case.

———. "54 Actions for 17 Tons of CO_2 Reduction." https://www.urgenda.nl/en/themas/climate-case/dutch-implementation-plan.

———. "A Moment of Hope: Urgenda Wins Historic Climate Case in Supreme Court of the Netherlands." December 20, 2019. https://www.urgenda.nl/en/home-en.

World Food Prize Foundation. "2022 World Food Prize Awarded to NASA Climate Scientist." May 5, 2022. https://www.worldfoodprize.org/index.cfm/87428/48752/2022_world_food_prize_awarded_to_nasa_climate_scientist.

World Water Atlas. "Dutch Water Legacy." https://www.worldwateratlas.org/en/events/shenzhen-design-week/dutch-water-legacy.

Yeo, Sophie. "Netherlands to Upgrade Flood Defences to Cope with Climate Change." *Climate Home News*, March 4, 2014. https://www.climatechangenews.com/2014/03/04/netherlands-to-upgrade-flood-defences-to-cope-with-climate-change.

Younger, Sally. "NASA Study: Rising Sea Level Could Exceed Estimates for U.S. Coasts." NASA Global Climate Change. November 15, 2022. https://climate.nasa.gov/news/3232/nasa-study-rising-sea-level-could-exceed-estimates-for-us-coasts.

CHAPTER EIGHT: A LINE RUNS THROUGH IT

Books

Benton-Banai, Edward. *The Mishomis Book: The Voice of the Ojibway*. Minneapolis: University of Minnesota Press, 2010.

Johnston, Basil. *Ojibway Heritage.* Bison Books. Kindle edition, pp. 29–30.

LaDuke, Winona. "Protecting the Culture and Genetics of Wild Rice." In *Original Instructions: Indigenous Teachings for a Sustainable Future,* p. 208. Edited by Melissa K. Nelson. Rochester, VT: Inner Traditions–Bear & Company, 2008. Kindle edition.

———. *To Be a Water Protector: The Rise of the Wiindigoo Slayers.* Halifax: Fernwood Publishing, 2020.

Lindstrom, Carole. *We Are Water Protectors.* New York: Roaring Book Press, 2020.

McKibben, Bill. *Falter.* New York: Henry Holt, 2019.

Thompson, Clifford, ed. "LaDuke, Winona." From *Current Biography Yearbook 2003.* New York: H. W. Wilson, 2003. http://web.sbu.edu/friedsam/laduke/winona_laduke.pdf.

Articles

Amnesty International. "Niger Delta Negligence." March 16, 2018. https://www.amnesty.org/en/latest/news/2018/03/niger-delta-oil-spills-decoders.

Averett, Nancy. "For Decades, the Ojibwe Tribe Shunned Scientists—until Their Partnership Became Vital." *The Nation,* February 15, 2023.

Beattie, Samantha. "Line 5 Pipeline between U.S. and Canada Could Cause 'Devastating Damage' to Great Lakes, Say Environmentalists." CBC News, August 3, 2021. https://www.cbc.ca/news/canada/toronto/line-five-environment-great-lakes-1.6120882.

Beaumont, Hilary. "Why Indigenous Women Are Risking Arrest to Fight Enbridge's Line 3 Pipeline through Minnesota." *Environmental Health News,* May 10, 2021. https://www.ehn.org/line-3-minnesota-2652795873.html.

Bibeau, Frank. "Press Release: First 'Rights of Nature' Enforcement Case Filed in Tribal Court to Enforce Treaty Guarantees Action Filed against Minnesota Department of Natural Resources to Stop Diversion of 5 Billion Gallons of Water for Enbridge 'Line 3' Pipeline." Center for Democratic and Environmental Rights. August 5, 2021. https://www.centerforenvironmentalrights.org/news/press-release-first-rights-of-nature-enforcement-case-filed-in-tribal-court-to-enforce-treaty-guarantees.

Bjorhus, Jennifer. "MPCA Reports More Drilling Mud Spills along Line Construction Route." *Star Tribune,* August 11, 2021. https://www.startribune.com/mpca-reports-more-drilling-mud-spills-along-line-3-construction-route/600086700.

Bluestem Prairie. "No Small Potatoes: Dept of Natural Resources Requires EAW for Pinelands to Spud Fields Project." September 11, 2015. https://www.bluestemprairie.com/bluestemprairie/2015/02/no-small-potatoes-dept-of-natural-resources-requires-eaw-for-pinelands-to-spud-fields-project.html.

Bouayad, Aurelien. "Wild Rice Protectors: An Ojibwe Odyssey." *Environmental Law Review* 22, no. 1 (2020): 25–42. https://doi.org/10.1177/1461452920912909.

Brown, Alleen. "Pipeline Protesters Face Corporate Counterinsurgency." *The Intercept,* July 7, 2021. https://theintercept.com/2021/07/07/line-3-pipeline-minnesota-counterinsurgency.

Brown, Alleen, and John McCracken. "Documents Show How a Pipeline Company Paid Minnesota Millions to Police Protests." *Grist,* February 9, 2023. https://grist.org/protest/enbridge-line-3-pipeline-minnesota-public-safety-escrow-account-invoices.

Bures, Frank. "Winona LaDuke's Last Battle." *Belt Magazine*, November 9, 2018. https://beltmag.com/winona-laduke-line-3-pipeline-future.

Canadian Association of Petroleum Producers. "Oil Sands." https://www.capp.ca/en/oil-natural-gas-you/oil-natural-gas-canada/oil-sands.

Dean, Dave. "75% of the World's Mining Companies Are Based in Canada." *Vice*, July 9, 2013. https://www.vice.com/en/article/wdb4j5/75-of-the-worlds-mining-companies-are-based-in-canada.

Decolonial Atlas. "Native Names for the Mississippi River." January 5, 2015. https://decolonialatlas.wordpress.com/2015/01/05/native-names-for-the-mississippi-river.

Dennis, Brady, and Steven Mufson. "The Company behind the Keystone XL Pipeline Ended the Project on Wednesday, Ending a Fight That Stretched More Than a Decade." *Washington Post*, June 9, 2021. https://www.washingtonpost.com/climate-environment/2021/06/09/keystone-pipeline-dead.

EarthRights International. "Line 3: *Tara Houska, Winona Laduke, Ahnacole Chapman, and Switchboard Trainers Network v. County of Hubbard, Corwin Aukes, and Mark Lohmeier*." https://earthrights.org/case/line-3-cases.

Enbridge. "Enbridge—Our History." https://www.enbridge.com/about-us/our-history.

Environmental Health News. "FBI's Deep Surveillance at Standing Rock Revealed in Court." March 18, 2024. https://www.ehn.org/fbi-s-deep-surveillance-at-standing-rock-revealed-in-court-2667534461.html.

Erdrich, Louise. "Not Just Another Pipeline." *New York Times*, December 28, 2020. https://www.nytimes.com/2020/12/28/opinion/minnesota-line-3-enbridge-pipeline.html.

Furst, Randy. "Judge Dismisses Charges against Activists Accused of Disrupting Enbridge Line 3." *Star Tribune*, September 18, 2023. https://www.startribune.com/judge-dismisses-charges-against-activists-accused-of-disrupting-enbridge-line-3/600305634.

Gordon, Matt. "I'm a Native American Proud to Be Working on Line 3." *Star Tribune*, September 2, 2021. https://www.startribune.com/im-a-native-american-proud-to-be-working-on-line-3/600093395.

Greenpeace Reports. "Enbridge's History of Spills Threatens Minnesota Waters." Report executive summary. November 14, 2018. https://www.greenpeace.org/usa/reports/dangerous-pipelines.

Haskard, Joel. "Young Anishinaabe at 8th Fire Solar Are Building the Clean Energy Economy." Clean Energy Resource Teams, July 2020. https://www.cleanenergyresourceteams.org/story/young-anishinaabe-8th-fire-solar-are-building-clean-energy-economy.

Healing Minnesota Stories. "DNR Evades Key Questions about Enbridge's Clearbrook Aquifer Breach." November 9, 2022. https://healingmnstories.wordpress.com/2022/11/09/dnr-evades-key-questions-about-enbridges-clearbrook-aquifer-breach.

———. "Final Payout from the Enbridge Line 3 Public Safety Escrow Account: $8.5 Million." June 10, 2022. https://healingmnstories.wordpress.com/2022/06/10/final-payout-from-the-enbridge-line-3-public-safety-escrow-account-8-5-million.

Healthy Materials Labs. "Hempcrete Revolution: How 'Pot's Benevolent Cousin' Is Staging Its Comeback Through Architecture." *Architizer*. https://architizer.com/blog/inspiration/stories/hempcrete-revolution-architecture-green-building.

Hofschneider, Anita. "Why More Than 60 Indigenous Nations Oppose the Line 5 Oil Pipeline." *Grist*, December 20, 2023. https://grist.org/indigenous/why-more-than-60-indigenous-nations-oppose-the-line-5-oil-pipeline.

Hosea, Leana. "Why Michigan Is Trying to Shut Down Canada's Enbridge Line 5 Pipeline." BBC, January 14, 2023. https://www.bbc.com/news/world-us-canada-63879493.

Hughlett, Mike. "Enbridge Says Aggressive Climate Policies Shortening Life of Its Pipelines." *Star Tribune*, December 5, 2021. https://www.startribune.com/enbridge-says-aggressive-climate-policies-shortening-life-of-its-pipelines/600124059.

———. "Minnesota's OK for Enbridge to Temporarily Move 5B Gallons of Water Sows Tensions." *Star Tribune*, June 29, 2021. https://www.startribune.com/minnesotas-ok-for-enbridge-to-temporarily-move-5b-gallons-of-water-sows-tension-line-3-pipeline/600073288.

Hughlett, Mike, and Brooks Johnson. "Enbridge Shells Out $750K to Law Enforcement for Line 3 Protest Costs." *Star Tribune*, April 24, 2021.

Independent Global News. "'Not Having It': Winona LaDuke on Mass Protest by Water Protectors to Halt Line 3 Pipeline in Minnesota." June 8, 2021. https://www.democracynow.org/2021/6/8/line_3_protests_treaty_people_gathering.

Karnowski, Steve. "U.S. Army Corps of Engineers Approves Key Line 3 Permit." MPR News, November 23, 2020. https://www.mprnews.org/story/2020/11/23/us-army-corps-of-engineers-approves-key-line-3-permit.

Kemble, Rebecca. "Winona LaDuke and Six Elder Women Arrested at the Shell River." *Wisconsin Citizens Media Cooperative*, July 20, 2021. https://wcmcoop.org/2021/07/20/winona-laduke-and-six-elder-women-arrested-at-the-shell-river.

Kolpack, Dave. "Police Say Nearly 250 Arrested in Minnesota Pipeline Protest." Associated Press, June 9, 2021. https://apnews.com/article/joe-biden-minnesota-arrests-business-1523d769d7a7e097503f4088c420065b.

Kraker, Dan, and Kirsti Marohn. "30 Years Later, Echoes of Largest Inland Oil Spill Remain in Line 3 Fight." MPR News, March 3, 2021. https://www.mprnews.org/story/2021/03/03/30-years-ago-grand-rapids-oil-spill#.

LaDuke, Winona. "Enbridge, Columbus, and the Last Tar Sands Pipeline." *Star Tribune*, October 9, 2021. https://www.startribune.com/line-3-opponents-can-savor-this-defeat/600105209.

———. "Happy Anniversary: The Largest Inland Oil Spill in U.S. History Happened in Minnesota." *Grand Rapids Herald-Review*, March 3, 2017. https://www.grandrapidsmn.com/opinion/happy-anniversary-the-largest-inland-oil-spill-in-u-s-history-happened-in-minnesota/article_2ade2706-004f-11e7-9023-2b31a01741a6.html.

———. "In Response: Northern Mayors, So-Called Minnesota 'Leaders' Still Beholden to Enbridge." *Duluth News Tribune*, September 16, 2021. https://www.duluthnewstribune.com/opinion/columns/in-response-northern-mayors-so-called-minnesota-leaders-still-beholden-to-enbridge.

———. "LaDuke: The Time for Women Leaders." *InForum*, March 29, 2023. https://www.inforum.com/opinion/columns/laduke-the-time-for-women-leaders.

———. "LaDuke: A Water Protector Comes Home to Reflect." *InForum*, October 13, 2021. https://www.inforum.com/opinion/columns/laduke-a-water-protector-comes-home-to-reflect.

Lake, Osprey Orielle, and Katherine Quaid. "Indigenous Women Lead the Movement to Stop Line 3 Pipeline: 'This Is Everything We Have.'" *Ms. Magazine*, May 24, 2021. https://msmagazine.com/2021/05/24/indigenous-women-stop-line-3-pipeline-enbridge.

"Law Enforcement Line 3 Protest Exercise: 'Operation River Crossing.'" Copy of manual *Operation Northern Lights: Multi-Agency Response Drill Series*, September 17, 2020. Contributed by *The Intercept*. https://www.documentcloud.org/documents/20519627-law-enforcement-line-3-protest-exercise-operation-river-crossing.

Lim, Audrea. "'The Next Standing Rock.'" Minnesota's Indigenous Water Protectors Stand Up to Line 3." *Progressive Magazine*, December 1, 2017. https://progressive.org/magazine/the-next-standing-rock-minnesotas-line-3-pipeline.

Little, Lorraine. "Counterpoint: Line 3 Is Needed, Vetted and Done." *Star Tribune*, October 13, 2021. https://www.startribune.com/counterpoint-line-3-is-needed-vetted-and-done/600106329.

Lovett, Tim. "Wild Rice Goes to Court as the Rights of Nature Movement Hits Minnesota." Hennepin County Bar Association. https://www.mnbar.org/hennepin-county-bar-association/resources/hennepin-lawyer/articles/2022/01/14/wild-rice-goes-to-court-as-the-rights-of-nature-movement-hits-minnesota.

Manoomin et al. v. Minnesota Dept of Natural Resources (DNR) et al. https://ecojurisprudence.org/wp-content/uploads/2022/06/manoomin-et-al-v-dnr-complaint-w-exhibits-8-4-21.pdf.

Marchese, David. "Winona LaDuke Feels That President Biden Has Betrayed Native Americans." *New York Times*, August 6, 2021. https://www.nytimes.com/interactive/2021/08/09/magazine/winona-laduke-interview.html.

Marohn, Kirsti. "MPCA: Line 3 Drilling Fluid Spilled into Wetlands." MPR News, August 10, 2021. https://www.mprnews.org/story/2021/08/10/mpca-line-3-drilling-fluid-spilled-into-wetlands.

———. "'We Held a Lot of Good Ground': Line 3 Protest and Prayer Camp Disbands near Mississippi River Crossing." MPR News, June 15, 2021. https://www.mprnews.org/story/2021/06/15/line-3-protest-and-prayer-camp-disbands-near-mississippi-river-crossing#.

McLaughlin, Shaymus. "MPCA Acknowledges 8 More Drilling Fluid Releases during Line 3 Construction, Including into Wetlands." *Bring Me the News*, July 29, 2021. https://bringmethenews.com/minnesota-news/mpca-acknowledges-8-more-drilling-fluid-releases-during-line-3-construction-including-into-wetlands.

Milgroom, Jessica. "Wild Rice and the Ojibwe." *Mnopedia*, July 20, 2020. https://www.mnopedia.org/thing/wild-rice-and-ojibwe.

Military Historical Society of Minnesota. "The Indian Wars of Minnesota." https://www.mnmilitarymuseum.org/files/1213/2249/9221/Indian_Wars_of_Minnesota.pdf.

MN350. "Line 3 Report: A Giant Step Backward." https://mn350.org/giant-step-backward.

Munteanu, Nina. "Henry Ford's Hemp Cars." *The Meaning of Water*, October 3, 2020. https://themeaningofwater.com/2020/10/03/henry-fords-hemp-cars.

Native Harvest Ojibwe Products. "Manoomin (Wild Rice): The Food That Grows on Water," part 1 of 3. https://nativeharvest.com/blogs/news/manoomin-wild-rice-the-food-that-grows-on-water-part-1-of-3.

Neef, Andrew. "Multi-Agency Task Force Prepares 'Rules of Engagement' for Line 3 Protests." *Unicorn Riot*, February 11, 2019. https://unicornriot.ninja/2019/multi_agency_task_force_prepares_rules_of_engagement_for_line_3_protests.

Parrish, Will, and Alleen Brown. "How Police, Private Security, and Energy Companies Are Preparing for a New Pipeline Standoff." *The Intercept*, January 30, 2019. https://theintercept.com/2019/01/30/enbridge-line-3-pipeline-minnesota.

Partlow, Joshua. "Pipeline Protestors Seize Minnesota Construction Site in Bid to Stop $4 Billion Project." *Washington Post*, June 8, 2021. https://www.washingtonpost.com/climate-environment/2021/06/07/pipeline-protest-line-3-minnesota.

Perret, Dorothée, and Oscar Tuazon. "Winona's Hemp." *The Vessel*. https://vessel-magazine.no/issues/2/re-acting-fibres/winonas-hemp.

Regan, Sheila. "Nancy Marie Beaulieu: Protector of Treaties." *Minnesota Women's Press*, November 30, 2021. https://www.womenspress.com/nancy-marie-beaulieu-protector-of-treaties.

Rehkamp, Patrick. "Measuring Enbridge Energy's Clout at the Minnesota Capitol." *Minneapolis/St. Paul Business Journal*, April 20, 2018. https://www.bizjournals.com/twincities/news/2018/04/20/measuring-enbridge-energys-clout-at-the-minnesota.html.

Robertson, Tom, and Dan Gunderson. "Part 1: The Spirits Spoke to Him." Minnesota Public Radio, August 20, 2003. http://news.minnesota.publicradio.org/features/2003/08/18_gundersond_spiritualitytwo.

Rougeau, Naomi. "Winona LaDuke Is Determined to Re-invent the Economy." *Mission Magazine*. https://www.missionmag.org/winona-laduke-is-determined-to-re-invent-the-economy.

Schillaci, William C. "The Montana Exceptions and a Hazardous Waste Case." *EHS Daily Advisor*, December 10, 2019. https://ehsdailyadvisor.blr.com/2019/12/the-montana-exceptions-and-a-hazardous-waste-case.

Searcey, Dionne, and Mira Rojanasakul. "Big Farms and Flawless Fries Are Gulping Water in the Land of 10,000 Lakes." *New York Times*, September 3, 2023. https://www.nytimes.com/interactive/2023/09/03/climate/minnesota-drought-potatoes.html.

Seven Fires Foundation. "Prophecy of the Seven Fires of the Anishinaabe." www.7fires.org.

Shapiro, Aaron. "How Minnesota's Arrowhead Transformed into a Tourism Mecca." *Racket*, January 25, 2023. https://racketmn.com/minnesota-arrowhead-north-shore-iron-range-lakes-tourism-history#.

Soderstrom, Mark. "Family Trees and Timber Rights: Albert E. Jenks, Americanization, and the Rise of Anthropology at the University of Minnesota." *Journal of the Gilded Age and Progressive Era* 3, no. 2 (April 2004): 176–204. https://www.jstor.org/stable/25144365.

"Sonic Weapon Used on Water Protectors." Video posted by Giniw Collective. https://www.facebook.com/watch/?v=525627551798397.

StopLine3.org. "Stop the Line 3 Pipeline." https://www.stopline3.org.

Tabuchi, Hiroko. "Biden Administration Backs Oil Sands Pipeline Project." *New York Times*, June 24, 2021. https://www.nytimes.com/2021/06/24/climate/line-3-pipeline-biden.html.

Tabuchi, Hiroko, Matt Furber, and Coral Davenport. "Police Make Mass Arrests at Protest against Oil Pipeline." *New York Times*, June 9, 2021. https://www.nytimes.com/2021/06/07/climate/line-3-pipeline-protest-native-americans.html.

Thompson, Darren. "Aitkin County Dismisses Line 3 Trespassing Charges against Winona LaDuke." *Native News Online*, April 22, 2023. https://nativenewsonline.net/environment/aitkin-county-dismisses-line-3-trespassing-charges-against-winona-laduke.

Treuer, Anton. https://antontreuer.com/about.

Tribal Treaties Database. "Treaty with the Chippewa, 1855." Oklahoma State University Libraries. https://treaties.okstate.edu/treaties/treaty-with-the-chippewa-1855-0685.

Tuttle, Robert. "Enbridge's Lake Pipeline Tunnel Faces Long Environmental Review." *Bloomberg Terminal*, June 23, 2021.

Village Earth. "The History of 'Competency' as a Tool to Control Native American Lands." https://villageearth.org/the-history-of-competency-as-a-tool-to-control-native-american-lands.

Walker, Richard Arlin. "Winona LaDuke Resigns from Honor the Earth." *Minnesota Reformer*, April 5, 2023. https://minnesotareformer.com/2023/04/05/winona-laduke-resigns-from-honor-the-earth.

Watch the Line Minnesota. "MPCA: Enbridge Polluted Water at 63% of Horizontal Drilling Locations." August 10, 2021. https://watchthelinemn.org/2021/08/10/mpca-enbridge-polluted-water-at-63-of-horizontal-drilling-locations.

WCCO Staff. "Enbridge to Pay Millions in Fines for Line 3 Water Quality Violations, Aquifer Breaches." CBS Minnesota. October 17, 2022. https://www.cbsnews.com/minnesota/news/enbridge-to-pay-millions-in-fines-for-line-3-water-quality-violations-aquifer-breaches.

Weisman, Alan. "Op-Ed: Will Biden Choose Fossil Fuel or Minnesota's Rivers, and a Cooler Planet, in the Fight against Line 3?" *Los Angeles Times*, June 30, 2021. https://www.latimes.com/opinion/story/2021-06-30/line-3-tar-sands-minnesota-winona-laduke-joe-biden-mississippi-river.

Woodside, John. "Risk of 'Death Spiral' for Enbridge Increases: Rate Hike Application." *Canada's National Observer*, February 22, 2023. https://www.nationalobserver.com/2023/02/22/news/risk-death-spiral-enbridge-increases-rate-hike-application.

Zoledziowski, Anya. "At Least 4 Oil Pipeline Workers Linked to Sex Trafficking in Minnesota." *Vice*, July 28, 2021. https://www.vice.com/en/article/g5gkpw/four-enbridge-pipeline-workers-linked-to-sex-trafficking-minnesota#.

CHAPTER NINE: THE PLAN AND THE SUN

Books

Bangladesh Planning Commission. *Bangladesh Delta Plan 2100: Bangladesh in the 21st Century.* Abridged version. Dhaka: General Economics Division, Ministry of Planning, Government of the People's Republic of Bangladesh, 2018.

———. *Bangladesh Delta Plan 2100: Volume 1, Strategy.* Dhaka: General Economics Division, Ministry of Planning, Government of the People's Republic of Bangladesh, 2018.

Islan, S. Nazrul. *A Review of Bangladesh Delta Plan 2100.* Dhaka: Eastern Academic, 2022.

Articles

Agence France-Presse. "Why Orange Hair Is Everywhere in Bangladesh." NDTV. October 21, 2019. https://www.ndtv.com/offbeat/why-orange-hair-is-everywhere-in-bangladesh-2120072.

Ahmed, Kaamil. "World Has Left Bangladesh to Shelter 1M Rohingya Refugees Alone, Says Minister." *The Guardian*, October 26, 2022. https://www.theguardian.com/global-development/2022/oct/26/world-has-left-bangladesh-to-shelter-1m-rohingya-refugees-alone-says-minister.

Al Hasnat, Mahadi. "World's Worst Air Pollution Slashes 7 Years Off Life Expectancy in Bangladesh." *Mongabay*, June 16, 2022. https://news.mongabay.com/2022/06/worlds-worst-air-pollution-slashes-7-years-off-life-expectancy-in-bangladesh.

Alam, Shamsul. "Bangladesh Delta Plan 2100—Making Growth Sustainable." *Financial Express*, November 3, 2018. https://thefinancialexpress.com.bd/views/reviews/bangladesh-delta-plan-2100-making-growth-sustainable-1541259766.

Anchondo, Carlos. "Another Insurer Abandons Coal, Oil Sands." *E&E News*, October 21, 2019. https://subscriber.politicopro.com/article/eenews/2019/10/21/another-insurer-abandons-coal-oil-sands-023230.

Aziz, Maksuda. "As Stronger Storms Hit, Farmers in Bangladesh Struggle to Pay Agricultural Loans." *Mongabay*, September 9, 2022. https://india.mongabay.com/2022/09/as-stronger-storms-hit-farmers-in-bangladesh-struggle-to-pay-agricultural-loans.

BDNews24. "Bangladesh-Netherlands Signs Twinning Arrangement: Extends Cooperation Scheme in Water Sector till 2007." February 13, 2006. https://bdnews24.com/bangladesh/bangladesh-netherlands-signs-twinning-arrangement-extends-cooperation-scheme-in-water-sector-till-2007.

Berwyn, Bob. "A Long-Sought Loss and Damage Deal Was Finalized at COP27. Now, the Hard Work Begins." *Inside Climate News*, March 1, 2023. https://insideclimatenews.org/news/01032023/cop27-loss-damage-deal-developing-nations.

Bilah, Masum Eyamin Sajid. "When Nations Subsidise a Billionaire's 'Adventure.'" *Business Standard*, August 27, 2019. https://www.tbsnews.net/analysis/when-nations-subsidise-billionaires-adventure.

Bkash. https://www.bkash.com.

Bremner, Matthew. "The Sinking Brothel." *Vice*, February 24, 2018. https://www.vice.com/en/article/4374zq/sex-workers-sinking-brothel-bangladesh-climate-change.

Business Standard. "Cordon Approach to Water Management Less Effective: Experts." March 10, 2022. https://www.tbsnews.net/bangladesh/cordon-approach-water-management-less-effective-experts-383017.

———. "First Coal Shipment for Rampal Power Plant Arrives from Indonesia." August 5, 2022. https://www.tbsnews.net/bangladesh/first-coal-shipment-rampal-power-plant-arrives-indonesia-471970.

———. "Shorter Trees and Fewer Leaves: How Farakka Is Transforming the Sundarbans." July 9, 2021. https://www.tbsnews.net/features/panorama/shorter-trees-and-fewer-leaves-how-farakka-transforming-sundarbans-272338#.

Cairns, Rebecca. "How Electric Tuk-Tuks Could Become a 'Virtual Power Plant' for This Country." CNN, April 10, 2023. https://www.cnn.com/2023/04/10/world/solshare-energy-bangladesh-climate-hnk-spc-intl/index.html.

Centre for Public Impact. "The Solar Home Systems Initiative in Bangladesh." October 20, 2017. https://www.centreforpublicimpact.org/case-study/solar-home-systems-bangladesh.

Char Development Settlement Project. "About Us." https://cdsp.org.bd/about.

Chowdhury, Mohammed Shah Nawaz. "Ecological Engineering with Oysters for Coastal Resilience: Habitat Suitability, Bioenergetics, and Ecosystem Services." Ph.D. thesis, 2019, Wageningen University, Wageningen, The Netherlands.

Chowdhury, Zakir Hossain. "Melting Himalayan Glaciers Alter Water Supplies Near and Far." *Undark*, August, 29, 2022.

Daily Star. "Rampal Electric Cost Nearly Doubles Due to Rise in Coal Price, Dollar Rate." February 18, 2023. https://www.thedailystar.net/environment/natural-resources/energy/news/rampal-electricity-cost-nearly-doubles-due-rise-coal-price-dollar-rate-3251481.

———. "United Nations Appoints Dr Nazrul Islam as Chief of Development Research." August 19, 2021. https://www.thedailystar.net/news/bangladesh/news/united-nations-appoints-dr-nazrul-islam-chief-development-research-2156236.

———. "Vessels Sinking: Sundarbans Ecology Braces for Disaster." April 17, 2018. https://www.thedailystar.net/country/vessels-sinking-sundarbans-ecology-braces-disaster-world-largest-mangrove-forest-1563724.

Dhaka Tribune. "Rights Activists' Open Letter to PM Hasina on Saving Farmlands from Development Projects." September 1, 2022. https://www.dhakatribune.com/bangladesh/293374/rights-activists%E2%80%99-open-letter-to-pm-hasina-on.

———. "SOLshare Showcases Breakthrough in EV Battery Technology at Dhaka Event." June 13, 2023. https://www.dhakatribune.com/bangladesh/285261/solshare-showcases-breakthrough-in-ev-battery#.

———. "The Sordid Tale of Banishanta Sex Workers." September 22, 2016. https://www.dhakatribune.com/bangladesh/5771/the-sordid-tale-of-banishanta-sex-workers.

———. "Study: Bangladesh's Power Sector Heading Towards Financial Disaster." May 18, 2020. https://www.dhakatribune.com/bangladesh/power-energy/210275/study-bangladesh%E2%80%99s-power-sector-heading-towards.

Dutch Water Sector. "Bangladesh Water Plan 2100." May 20, 2019. https://www.dutchwatersector.com/news/bangladesh-delta-plan-2100.

———. "Unlocking the Potential of the Coastal Zone in Bangladesh." July 23, 2020. https://www.dutchwatersector.com/news/unlocking-the-potential-of-the-coastal-zone-in-bangladesh.

Earth Observatory. "The Braided Brahmaputra." https://earthobservatory.nasa.gov/images/147591/the-braided-brahmaputra.

Elles, David. "World Bank President, Dogged by Climate Questions, Will Step Down Early." *New York Times*, February 15, 2023. https://www.nytimes.com/2023/02/15/climate/david-malpass-world-bank.html.

Faruque, Mohammad, and Rebecca Tan. "Massive Fire at Rohingya Refugee Camp in Bangladesh Displaces 12,000." *Washington Post*, March 6, 2023. https://www.washingtonpost.com/world/2023/03/06/rohingya-refugee-camp-fire-bangladesh.

Food and Agriculture Organization. "Bangladesh Environment Conservation Act, 1995 (Act No. 1 of 1995)." Unofficial English version. Bangla text of the act was published in the *Bangladesh Gazette*, extraordinary issue of 16-2-1995 and amended by Act Nos. 12 of 2000 and 9 of 2002.

Frayer, Laura. "Facing Floods: What the World Can Learn from Bangladesh's Climate Solutions." NPR, March 26, 2023. https://www.npr.org/sections/goatsandsoda/2023/03/26/1165779335/facing-floods-what-the-world-can-learn-from-bangladeshs-climate-solutions.

Gopal, Brij. "Future of Wetlands in Tropical and Subtropical Asia, Especially in the Face of Climate Change." *Aquatic Sciences* 75 (2013): 39–61. https://doi.org/10.1007/s00027-011-0247-y.

Grameen Shakti. https://www.gshakti.org.

Groh, Sebastian. "An Energy Crisis Remedy: Let Us Use the Solar Panels on Our Rooftops." *Business Standard*, August 7, 2022. https://www.tbsnews.net/thoughts/energy-crisis-remedy-let-us-use-solar-panels-our-rooftops-472978.

Hossain, MdShahadat, C. Kwei Lin, and M. Zakir Hussain. "Goodbye Chakaria Sunderban: The Oldest Mangrove Forest." *Wetland Science and Practice* 18, no. 3 (September 2001): 19–22. https://doi.org/10.1672/0732-9393(2001)018[0019:GCSTOM]2.0.CO;2.

IIX Global. "Bangladesh-Based Clean Energy Firm SOLshare Successfully Closes US$1.1M Financing Round with Support from IIX Impact Partners." June 3, 2020. https://iixglobal.com/solsharedealclose.

Imtiaz, Aysha. "The Unlikely Protector against Bangladesh's Rising Seas." BBC, August 31, 2021. https://www.bbc.com/future/article/20210827-the-unlikely-protector-against-rising-seas-in-bangladesh.

Islam, Rafiqul. "To Help Climate Migrants, Bangladesh Takes Back Land from the Sea." Reuters, September 9, 2015. https://www.reuters.com/article/idUSKCN0R90TY.

Islam, S. Nazrul. "Why Western Approaches Can't Prevent River Erosion in Bangladesh." *Daily Star*, August 26, 2023. https://www.thedailystar.net/opinion/views/news/why-western-approaches-cant-prevent-river-erosion-bangladesh-3403271#.

Jarvis, Oliver. "COP26: Could Oysters Help to Save Bangladesh from Rising Seas?" Al Jazeera, November 11, 2021. https://www.aljazeera.com/program/newsfeed/2021/11/11/cop26-could-oysters-help-to-save-bangladesh-from-rising-seas.

Jerryson, Michael. "Buddhist Inspired Genocide." Berkeley Center for Religion, Peace, and World Affairs, October 13, 2017. https://berkleycenter.georgetown.edu/responses/buddhist-inspired-genocide.

Kammen, Daniel. "Clean Energy in Bangladesh: Innovating for a Clean Energy Economy in Bangladesh. *Bangladesh Notes*, Spring 2019. Institute for South Asia Studies University of California at Berkeley. https://issuu.com/csas_ucberkeley/docs/kammen_sarn-web.

Kaphle, Anup. "An Interview with Bangladeshi Prime Minister Sheikh Hasina." *Washington Post*, October 11, 2011. https://www.washingtonpost.com/world/an-interview-with-bangladesh-pm-sheikh-hasina/2011/10/10/gIQAXAQRcL_story.html.

Khan, Shahidur Rahman. "Sandwip-Urir Char-Noakhali Cross Dam for Long-Term Food Security." *Daily Star*, April 26, 2008. https://www.thedailystar.net/news-detail-33780.

Lau, Chris, and Laura Paddison. "Bangladesh and Myanmar Brace for the Worst as Cyclone Mocha Makes Landfall." CNN, May 14, 2023. https://www.cnn.com/2023/05/13/asia/cyclone-mocha-aid-agencies-myanmar-bangladesh-climate-intl-hnk/index.html.

Mahmud, Faisal. "Bangladesh in Hot Seat over Adani's Power Deal." Al Jazeera, March 30, 2023. https://www.aljazeera.com/economy/2023/3/30/bangladesh-in-a-hot-seat-over-adanis-power-deal.

———. "PM Hasina Opens Bangladesh's Longest Bridge over River Padma." Al Jazeera, June 25, 2022. https://www.aljazeera.com/news/2022/6/25/bangladesh-unveils-its-longest-bridge-over-river-padma.

McBride, James, Noah Berman, and Andrew Chatzky. "China's Massive Belt and Road Initiative." Council on Foreign Relations. February 2, 2023. https://www.cfr.org/backgrounder/chinas-massive-belt-and-road-initiative.

McVeigh, Karen, and Dinakar Peri. "Fatal Elephant Attacks on Rohingya Refugees Push Bangladesh to Act." *The Guardian*, May 9, 2018. https://www.theguardian.com/global-development/2018/may/09/fatal-elephant-attacks-on-rohingya-refugees-push-bangladesh-to-act.

New Age. "Hasina, Modi Open Rampal Power Plant, Built Ignoring Protests." September 6, 2022. https://www.newagebd.net/article/180356/hasina-modi-open-rampal-power-plant-built-ignoring-protests.

Nicholas, Simon. "Rampal Coal Plant Inauguration Won't Solve Bangladesh's Power Woes." Institute for Energy Economics and Financial Analysis. September 7, 2022. https://ieefa.org/resources/rampal-coal-plant-inauguration-wont-solve-bangladeshs-power-woes.

Nicholls, Robert J., Daniel Lincke, Jochen Hinkel, Sally Brown, et al. "A Global Analysis of Subsidence, Relative Sea-Level Change and Coastal Flood Exposure." *Nature Climate Change* 11 (April 2021): 338–342. https://doi.org/10.1038/s41558-021-00993-z.

NWO (Dutch Research Council). "Living Polders in Bangladesh." September 22, 2020. https://www.nwo.nl/en/cases/living-polders-bangladesh.

Oxfam. "Worst Monsoon Rains in over a Century Submerge Most of Northeast Bangladesh and Devastate the Lives of over 4 Million People." June 23, 2022. https://reliefweb.int/report/bangladesh/worst-monsoon-rains-over-century-submerge-most-northeast-bangladesh-and-devastate-lives-over-4-million-people?_gl=1*1ygbxa4*_ga*MTU1NDU5NzA5Ni4xNzE2OTE4NjI3*_ga_E60ZNX2F68*MTcxNjkxODYyNy4xLjAuMTcxNjkxODYyNy42MC4wLjA.

Paprocki, Kasia. "Threatening Dystopias: Development and Adaptation Regimes in Bangladesh." *Annals of the American Association of Geographers* 108, no. 4 (2018): 1–19. https://doi.org/10.1080/24694452.2017.1406330.

Phys.org. "Bangladesh 'Extremely Worried' over Low Male Tiger Population." July 30, 2019. https://phys.org/news/2019-07-bangladesh-extremely-male-tiger-population.html.

Pitol, M. N. S. "Trends of Sundarbans Mangroves Biodiversity Declination in Bangladesh." *Academia Letters*, 2022: article 5195.

Pollee Unnayon Prokolpo. "Bangladesh Poribesh Andolon (BAPA)." http://www.pupbd.org/Membership/Bangladesh_Poribesh_Andolon.

Rahman, Md. Mamunur, Nahid Nasrin, Swarup Chakraborty, Asif Siddique, et al. "Renewable Energy: The Future of Bangladesh." *International Journal of Engineering Research and Application* 7, no. 6 (June 2017): 23–29.

Raju, Mohammed Norul Alam, and Faima Rahman. "Why Saving the Sundarbans Is So Urgent." *Daily Star*, January 2, 2020. https://www.thedailystar.net/opinion/environment/news/why-saving-the-sundarbans-so-urgent-1848145.

Ratcliffe, Rebecca, Liz Ford, Lydia McMullan, Pablo Gutiérrez, and Garry Blight. "Cox's Bazar Refugee Camps: Where Social Distancing Is Impossible." *The Guardian*, June 29, 2020. https://www.theguardian.com/world/ng-interactive/2020/jun/29/not-fit-for-a-human-coronavirus-in-coxs-bazar-refugee-camps.

Rezwan. "Bangladesh Reassesses Its Belt and Road Initiative Strategy with China as the US Offers a New Alternative." *Global Voices*, February 28, 2023. https://globalvoices.org/2023/02/28/bangladesh-reassesses-its-belt-and-road-initiative-strategy-with-china-as-the-us-offers-a-new-alternative.

Rokon, Sheikh. "Another Shipping Disaster in the Sundarbans." *Dialogue Earth*, January 16, 2017. https://dialogue.earth/en/pollution/another-shipping-disaster-in-the-sundarbans.

Roome, John. "Implementing Bangladesh Delta Plan 2100: Key to Boost Economic Growth." *End Poverty in South Asia*, June 9, 2021. World Bank blog. https://blogs.worldbank.org/en/endpovertyinsouthasia/implementing-bangladesh-delta-plan-2100-key-boost-economic-growth.

Roskamp, Hanny. "How Oysters Build Strong Coastlines." *Wageningen Climate Solutions*, September 2019. https://magazines.wur.nl/climate-solutions-en/oysters.

Sarkar, Jayita. "Rohingyas and the Unfinished Business of Partition." *The Diplomat*, January 16, 2018. https://thediplomat.com/2018/01/rohingyas-and-the-unfinished-business-of-partition.

Sarkar, Samir. "Which Animals Are Found in Sundarban National Park?" *Sundarban Wildlife Expedition*, July 18, 2021. https://www.sundarbanwildlifetourism.com/blog/which-animals-are-found-in-sundarban-national-park.

Sengupta, Somini, Jacqueline Williams, and Aruna Chandrasekhar. "How One Billionaire Could Keep Three Countries Hooked on Coal for Decades." *New York Times*, August 15, 2019. https://www.nytimes.com/2019/08/15/climate/coal-adani-india-australia.html.

Shihab, A. K. M. "Environmental Movement: Save Sundarbans, Bangladesh." Libcom.org, March 25, 2018. https://libcom.org/article/environmental-movement-save-sundarbans-bangladesh.

Siddique, Abu. "Sand Mining a Boon for Illegal Industry at Expense of Bangladesh's Environment." *Mongabay*, August 16, 2022. https://news.mongabay.com/2022/08/sand-mining-a-boon-for-illegal-industry-at-expense-of-bangladeshs-environment.

Simon, Ellen. "Battling for Bangladesh's Waters and Its People | Sharif Jamil, Buriganga Riverkeeper." *Waterkeeper* 15, no. 1 (2019). https://waterkeeper.org/magazines/volume-15-issue-1/battling-for-bangladeshs-waters-and-its-people-sharif-jamil-buriganga-riverkeeper/.

SOLshare. https://solshare.com.

———. *Annual Report*. 2022. Dhaka: ME Solshare Ltd.

———. "The Journey to 100 Solar Peer-to-Peer Microgrids." September 19, 2022. https://www.youtube.com/watch?v=FfTy0yOSGqU.

Sunderban National Park. "Birds in Sunderban National Park." https://www.sunderbannationalpark.in/birds-in-sunderban.html.

United Nations. "Bangladesh: Tigers' Sundarbans Forest Habitat Threatened by Heedless Industrialisation, Says UN Expert." Press release. July 31, 2018. https://www.ohchr.org/en/press-releases/2018/07/bangladesh-tigers-sundarbans-forest-habitat-threatened-heedless.

Westerman, Ashley. "Should Rivers Have Same Legal Rights as Humans? A Growing Number of Voices Say Yes." NPR, August 3, 2019. https://www.npr.org/2019/08/03/740604142/should-rivers-have-same-legal-rights-as-humans-a-growing-number-of-voices-say-yes.

World Heritage Convention. "World Heritage Centre and IUCN Call for Relocation of Rampal Power Plant, a Serious Threat to the Sundarbans." October 18, 2016. https://whc.unesco.org/en/news/1573.

Zaman, Asif M. "Dhaka and Her Rivers: A Beautiful Relationship Gone Sour." *Daily Star*, August 7, 2017. https://www.thedailystar.net/opinion/environment/dhaka-and-her-rivers-1444537.

CHAPTER TEN: PIVOTING

Books

Smil, Vaclav. *Enriching the Earth.* Cambridge, MA: MIT Press, 1994.

Articles

Adee, E. A., E. D. Nafziger, and L. E. Paul. "Optimizing Plant Population and Nitrogen Rate for High-Oil Corn." *Crop Management* 6, no. 1 (2007). https://doi.org/10.1094/CM-2007-0831-01-RS.

Arkansas Division of Agriculture. "Rice Expo 2011—Dr. Trent Roberts." August 1, 2011. https://www.youtube.com/watch?v=0zqiywvoA5c.

Barnard, Michael. "CleanTech Talk 1/2: CEO Karsten Temme of Pivot Bio on Microbes Slashing Agriculture GHG Emissions." *Clean Technica*, August 21, 2021. https://cleantechnica.com/2021/08/21/cleantech-talk-1-2-ceo-karsten-temme-of-pivotbio-on-microbes-slashing-agriculture-ghg-emissions.

———. "CleanTech Talk 2/2: Pivot Bio's Genetically Engineered Microbes Displace 20–25% of Fertilizer Today, 100% by 2030." *Clean Technica*, August 22, 2021. https://cleantechnica.com/2021/08/22/cleantech-talk-2-2-pivotbios-genetically-engineered-microbes-displace-20-25-of-fertilizer-today-100-by-2030.

Bloch, Sara E., Rosemary Clark, Shayin S. Gottlieb, L. Kent Wood, Neal Shah, An-Ming Mak, James G. Lorigan, et al. "Biological Nitrogen Fixation in Maize: Optimizing Nitrogenase Expression in a Root-Associated Diazotroph." *Journal of Experimental Botany* 71, no. 15 (July 25, 2020): 4591–4603. https://doi.org/10.1093/jxb/eraa176.

Business Wire. "Pivot Bio Joins AIM for Climate as Innovation Sprint Partner." November 11, 2022. https://www.businesswire.com/news/home/20221111005122/en/Pivot-Bio-Joins-AIM-for-Climate-as-Innovation-Sprint-Partner.

Chrobak, Ula. "The World's Forgotten Greenhouse Gas." BBC, June 3, 2021. https://www.bbc.com/future/article/20210603-nitrous-oxide-the-worlds-forgotten-greenhouse-gas.

CNBC.com Staff. "12. Pivot Bio." CNBC, May 9, 2023. https://www.cnbc.com/2023/05/09/pivot-bio-disruptor-50.html.

Currie, Mason, Daniel E. Kaiser, and Jeffrey A. Vetsch. "Can ProveN Reduce Corn Nitrogen Requirement in Minnesota?" Presented at the annual international meeting of the American Society of Agronomy, the Crop Science Society of America, and the Soil Science Society of America. Salt Lake City, November 9, 2021. https://scisoc.confex.com/scisoc/2021am/meetingapp.cgi/Paper/137782.

Dakota Farmer. "Universities Test Product in Field Trials." https://editions.mydigitalpublication.com/publication/?i=813800&article_id=4716442&view=articleBrowser.

Dinneen, James. "Nitrogen-Producing Bacteria Slash Fertiliser Use on Farms." *New Scientist*, June 21, 2023. https://www.newscientist.com/article/2379100-nitrogen-producing-bacteria-slash-fertiliser-use-on-farms.

Discovered in Berkeley. "How a Berkeley Startup Is Changing Farming with a New Kind of Fertilizer." September 12, 2022. https://www.discoveredinberkeley.com/2022/09/12/how-a-berkeley-startup-is-changing-farming-with-a-new-kind-of-fertilizer.

Feldman, Amy. "Pivot Bio Nears $2 Billion Valuation as It Raises a Whopping $430 Million to Replace Synthetic Fertilizers on Corn and Wheat." *Forbes*, July 19, 2021. https://www.forbes.com/sites/amyfeldman/2021/07/19/pivot-bio-nears-2-billion-valuation-as-it-raises-whopping-430-million-to-replace-synthetic-fertilizers-on-corn-and-wheat-sustainability/?sh=6b0b0c782273.

Genetics Science Learning Center. "Evolution of Corn." Learn.Genetics. August 1, 2017. https://learn.genetics.utah.edu/content/evolution/corn.

Global Impact Venturing. "Investors Plant $430M in Pivot Bio." July 29, 2021. https://globalventuring.com/corporate/investors-plant-430m-in-pivot-bio.

Globe News Wire. "Pivot Bio Launches the First-Ever On-Seed Nitrogen." August 30, 2022. https://www.globenewswire.com/news-release/2022/08/30/2506837/0/en/Pivot-Bio-Launches-the-First-Ever-On-Seed-Nitrogen.html.

Gordon, Paul. "How Reporter Yanqi Xu Connected Nitrate Pollution to Pediatric Cancer." *Association of Health Care Journalists*, January 24, 2024. https://healthjournalism.org/blog/2024/01/how-reporter-yanqi-xu-connected-nitrate-pollution-to-pediatric-cancer.

Guidehouse Insights. "Green Ammonia and the Electrification of the Haber-Bosch Process Reduce Carbon Emissions." April 6, 2021. https://guidehouseinsights.com/news-and-views/green-ammonia-and-the-electrification-of-the-haber-bosch-process-reduce-carbon-emissions.

Grunwald, Michael. "Chemical Fertilizer Is a Climate Disaster. Can High-Tech Biology Fix It?" *Canary Media*, August 29, 2023. https://www.canarymedia.com/articles/food-and-farms/chemical-fertilizer-is-a-climate-disaster-can-high-tech-biology-fix-it.

Hanson, Matthew, and Yanqi Xu. "Nebraska's Nitrate Problem Is Serious, Experts Say. Can We Solve It?" *Nebraska Public Media*. https://www.google.com/search?client=safari&rls=en&q=Nebraska%E2%80%99s+nitrate+problem+is+serious%2C+experts+say.+Can+we+solve+it%3F&ie=UTF-8&oe=UTF-8.

Hill, David. "Census of Agriculture: Upstate Farming Shrinking but Evolving." *Daily Sentinel*, April 20, 2019. https://www.romesentinel.com/news/census-of-agriculture-upstate-farming-shrinking-but-evolving/article_f2ce5940-8257-50fe-9dc4-219812ce83ad.html.

Hopkin, Karen. "Bite Me: The Mutation That Made Corn Kernels Consumable." *Scientific American*, August 11, 2015. https://www.scientificamerican.com/podcast/episode/bite-me-the-mutation-that-made-corn-kernels-consumable.

Hough, Cassandra. "'Lightning' in a Bottle Could Extract Nitrogen from the Air, Offering Green Ammonia to Farmers." ABC News. https://www.abc.net.au/news/2022-11-13/lightning-in-a-bottle/101636690.

Huber, Bridget. "Report: Fertilizer Responsible for More Than 20 Percent of Total Agricultural Emissions." *Food and Environment Reporting Network*, November 1, 2021. https://thefern.org/ag_insider/report-fertilizer-responsible-for-more-than-20-percent-of-total-agricultural-emissions.

Jenkins, Jason. "Do N-Fixing Biological Products Work?" *Progressive Farmer*, May 3, 2023. https://www.dtnpf.com/agriculture/web/ag/crops/article/2023/05/03/university-research-casts-skepticism.

Jett, Tyler. "Microbe Fertilizer Companies Look to Expand on Iowa Farms after Big 2022. Do Their Products Work?" *Des Moines Register*, October 9, 2022. https://www.desmoinesregister.com/story/money/business/2022/10/09/microbe-fertilizer-startups-look-to-expand-iowa-farms/8086016001.

Johnson, Nathanael. "Death by Fertilizer." *Grist*, October 2, 2018. https://grist.org/article/billionaires-and-bacteria-are-racing-to-save-us-from-death-by-fertilizer.

Kaplan, Sarah. "One of the Most Potent Greenhouse Gases Is Rising Faster Than Ever." *Washington Post*, September 12, 2024.

Marston, Jennifer. "Pivot Bio Pilot Replaces Synthetic Nitrogen on Nearly 1M Acres of Farmland." *Ag Funder News*, April 19, 2023. https://agfundernews.com/breaking-pivot-bio-pilot-replaces-synthetic-nitrogen-on-nearly-1m-acres-of-farmland.

McCullough, Currey. "Pivot Bio's PROVEN Has Been Everywhere!" RFD-TV, October 5, 2022. https://www.rfdtv.com/pivot-bios-proven-has-been-everywhere.

Molteni, Megan. "Farmers Can Now Buy Designer Microbes to Replace Fertilizer." *Wired*, October 2, 2018. https://www.wired.com/story/farmers-can-now-buy-designer-microbes-to-replace-fertilizer.

Nature Biotechnology. "Can Microbes Save the Planet?" Editorial. Vol. 41 (June 2023): 735. https://doi.org/10.1038/s41587-023-01837-1.

Nebraska On-Farm Research Network. "2021 On-Farm Research Results." University of Nebraska–Lincoln. https://on-farm-research.unl.edu/pdfs/research/result-publications/2021research-results.pdf.

Ohio Ag Net & Ohio's Country Journal. "Atley Reduces Passes with Pivot Bio ProveN Products." https://www.youtube.com/watch?v=tydwfruWqUA.

Performance Premixes. https://www.performancepremixes.com.

Pivot Bio. https://www.pivotbio.com.

———. "Pivot Bio Commitment—AIM for Climate." https://youtu.be/T8NTQpXF1TM?si=4z8jbi8G-nvUbbMi.

———. "Pivot Bio Virtual Field Day." June 18, 2020. https://www.youtube.com/live/t7pr97Du4uQ?si=nOE31zcqXE62pR9n.

———. "Safety Data Sheet." For Proven 40. November 23, 2023. https://cdn.sanity.io/files/cju0suru/prod/ddf655146da23567713daf62708b9c46048e63d1.pdf.

———. "Three Land-Grant Universities Study Pivot Bio Performance." August 4, 2023. https://www.pivotbio.com/blog/performance.

Precision Technology Institute. *2020 PTI Farm Research Summary*. Pontiac, IL. https://irp-cdn.multiscreensite.com/25c4dd71/files/uploaded/PTI%20Yield%20Summary%20Book%20NACHURS%20EDIT.pdf.

———. *2021 PTI Farm Research Summary*. Pontiac, IL. https://www.precisionagriservices.com/PDFs/InsidePTI-Results/PTI_2021%20Summary%20Results.pdf.

Preferred Popcorn. https://www.preferredpopcorn.com.

Proceedings: North Central Extension–Industry Soil Fertility Conference. 51st annual conference. Des Moines, 2021.

Ramachadran, Vijaya. "The West Needs to Come to Grips with African Fertilizer Needs." *Breakthrough Journal*, April 29, 2024. https://thebreakthrough.org/journal/no-20-spring-2024/the-west-needs-to-come-to-grips-with-african-fertilizer-needs.

Sawyer, John E., and Emerson D. Nafziger. "Regional Approach to Making Nitrogen Fertilizer Rate Decisions for Corn." *Proceedings of the 2006 Wisconsin Fertilizer, Aglime and Pest Management Conference*, p. 68. https://soilsextension.webhosting.cals.wisc.edu/wp-content/uploads/sites/68/2014/02/2006_wfapm_proc.pdf.

Sound Agriculture. "Meet Brandon Hunnicutt: 5th Generation Farmer Embracing Innovation to Improve Sustainability." June 10, 2021. https://www.sound.ag/blog/meet-brandon-hunnicutt-5th-generation-farmer-embracing-innovation-to-improve-sustainability.

———. "Source: How It Works." https://www.sound.ag/source.

Statista. "Leading Fertilizer Exporting Countries Worldwide in 2022, Based on Value." September 2023. https://www.statista.com/statistics/1278057/export-value-fertilizers-worldwide-by-country.

Thompson, Laura, Taylor Lexow, Sarah Sivits, Aaron Nygren, and Laila Puntel. "Farmer Focus: Nebraska Growers Put Pivot Bio to the Test via On-Farm Research Studies." CropWatch, April 7, 2022. Institute of Agriculture and Natural Resources, University of Nebraska–Lincoln. https://cropwatch.unl.edu/2022/farmer-focus-nebraska-growers-put-pivot-bio-test-farm-research-studies.

University of Cambridge. "Carbon Emissions from Fertilisers Could Be Reduced by as Much as 80% by 2050." February 9, 2023. https://www.cam.ac.uk/research/news/carbon-emissions-from-fertilisers-could-be-reduced-by-as-much-as-80-by-2050.

University of Nebraska–Lincoln. "Study Confirms That Common Fertilizer Compound Can Release Uranium into Groundwater." *SciTechDaily*, April 1, 2023. https://scitechdaily.com/study-confirms-that-common-fertilizer-compound-can-release-uranium-into-groundwater.

US Food and Drug Administration. "GMO Crops, Animal Food, and Beyond." https://www.fda.gov/food/agricultural-biotechnology/gmo-crops-animal-food-and-beyond.

Vyn, Tony. "Can New Microbes Lower Nitrogen Rates in Corn?" *Pest and Crop Newsletter*, January 25, 2022. Purdue University Entomology Extension. https://extension.entm.purdue.edu/newsletters/pestandcrop/article/can-new-microbes-lower-nitrogen-rates-in-corn.

Wang, Huai, Anthony J. Studer, Qiong Zhao, et al. "Evidence That the Origin of Naked Kernels during Maize Domestication Was Caused by a Single Amino Acid Substitution in *TGA1*." *Genetics* 200, no. 3 (July 2015): 965–974. https://doi.org/10.1534/genetics.115.175752.

Wen, Amy, Keira L. Havens, Sarah E. Bloch, Neal Shah, et al. "Enabling Biological Nitrogen Fixation for Cereal Crops in Fertilized Fields." *ACS Synthetic Biology* 10, no. 12 (2021): 3264–3277. https://doi.org/10.1021/acssynbio.1c00049.

Wood, Jamin, and Bernardino Virdis. "Why Green Ammonia Might Not Be That Green." *Renew Economy*, April 28, 2023. https://reneweconomy.com.au/why-green-ammonia-might-not-be-that-green.

Xu, Yanqi. "Our Dirty Water: Nebraska's Nitrate Problem Is Growing Worse. It's Likely Harming Our Kids." *Flatwater Free Press*, October 27, 2022. https://flatwaterfreepress.org/our-dirty-water-nebraska-water-nitrates.

CHAPTER ELEVEN: XNOIS

Books

Bowen, Thomas. *Seri Prehistory: The Archaeology of the Central Coast of Sonora, Mexico.* Anthropological Papers of the University of Arizona No. 27. Tucson: University of Arizona Press, 1976.

Felger, Richard Stephen, and Mary Beck Moser. *People of the Desert and Sea: Ethnobotany of the Seri Indians.* Century Collection. Tucson: University of Arizona Press, 1985. Kindle edition.

McGee, W. J. *The Seri Indians.* Seventeenth Annual Report of the Bureau of American Ethnology to the Secretary of the Smithsonian Institution, 1895–96. Washington, DC: Government Printing Office, 1898. Ebook. https://www.gutenberg.org/files/49403/49403-h/49403-h.htm.

Mellado Moreno, Alberto, Erika Barnett, Servando López Monroy, Laura Smith Monti, et al. *Renacimiento de Xnois.* Tucson: Borderlands Restoration Network, 2021.

Moser, Mary Beck, and Stephen A. Marlett. *Comcaác quih yaza quih hant ihiip hac: Diccionario seri–español–inglés.* Hermosillo, Son: Editorial UniSon: Plaza y Valdés Editores, 2010 / 2a edición.

Nabhan, Gary Paul. *The Desert Smells like Rain: A Naturalist in O'odham Country.* Tucson: University of Arizona Press, 2002.

———. *Gathering the Desert.* Tucson: University of Arizona Press, 1986.

———. *Growing Food in a Hotter, Drier Land.* White River Junction, VT: Chelsea Green, 2013.

Articles

Abad, Ana Fernández. "'Zostera Marina,' a Michelin-Starred Sea Grain That Could Point to the Future of Hydroponic Crops." *El País,* May 11, 2023. https://english.elpais.com/science-tech/2023-05-11/zostera-marina-a-michelin-starred-sea-grain-that-could-point-to-the-future-of-hydroponic-crops.html.

Adler, Jennifer. "Seagrass Is a Vital Weapon Against Climate Change, but We're Killing It." *HuffPost,* February 20, 2021. https://www.huffpost.com/entry/seagrass-ocean-climate-change-pollution-florida_n_602ced75c5b6cc8bbf3819ff.

Andalusia.com. "Aponiente Restaurant El Puerto de Santa Maria." Review from the *Michelin Guide to Spain and Portugal 2018.* https://www.andalucia.com/province/cadiz/puertosantamaria/restaurant-aponiente.

Arellano, Astrid. "Erika Barnett, a Seri Woman, Created an Organic Garden 3 Years Ago in Punta Chueca, Today She Helps Them in the Face of the Food Crisis." *Proyecto Puente,* April 30, 2020. https://proyectopuente.com.mx/2020/04/30/erika-barnett-mujer-indigena-cmiique-crea-huerto-organico-en-punta-chueca-para-enfrentar-crisis-alimentaria-durante-contingencia.

———. "The Grain of the Sea: The Comcaac Nation Preserves Seagrasses and Awakens the Ancestral Food They Obtain from Them." *Mongabay,* June 6, 2022. https://es.mongabay.com/2022/06/nacion-comcaac-conserva-pastos-marinos-en-mexico.

———. "Indigenous Comcaac Serve Up an Oceanic Grain to Preserve Seagrass Meadows." *Mongabay,* March 10, 2023. https://news.mongabay.com/2023/03

/indigenous-comcaac-serve-up-an-oceanic-grain-to-preserve-seagrass-meadows.

Balthazar, Deborah. "The Sonoran Desert Toad Can Alter Your Mind—It's Not the Only Animal." *Science News*, May 14, 2023. https://www.sciencenews.org/article/sonoran-desert-toad-psychedelic-animal-chemistry.

Barrio Bread. https://www.barriobread.com.

Blust, Kendal. "Eelgrass Is an Ancestral Food in Sonora's Comcaac Community. Now It's Gaining Global Attention." *Fronteras*, May 18, 2022. https://fronterasdesk.org/content/1780626/eelgrass-ancestral-food-sonoras-comcaac-community-now-its-gaining-global-attention.

Borderlands Restoration Network. https://www.borderlandsrestoration.org.

Brusca, Richard. "A Brief Geologic History of Northwestern Mexico." January 27, 2022. http://rickbrusca.com/http___www.rickbrusca.com_index.html/Sea_of_Cortez_files/Geology%20of%20NW%20Mexico.pdf.

Bugel, Safi. "Sunscreen Chemicals Accumulating in Mediterranean Seagrass, Finds Study." *The Guardian*, April 12, 2022. https://www.theguardian.com/environment/2022/apr/12/sunscreen-chemicals-accumulating-in-mediterranean-seagrass-finds-study.

Cañas, Jesús A. "Spain's Disappearing Reserves of White Gold." *El País*, September 5, 2016. https://english.elpais.com/elpais/2016/08/29/inenglish/1472461713_166345.html.

Cosas de Comé. "Ángel León's Magic Salt." January 30, 2019. https://cadiz.cosasdecome.es/la-sal-magica-angel-leon/Aponiente. https://www.aponiente.com.

Courage, Katherine Harmon. "Prairies of the Sea." *Smithsonian Magazine*, December 2020. https://www.smithsonianmag.com/science-nature/seagrass-ocean-secret-weapon-climate-change-180976235.

Curt Bergfors Food Planet Prize. "Aponiente's Sea Pantry Wants to Feed the World Through Sea Grains." https://roadsandkingdoms.com/2023/a-sea-grain-to-feed-the-world.

De Greef, Kimon. "The Pied Piper of Psychedelic Toads." *New Yorker*, March 21, 2022. https://www.newyorker.com/magazine/2022/03/28/the-pied-piper-of-psychedelic-toads.

Felger, Richard, and Mary Beck Moser. "Eelgrass (*Zostera marina L.*) in the Gulf of California: Discovery of Its Nutritional Value by the Seri Indians." *Science* 181, no. 4097 (July 27, 1973): 355–356. http://doi.org/10.1126/science.181.4097.355.

Flavelle, Christopher. "Biden Administration Proposes Evenly Cutting Water Allotments from Colorado River." *New York Times*, April 11, 2023. https://www.nytimes.com/2023/04/11/climate/colorado-river-water-cuts-drought.html.

Fund for the Stewardship and Recovery of the Salt Marshes (SALARTE). "Crónicas desde el Golfo de California—Desierto de Sonra" [Chronicles from the Gulf of California—Sonoran Desert]. https://salarte.org/noticias/cronicas-del-golfo-de-california-desierto-de-sonora.

Gary Nabhan. https://www.garynabhan.com.

———. "Desert Is a Homeland That Has Migrated." March 3, 2006. https://www.garynabhan.com/news/2006/03/desert-is-a-homeland-that-has-migrated.

Goulding, Matt. "Seeding the Ocean: Inside a Michelin-Starred Chef's Revolutionary Quest to Harvest Rice from the Sea." *Time*, January 9, 2021. https://time.com/5926780/chef-angel-leon-sea-rice.

Hoeffner, Regidor Francisco Fonseca, Oscar Andrade, Laura Monti, Rafael Cabanillas, and Gary Nabhan. "Water, Energy and Food Crisis Threatening the Coastal Desert Villages of the Comcaac (Seri) Community." Gary Nabhan webpage. March 12, 2021. https://www.garynabhan.com/news/2021/03/water-energy-and-food-crises-that-threaten.

Jones, Nicola. "Why the Market for 'Blue Carbon' Credits May Be Poised to Take Off." *Yale Environment 360*, April 13, 2021. https://e360.yale.edu/features/why-the-market-for-blue-carbon-credits-may-be-poised-to-take-off.

Judkis, Maura. "An Ocean-Obsessed Spanish Chef Brings Plankton to the Plate—and Makes It Glow." *Washington Post*, November 4, 2016. https://www.washingtonpost.com/news/food/wp/2016/11/04/an-ocean-obsessed-spanish-chef-brings-plankton-to-the-plate-and-makes-it-glow.

Kassam, Ashifa. "The Rice of the Sea: How a Tiny Grain Could Change the Way Humanity Eats." *The Guardian*, April 9, 2021. https://www.theguardian.com/environment/2021/apr/09/sea-rice-eelgrass-marine-grain-chef-angel-leon-marsh-climate-crisis.

Lengamer.org. "General Information about the Culture of the Comcaac and Their Language." http://lengamer.org/asoc/language.php?language=seri.

———. "Lists of Species Found in the Seri Zone." http://lengamer.org/admin/language_folders/seri/user_uploaded_files/links/File/Listados/Listado.html.

López Calderón, Jorge, Rafael Riosmena Rodríguez, Jorge Torre, and Alf Meling López. "El pasto marino en el Golfo de California: estado actual y amenazas." CONABIO: *Biodiversitas* 97: 20–15. https://cobi.org.mx/wp-content/uploads/2013/02/biodiversitas.pdf.

Merello, Patricia. "Resurrect Historic Salt Flats, a Treasure of the Bay of Cádiz That Is More Than 300 Years Old." Lavozdeslsur.es. July 2, 2023. https://www.lavozdelsur.es/la-voz-seleccion/reportajes/resucitar-salinas-tesoro-de-la-bahia-de-cadiz-con-mas-de-300-anos_297908_102.html.

Montojo, Marta. "A Sea Grain to Feed the World." *Roads and Kingdoms*, May 22, 2023. https://roadsandkingdoms.com/2023/a-sea-grain-to-feed-the-world.

Morales Vera, Thor Edmundo. "Las Aves de Los *Comcáac* (Sonora, México)." Bachelor of biology thesis, Universidad Veracruzana, 2006. http://www.lengamer.org/admin/language_folders/seri/user_uploaded_files/links/File/bibliografia_seri/Estudios_cientificos_files/Morales_Tesis._Aves.pdf.

Moser, Edward. "Two Seri Myths." Summer Institute of Linguistics. *Tlalocán Revista de Fuentes para el Conocimiento de las Culturas indígenas de México* 5, no. 4 (1968): 364–467. https://repositorio.unam.mx/contenidos/60245.

Moser, Mary B., and Stephen A. Marlett. "Seri Dictionary: Earth, Sea, Sky, Time and Weather." *Work Papers of the Summer Institute of Linguistics, University of North Dakota Session* 42 (1998): article 2. https://doi.org/10.31356/silwp.vol42.02.

Nabhan, Gary Paul, Erin C. Riordan, Laura Monti, Amadeo M. Rea, et al. "An Aridamerican Model for Agriculture in a Hotter, Water Scarce World." *Plants, People, Planet* 2, no. 6 (November 2020): 581–687. https://doi.org/10.1002/ppp3.10129.

Native Seed Search. https://www.nativeseeds.org.

Newswise. "Climate and Agriculture in the Mediterranean: Less Water Resource, More Irrigation Demand." November 16, 2021. https://www.newswise.com/articles

/climate-and-agriculture-in-the-mediterranean-less-water-resource-more-irrigation-demand.

Niazi, Kamran Ali Khan, and Marta Victoria. "Comparative Analysis of Photovoltaic Configurations for Agrivoltaic Systems in Europe." *Progress in Photovoltaics: Research and Applications* 31, no. 11 (2023): 1101–1113. https://www.doi.org/10.1002/pip.3727.

Nielsen, Tina. "Ángel León's Discovery of Sea Grains Opens New Frontier in Food and Sustainability." *Foodservice Consultant*, January 21, 2021. https://www.fcsi.org/foodservice-consultant/eame/angel-leons-discovery-sea-grains-opens-new-frontier-food-sustainability.

Olsen, Jeanine, and Wytze Stam. "Genome of the Flowering Plant That Returned to the Sea." *University of Groningen News*, January 26, 2016. https://www.rug.nl/about-ug/latest-news/news/archief2016/nieuwsberichten/het-genoom-van-de-bloeiende-plant-die-naar-zee-ging?lang=en.

Paddison, Laura. "How Replanted Seagrass Is Restoring the Ocean." *Yes! Solutions Journalism*, August 2, 2021. https://www.yesmagazine.org/environment/2021/08/02/replanted-seagrass-is-restoring-the-ocean.

Page, Tom. "Call to Earth: Special Report." CNN, https://www.cnn.com/interactive/2021/10/world/future-of-food-in-the-past-c2e-spc.

Parque de los Toruños. "El parque." http://www.juntadeandalucia.es/avra/opencms/parque-torunos/contenido/1-Conoce-el-parque/parque.html?ulSelected=md-js-menu_submenu-1296092944&liSelected=El%20Parque.

Pearlstein, S. L., R. S. Felger, E. P. Glenn, J. Harrington, K. A. Al-Ghanem, and S. G. Nelson. "Nipa (*Distichlis palmeri*): A Perennial Grain Crop for Saltwater Irrigation." *Journal of Arid Environments* 82 (2012): 60–70.

Restauración News. "Aponiente, the Most Sustainable Restaurant in the World According to the World's 50 Best Restaurants." July 19, 2022. https://www.theworlds50best.com/discovery/Establishments/Spain/El-Puerto-de-Santa-Mar%C3%ADa/Aponiente.html.

Southwest Word Fiesta. "In Memory: Richard Felger." November 6, 2020. https://swwordfiesta.org/in-memory-richard-felger.

Toscano. "Agrovoltaics: Everything You Need to Know." https://toscano.es/2023/01/25/agrovoltaica-beneficios.

University of Arizona Press. "Mary Beck Moser." https://uapress.arizona.edu/author/mary-beck-moser.

University of Groningen. "Consortium with Laura Govers Wins Tender from Dutch Ministry of Infrastructure and Water Management for Scaling Up Seagrass Restoration." October 26, 2022. https://www.rug.nl/fse/news/consortium-with-laura-govers-wins-tender-from-dutch-ministry-of-infrastructure-and-waterstaat-fo?lang=en.

Urquijo, Gema, et al. *Aponiente: Press Book 2022.* El Puerto de Santa María, Andalucía. https://www.aponiente.com/wp-content/uploads/2022/07/PRESS-BOOK-APONIENTE-2022_ENGLISH-20220718.pdf.

Utopicfood. "Chef Ángel León Talks: Love for the Ocean and Planktons—Can Chefs Save the World?" August 11, 2020. https://www.youtube.com/watch?v=jw06TP5ylwk.

World Wildlife Foundation. "Planting Hope: Seagrass." https://www.wwf.org.uk/what-we-do/planting-hope-how-seagrass-can-tackle-climate-change.

CHAPTER TWELVE: UMAMI

Books

Brecher, Jeremy, Jill Cutler, and Brendan Smith, eds. *In the Name of Democracy: American War Crimes in Iraq and Beyond.* New York: Henry Holt, 2005. Kindle edition.

Doumeizel, Vincent. *The Seaweed Revolution: How Seaweed Has Shaped Our Past and Can Save Our Future.* Translated by Charlotte Coombe. London: Legend Press, 2023.

Flavin, Katie, Nick Flavin, and Bill Flahive. *Kelp Farming Manual: A Guide to the Processes, Techniques, and Equipment for Farming Kelp in New England Waters.* Saco, ME: Ocean Approved, 2013. https://maineaqua.org/wp-content/uploads/2020/06/OceanApproved_KelpManualLowRez.pdf.

Smith, Bren. *Eat Like a Fish.* New York: Alfred Knopf, 2019.

Stephenson, W. A. *Seaweed in Agriculture and Horticulture.* London: Faber and Faber, 1968.

Articles

AgriSea New Zealand. "New Zealand's Leading Biostimulants Solution Provider." https://agrisea.co.nz.

Ajinomoto. "What Is Umami?" https://www.ajinomoto.com/umami/5-facts.

AKUA. https://akua.co.

Allen, J. D. "Scallops Dying Off in Long Island Are 'a Cautionary Tale' for New England." WBUR, January 23, 2023. https://www.wbur.org/news/2023/01/23/scallop-death-massachusetts-climate-change#.

Atlantic Sea Farms. https://atlanticseafarms.com.

Baker, Chelsey A., Adrian P. Martin, Andrew Yool, and Ekaterina Popova. "Biological Carbon Pump Sequestration Efficiency in the North Atlantic: A Leaky or a Long-Term Sink?" *Global Biogeochemical Cycles,* May 4, 2022. https://doi.org/10.1029/2021GB007286.

Bauck, Whitney. "It's the 'Swiss Army Knife of the Sea.' But Can Kelp Survive Rising Marine Heat?" *The Guardian,* August 12, 2023. https://www.theguardian.com/environment/2023/aug/12/kelp-farming-ocean-temperatures-climate-crisis.

Bioneers. "The Blue Economy: Too Good Not to Be True | Bren Smith." September 21, 2021. https://bioneers.org/the-blue-economy-too-good-not-to-be-true-bren-smith.

Birnbaum, Michael. "Plastic Pollution in the Ocean Is Doubling Every 6 Years." *Washington Post,* March 8, 2023. https://www.washingtonpost.com/climate-environment/2023/03/08/ocean-plastics-pollution-study.

Boyd, P. W., L. T. Bach, C. L. Hurd, et al. "Potential Negative Effects of Ocean Afforestation on Offshore Ecosystems." *Nature Ecology and Evolution* 6 (April 2022): 675–683. https://doi.org/10.1038/s41559-022-01722-1.

Buckminster Fuller Institute. "GreenWave Wins the 2015 Fuller Challenge." October 21, 2015. https://www.bfi.org/2015/10/21/greenwave-wins-the-2015-fuller-challenge.

Charles, Dan. "Carrageenan Backlash: Food Firms Are Ousting a Popular Additive." NPR, December 12, 2016. https://www.npr.org/sections/thesalt/2016/12/12/504558025/carrageenan-backlash-why-food-firms-are-ousting-a-popular-additive.

CSRWire. "Swedish Philanthropist Awards World's Biggest Environmental Prize." Curt Bergfors Food Planet Prize. November 18, 2021. https://www.csrwire.com/press_releases/731631-swedish-philanthropist-awards-worlds-biggest-environmental-prize.

Curt Bergfors Food Planet Prize. "GreenWave—Taking Ocean Farming to New Depths." https://foodplanetprize.org/initiatives/greenwave-taking-ocean-farming-to-new-depths.

Dillehay, Tom D., C. Ramírez, M. Pino, et al. "Monte Verde: Seaweed, Food, Medicine, and the Peopling of South America." *Science* 320, no. 5877 (May 9, 2008): 784-786. https://doi.org/10.1126/science.1156533.

Duarte, Carlos M., A. Bruhn, and D. A. Krause-Jensen. "A Seaweed Aquaculture Imperative to Meet Global Sustainability Targets." *Nature Sustainability* 5 (2022): 185–193. https://doi.org/10.1038/s41893-021-00773-9.

Duarte, Carlos M., Jean-Pierre Gattuso, Kasper Hancke, Hege Gundersen, et al. "Global Estimates of the Extent and Production of Macroalgal Forests." *Global Ecology and Biogeography*, May 5, 2022. https://doi.org/10.1111/geb.13515.

Eilperin, Juliet. "Can Alaska's Kelp Farms Transform Its Economy?" *Washington Post*, July 31, 2023. https://www.washingtonpost.com/climate-solutions/interactive/2023/alaska-kelp-farming.

Fantom, Lynn. "Can Small Seaweed Farms Help Kelp Scale Up?" *Civil Eats*, March 16, 2022. https://civileats.com/2022/03/16/can-small-seaweed-farms-help-kelp-scale-up.

Fletcher, Rob. "The 10 Most Promising Emerging Applications for Seaweed." *Fish Site*, August 21, 2023. https://thefishsite.com/articles/the-10-most-promising-emerging-applications-for-seaweed-world-bank-hatch-inovation-services.

———. "The Case for Seaweed Subsidies." *Fish Site*, April 20, 2022. https://thefishsite.com/articles/the-case-for-seaweed-subsidies-bren-smith-greenwave.

Fox, Maggie. "Ancient Seaweed Chews Confirm Age of Chilean Site." Reuters, May 8, 2008. https://www.reuters.com/article/idUSN08390999.

Gallagher, John Barry. "Kelp Won't Help: Why Seaweed May Not Be a Silver Bullet for Carbon Storage After All." *The Conversation*, March 10, 2022. https://theconversation.com/kelp-wont-help-why-seaweed-may-not-be-a-silver-bullet-for-carbon-storage-after-all-178018.

Gobler, Christopher J., et al. "Rebuilding a Collapsed Bivalve Population, Restoring Seagrass Meadows, and Eradicating Harmful Algal Blooms in a Temperate Lagoon Using Spawner Sanctuaries." *Frontiers in Marine Science* 9 (August 30, 2022). https://doi.org/10.3389/fmars.2022.911731.

Graham, Charlie. "3-D Ocean Farming." March 30, 2016. https://charliegrahamfood.weebly.com/blog/3-d-ocean-farming.

Grand View Research. "Biostimulants Market Size, Share and Trends Analysis Report by Active Ingredients (Acid Based, Microbial), by Crop Type, by Application (Foliar, Soil Treatment), by Region, and Segment Forecasts, 2023–2030." 2021. https://www.grandviewresearch.com/industry-analysis/biostimulants-market.

GreenBiz. "Sea Briganti." https://www.greenbiz.com/sea-briganti.

GreenWave. "Advanced Farmer Training: Hands-On Kelp Harvest, Co-ops, & Buyer Relationships." https://www.greenwave.org/blog-who-farms-matters/kodiak-harvest-training-2023.

———. "Seaweed-Powered Agriculture: Biostimulants Bridge Land and Sea." May 30, 2023. https://www.greenwave.org/blog-who-farms-matters/agrisea-biostimulant.

———. "Suzie Flores | Connecticut." https://www.greenwave.org/blog-who-farms-matters/suzie-flores.

GreenWave NZ. "What Is GreenWave?" https://smartmaoriaquaculture.co.nz/wp-content/uploads/2021/03/GreenWave-Summary-Slidedeck-for-Iwi-Aquaculture-Hui-2021.pdf.

Griffin, James, and Briana Warner. *Pursuit of Sea Vegetable Market Expansion: Consumer Preferences and Product Innovation.* Island Institute, September 2017. https://www.islandinstitute.org/wp-content/uploads/2020/09/Edible-Seaweed-Summary-Report-9-15-17.pdf.

Guillaume, Tena. "From Algae to Land Plants." *Nature Plants* 6, no. 594 (June 11, 2020). https://doi.org/10.1038/s41477-020-0712-5.

Hall, Georgia. "Lobstermen Face Hypoxia in Outer Cape Waters." *Inside Climate News,* September 4, 2023. https://provincetownindependent.org/local-journalism-project/next-generation/2023/08/16/lobstermen-face-hypoxia-in-outer-cape-waters.

Held, Lisa Elaine. "Kelp Is Not the New Kale; It's a Crop with Bigger Challenges, and Possibilities." *Food Print,* June 17, 2020. https://foodprint.org/blog/kelp-is-not-the-new-kale-its-a-crop-with-bigger-challenges-and-possibilities.

International Monetary Fund. "Climate Change—Fossil Fuel Subsidies." https://www.imf.org/en/Topics/climate-change/energy-subsidies.

Jones, Nicola. "Kelp Gets on the Carbon-Credit Bandwagon." *Hakai Magazine,* February 1, 2022. https://hakaimagazine.com/features/kelp-gets-on-the-carbon-credit-bandwagon.

Kang, Huiyu, Zhengyong Yang, and Zhiyi Zhang. "The Competitiveness of China's Seaweed Products in the International Market from 2002 to 2017." *Aquaculture and Fisheries* 8, no. 5 (September 2023): 579–586. https://doi.org/10.1016/j.aaf.2021.10.003.

Laird, Karen. "Montachem International Partners with Seaweed-Based Resin Maker Loliware." *Sustainable Plastics,* September 5, 2023. https://www.sustainableplastics.com/news/montachem-international-partners-seaweed-based-resin-maker-loliware.

Leone, Brad. "Bren Smith Is Changing the Way Seaweed Is Grown—and Eaten." *Bon Appétit,* October 11, 2013. https://www.bonappetit.com/test-kitchen/ingredients/article/seaweed-farming.

Lieb, Theresa. "Kelp Boom Hinges on Supply Chain and Carbon Market Investments." *GreenBiz,* April 21, 2022. https://www.greenbiz.com/article/kelp-boom-hinges-supply-chain-and-carbon-market-investments.

Loewe, Emma. "The Kelp Business Is Booming. How Big Is Too Big?" *Modern Farmer,* August 15, 2023. https://modernfarmer.com/2023/08/kelp-business-booming.

Loliware. "Introducing AI-Driven Seaweed Biomaterials." https://www.loliware.com.

Lower Carbon Capital. "Running Tide. Kelp Farming Carbon-Sinking Robots." https://lowercarboncapital.com/company/runningtide.

Maine Aquaculture Association. "The Faces of Maine's Working Waterfront: Atlantic Sea Farms." January 10, 2020. https://www.youtube.com/watch?v=U2SIH6ZECgs.

McKinley Research Group. *Alaska Seaweed Market Assessment.* Prepared for Alaska Fisheries Development Foundation, August 2021. https://mckinleyresearch.com/project/alaska-seaweed-market-assessment-2.

National Oceanic and Atmospheric Administration. "Science to Support Sustainable Shellfish and Seaweed Aquaculture Development in Alaska State Waters." August 11,

2022. https://www.smithsonianmag.com/innovation/facing-warming-waters-fishermen-are-taking-up-ocean-farming-180978091.

Nelson, Maggie. "Alaska's Kelp Farming Industry Hits Tricky Hurdle, Despite High Global Demand." KUCB, September 23, 2022. https://www.kucb.org/regional/2022-09-23/alaskas-kelp-farming-industry-hits-tricky-hurdle-despite-high-global-demand.

Notpla. https://www.notpla.com.

Nu, Jennifer. "Kodiak Kelp." *Edible Alaska*, May 23, 2019. https://ediblealaska.ediblecommunities.com/food-thought/kodiak-kelp.

Ocean Approved. http://www.oceanapproved.com.

Ocean Visions. "Macroalgae Cultivation and Carbon Sequestration." https://www2.oceanvisions.org/roadmaps/macroalgae-cultivation-carbon-sequestration.

PR Newswire. "Loliware Launches New Seaweed Resin to Replace Plastics at International Materials Conference." May 16, 2023. https://www.prnewswire.com/news-releases/loliware-launches-new-seaweed-resin-to-replace-plastics-at-international-materials-conference-301825694.html.

Richardson, Jeff. "Kelp Farms Could Help Reduce Coastal Marine Pollution." University of Alaska Fairbanks, January 18, 2023. https://www.uaf.edu/news/kelp-farms-could-help-reduce-coastal-marine-pollution.php.

Schiffman, Richard. "Can We Save the Oceans by Farming Them?" *Yale Environment 360*, October 6, 2016. https://e360.yale.edu/features/new_breed_of_ocean_farmer_aims_to_revive_global_seas.

Scranton, Shane. "Running Tide's Carbon Buoy: Engineering in Partnership with Nature." Running Tide, March 14, 2024. https://www.runningtide.com/blog-post/running-tides-carbon-buoy-engineering-in-partnership-with-nature.

Seafood New Zealand. "Tane and Clare Bradley." November 10, 2022. https://www.seafood.co.nz/news-and-events/news/detail/tane-and-clare-bradley.

Sellinger, Hannah. "Why Isn't Kelp Catching On?" *New York Times*, January 22, 2021. https://www.nytimes.com/2021/01/16/style/self-care/why-isnt-kelp-catching-on.html.

Shinnecock Kelp Farmers. "10,000+ Years Experience & We're Just Getting Started." https://www.shinnecockkelpfarmers.com.

Smith, Brendan [Bren]. "3-D Ocean Farming: Saving Our Seas." Kickstarter, August 26, 2014. https://www.kickstarter.com/projects/greenwave/3-d-ocean-farming-saving-our-seas.

———. "The Coming Green Wave: Ocean Farming to Fight Climate Change." *The Atlantic*, November 23, 2011. https://www.theatlantic.com/international/archive/2011/11/the-coming-green-wave-ocean-farming-to-fight-climate-change/248750.

———. "Ecological Redemption: Ocean Farming in the Era of Climate Change." Schumacher Center. https://centerforneweconomics.org/publications/ecological-redemption-ocean-farming-in-the-era-of-climate-change.

———. "A Green Art Manifesto." *Common Dreams*, December 19, 2009. https://www.commondreams.org/views/2009/12/19/green-art-manifesto.

———. "Who Ocean Farms Matters: Women Are Building the New Blue Economy." *Medium*, May 28, 2019. https://medium.com/@bren_7821/who-ocean-farms-matters-women-are-leading-the-new-blue-economy-4cff4941ac88.

Stańska, Katarzyna, and Antonii Krzeski. "The Umami Taste: From Discovery to Clinical Use." *Polish Journal of Otolaryngology* 70, no. 4 (June 30, 2016): 10–15. http://doi.org/10.5604/00306657.1199991.

Stekoll, Michael. "Development of Scalable Coastal and Offshore Macroalgal Farming." ARPA-E. https://arpa-e.energy.gov/sites/default/files/2021-07/UAF%20Integrated%20Cultivation%20-%202021%20Annual%20Review%20UAF%20CAT%201.pdf.

Stony Brook University News. "Kelp Mitigates Ocean Acidification, a Key to the Health and Abundance of Important Shellfish." May 25, 2022. https://news.stonybrook.edu/university/sbu-study-shows-kelp-can-reduce-ocean-acidification-and-protect-bivalves.

Temple, James. "Running Tide Is Facing Scientist Departures and Growing Concerns over Seaweed Sinking for Carbon Removal." *MIT Technology Review*, June 16, 2022. https://www.technologyreview.com/2022/06/16/1053758/running-tide-seaweed-kelp-scientist-departures-ecological-concerns-climate-carbon-removal.

United States Senate Committee on the Budget. "Sen. Whitehouse on Fossil Fuel Subsidies: 'We Are Subsidizing the Danger.'" May 3, 2023. https://www.budget.senate.gov/chairman/newsroom/press/sen-whitehouse-on-fossil-fuel-subsidies-we-are-subsidizing-the-danger-.

Virgil. *Aeneid.* Google Books. https://www.google.com/books/edition/The_Collected_Works_of_Virgil/w6HiEAAAQBAJ?hl=en&gbpv=1&dq=Aeneid,+Virgil+%C2%A0%22sea-weed%22+worthless&pg=PT412&printsec=frontcover.

Vittek, Shelby. "Facing Warming Waters, Fishermen Are Taking Up Ocean Farming." *Smithsonian*, June 30, 2021. https://www.smithsonianmag.com/innovation/facing-warming-waters-fishermen-are-taking-up-ocean-farming-180978091.

Wills, Matthew. "Burning Kelp for War." *JSTOR Daily*, January 13, 2022. https://daily.jstor.org/burning-kelp-for-war.

Wist, Allie. "This Farmer Thinks Kelp Will Help Save the World." *Saveur*, June 12, 2018. https://www.saveur.com/bren-smith-kelp-farmer.

World Bank. *Global Seaweed: New and Emerging Markets Report, 2023.* International Bank for Reconstruction and Development, 2023. http://documents.worldbank.org/curated/en/099081423104548226/P175786073c14c01609fe409c202ddf12d0.

World Bank Group. *Seaweed Aquaculture for Food Security, Income Generation and Environmental Health in Tropical Developing Countries.* World Bank Group, 2016. https://documents1.worldbank.org/curated/en/947831469090666344/pdf/107147-WP-REVISED-Seaweed-Aquaculture-Web.pdf.

Zhang, Jingsi, Çağrı Akyol, and Erik Meers. "Nutrient Recovery and Recycling from Fishery Waste and By-products." *Journal of Environmental Management* 348 (December 15, 2023): 119266. https://doi.org/10.1016/j.jenvman.2023.119266.

CHAPTER THIRTEEN: A TALE OF THREE CITIES

Articles

Aguirre-Torres, Luis. "Inventing a Sustainable Future for Latin America." The Whitehouse, President Barack Obama, blog, July 27, 2012. https://obamawhitehouse.archives.gov/blog/2012/07/27/inventing-sustainable-future-latin-america.

———. LinkedIn profile. https://www.linkedin.com/in/luis-aguirre-torres-80483b285.

Andean Forests. "San Sebastian Reserve: Special Forest Reserve." https://www.andeanforests.org/colombia/reserva-san-sebastian-de-la-castallana.

Baldwin, Eric. "CopenHill: The Story of BIG's Iconic Waste-to-Energy Plant." *ArchDaily*, October 7, 2019. https://www.archdaily.com/925966/copenhill-the-story-of-bigs-iconic-waste-to-energy-plant.

BBC. "Colombia's Medellin Named 'Most Innovative City.'" March 1, 2023. https://www.bbc.com/news/world-latin-america-21638308.

"BIG." Bjarke Ingels Group. https://big.dk.

Bliss-Herrera, Alexandra Clarke, and Peter Myers. "Medellín, Colombia: A Case Study for Healthy Cities." https://healthymedellin.weebly.com/comuna-13.html.

BlocPower. "City of Ithaca, NY Signs Contract with BlocPower to 'Green' Entire City, First Large-Scale City Electrification Initiative in the U.S." Press release. November 4, 2021. https://www.blocpower.io/posts/ithaca-contract-announcement#.

Butler, Matt. "The End of the Svante Myrick Era." *Ithaca Voice*, February 8, 2022. https://ithacavoice.org/2022/02/the-end-of-the-svante-myrick-era.

C40 Cities. "Mayors Warn Vehicle Manufacturers: 'The Health of Our Children Is More Important Than the Health of Your Profits.'" Press release. October 29, 2018. https://www.c40.org/news/who-air-quality.

Canary Media. "Ithaca's Novel Plan to Eliminate Carbon from Buildings." *Carbon Copy* podcast, November 30, 2021. https://www.canarymedia.com/podcasts/the-carbon-copy/ithacas-novel-plan-to-eliminate-carbon-from-buildings.

Cirino, Adriano. "Up Hill." *Plough*, November 17, 2023. https://www.plough.com/en/topics/justice/social-justice/economic-justice/up-hill.

City of Ithaca, New York. "Green New Deal." http://www.cityofithaca.org/642/Green-New-Deal.

Collins, Jessica. "Makoko Floating School, Beacon of Hope for the Lagos 'Waterworld'—a History of Cities in 50 Buildings, Day 48." *The Guardian*, June 2, 2015. https://www.theguardian.com/cities/2015/jun/02/makoko-floating-school-lagos-waterworld-history-cities-50-buildings.

Comisión de la Verdad. *Hay Futuro Si Hay Verdad*. https://www.comisiondelaverdad.co/hay-futuro-si-hay-verdad.

Cornell University. "Courtney Ann Roby." https://classics.cornell.edu/courtney-ann-roby.

Cruz Riaño, Íngrid. "Los corredores verdes le dan vida a Medellín." *Centrópolis*, July 22, 2019. https://www.centropolismedellin.com/os-corredores-verdes-le-dan-vida-medellin.

El País. "Medellín, Designated the Most Innovative City in the World by City of the Year." March 1, 2023. https://elpais.com/economia/2013/03/01/agencias/1362152298_236500.html.

Fixsen, Anna. "Bjarke Ingels Unveils an Ambitious Plan for Floating Cities at the U.N." *Architectural Digest*, April 4, 2019. https://www.architecturaldigest.com/story/bjarke-ingels-plan-floating-cities-un.

Flavelle, Christopher, Anne Barnard, Brad Plumer, and Michael Kimmelman. "Overlapping Disasters Expose Harsh Climate Reality: The U.S. Is Not Ready." *New York Times*, September 2, 2021. https://www.nytimes.com/2021/09/02/climate/new-york-rain-floods-climate-change.html.

Gibson, Rachel. "BIG Unveils Oceanix City Concept for Floating Villages That Can Withstand Hurricanes." *Dezeen*, April 4, 2019. https://www.dezeen.com/2019/04/04/oceanix-city-floating-big-mit-united-nations.

Guarino, Ben. "As Seas Rise, the U.N. Explores a Bold Plan: Floating Cities." *Washington Post*, April 5, 2019. https://www.washingtonpost.com/science/2019/04/05/seas-rise-un-explores-bold-plan-floating-cities.

Haasnoot, Marjolijn, Judy Lawrence, and Alexandre K. Magnan. "Pathways to Coastal Retreat." *Science* 372, no. 6548 (June 18, 2021): 1287–1290. https://doi.org/10.1126/science.abi6594.

Harding, Tanner. "Climate Czar: Luis Aguirre-Torres Leads Ithaca Toward a Greener Future." *Ithaca Times*, June 24, 2021. https://www.ithaca.com/news/ithaca/climate-czar-luis-aguirre-torres-leads-ithaca-toward-a-greener-future/article_24adef9e-d44c-11eb-a4ba-1f84fbb3ac4e.html.

Harris, Lee. "The Wall Street Bet behind Ithaca's Green New Deal." *New York Focus*, January 23, 2023. https://nysfocus.com/2023/01/23/ithaca-blocpower-goldman-sachs-green-new-deal.

Herbert-Read, J. E., A. Thornton, D. J. Amon, et al. "A Global Horizon Scan of Issues Impacting Marine and Coastal Biodiversity Conservation." *Nature Ecology and Evolution* 6 (July 7, 2022): 1262–1270. https://doi.org/10.1038/s41559-022-01812-0.

Ithaca Voice. "An Open Letter to the City of Ithaca Concerning Social Justice and the Green New Deal." October 13, 2022. https://ithacavoice.org/2022/10/op-ed-an-open-letter-to-the-city-of-ithaca-concerning-social-justice-and-the-green-new-deal.

Janke, Krista A. "Investment Key to Partnership with Ithaca, N.Y. to Lead City to Full Decarbonization by 2030." Kresge Foundation, July 14, 2022. https://kresge.org/news-views/kresge-commits-3-million-via-guarantee-into-climate-tech-firm-blocpower.

Jeffries, Scott. "How Much Money Is in the World Right Now?" GOBankingRates, March 22, 2024. https://www.gobankingrates.com/money/economy/how-much-money-is-in-the-world.

Jennewein, Chris. "USC: Rising Sea Levels Will Force 13 Million Americans to Relocate Inland." *Times of San Diego*, January 24, 2020. https://timesofsandiego.com/tech/2020/01/24/usc-rising-sea-levels-will-force-13-million-americans-to-relocate-inland.

Jonas, Natalie. "Can Floating Cities Save Us from Rising Sea Levels?" *Salon*, December 18, 2022. https://www.salon.com/2022/12/18/can-floating-cities-save-us-from-rising-sea-levels.

Jordan, Jimmy. "Aguirre-Torres Details Resignation from Ithaca Sustainability Leadership, Fears over Green New Deal." *Ithaca Voice*, November 7, 2022. https://ithacavoice.org/2022/11/aguirre-torres-details-resignation-from-ithaca-sustainability-leadership-fears-over-green-new-deal.

———. "Ithaca Is Electric: Stage Set for Decarbonization of City's Buildings." *Ithaca Voice*, July 14, 2022. https://ithacavoice.org/2022/07/ithaca-is-electric-stage-set-for-decarbonization-of-citys-buildings.

Kolbert, Elizabeth. "The Siege of Miami." *New Yorker*, December 13, 2015. https://www.newyorker.com/magazine/2015/12/21/the-siege-of-miami.

Kulp, S. A., and B. H. Strauss. "New Elevation Data Triple Estimates of Global Vulnerability to Sea-Level Rise and Coastal Flooding." *Nature Communications* 10, no. 4844 (October 29, 2019). https://doi.org/10.1038/s41467-019-12808-z.

LeRoy, Sverre, and Richard Wiles. *High Tide Tax: The Price to Protect Coastal Communities from Rising Seas.* Center for Climate Integrity. June 2019. https://climatecosts2040.org/files/ClimateCosts2040_Report-v5.pdf.

McCoy, Martha. "Ithaca Races against the Clock to Decarbonize All Buildings by 2030—Episode 144 of *Local Energy Rules.*" Institute for Local Self-Reliance podcast. November 24, 2021. https://ilsr.org/articles/ithaca-new-york-building-decarbonization-ler144.

Medellin Times. "Medellín Was Chosen as Discovery City 2019." June 25, 2019. https://medellintimes.com/medellin-was-chosen-as-discovery-city-2019.

Mimura, N. "Rising Seas and Subsiding Cities." *Nature Climate Change* 11 (April 2021): 296–297. https://rdcu.be/dI7TQ.

Oceanix. "Explore Oceanix Busan." https://oceanix.com/busan.

Ortiz, Heidy Yohana Tamayo. "The Atrocious Crime Wave in Bello, Where Human Remains Would Go to Pigs." *El Tiempo*, July 17, 2019. https://www.google.com/search?client=safari&rls=en&q=Comentar+++MEDELL%C3%8DN+La+atroz+ola+criminal+en+Bello%2C+donde+restos+humanos+ir%C3%ADan+a+marranera&ie=UTF-8&oe=UTF-8.

Patowary, Kaushik. "The Floating Houses of IJburg, Amsterdam." *Amusing Planet*, June 8, 2015. https://www.amusingplanet.com/2015/06/the-floating-houses-of-ijburg-amsterdam.html.

Planete Energies. "Medellín: Public Transportation for Social Change." March 3, 2018. https://www.planete-energies.com/en/media/article/medellin-public-transportation-social-change.

Pressley, Linda. "The Dump That Holds the Secrets of the Disappeared." BBC, December 30, 2014. https://www.bbc.com/news/magazine-30573931.

Prime Movers Lab. "Featured Founder: Oceanix's Marc Collins Chen and Itai Madamombe." *Medium*, October 18, 2021. https://medium.com/prime-movers-lab/featured-founder-oceanixs-marc-collins-chen-and-itai-madamombe-de3d352d5cf0.

———. "Ocean Technology." June 24, 2021. https://www.youtube.com/watch?v=6El0dSz7zxw.

Revkin, Andrew. "Floating Cities Could Ease the World's Housing Crunch, the UN Says." *National Geographic*, April 5, 2019. https://www.nationalgeographic.com/environment/article/floating-cities-could-ease-global-housing-crunch-says-un?hootPostID=58e232e2b5d4cdaa629663b0ef252c7f&loggedin=true&rnd=1717121078049.

Rodin, Judith. "The Transformation of Medellín Provides a Model for Cities Worldwide." *The Guardian*, April 10, 2014. https://www.theguardian.com/cities/2014/apr/10/medellin-transformation-model-cities-worldwide-resilience.

Root, Tik. "Ithaca, New York Votes to Decarbonize Every Building in Climate Change Fight." *Washington Post*, November 3, 2021. https://www.washingtonpost.com/climate-solutions/2021/11/03/ithaca-new-york-decarbonize-electrify.

Rubio, Isabel. "Oceanix Busan: The Dream of Building a Floating City to Tackle Climate Change." *El País*, June 17, 2023. https://english.elpais.com/science-tech

/2023-06-17/oceanix-busan-the-dream-of-building-a-floating-city-to-tackle-climate-change.html.

Stewart, Stanley. "How Medellin Went from Murder Capital to Hipster Holiday Destination." *The Telegraph*, January 4, 2018. https://www.telegraph.co.uk/travel/destinations/south-america/colombia/articles/medellin-murder-capital-to-hipster-destination.

Stone, Jon. "Richard Buckminster Fuller's Triton Floating City Project." Behance. https://www.behance.net/gallery/2971307/Richard-Buckminster-Fullers-Triton-City-project?tracking_source=search_projects%7Cjon+stone+triton.

Takemura, Alison F. "Chart: Here's How the US Could Get Heat Pumps in Every Home by 2050." *Canary Media*, June 23, 2023. https://www.canarymedia.com/articles/heat-pumps/chart-heres-how-the-us-could-get-heat-pumps-in-every-home-by-2050.

UN Environment Programme. "Medellín Shows How Nature Based Solutions Can Keep People and Planet Cool." July 17, 2019. https://www.unep.org/news-and-stories/story/medellin-shows-how-nature-based-solutions-can-keep-people-and-planet-cool.

UN-Habitat. "UN-Habitat and Partners Unveil OCEANIX Busan, the World's First Prototype Floating City." https://unhabitat.org/news/27-apr-2022/un-habitat-and-partners-unveil-oceanix-busan-the-worlds-first-prototype-floating.

UN-Habitat and Oceanix. "Busan, UN-Habitat and OCEANIX Set to Build the World's First Sustainable Floating City Prototype as Sea Levels Rise." November 18, 2021. https://unhabitat.org/news/18-nov-2021/busan-un-habitat-and-oceanix-set-to-build-the-worlds-first-sustainable-floating.

Urban Green Council. "Decoding New York State's All-Electric New Buildings Law." Urban Green, May 3, 2023. https://www.urbangreencouncil.org/decoding-new-york-states-all-electric-new-buildings-law.

Urban Rigger. https://urbanrigger.com.

Wainwright, Oliver. "Seasteading—a Vanity Project for the Rich or the Future of Humanity?" *The Guardian*, June 24, 2020. https://www.theguardian.com/environment/2020/jun/24/seasteading-a-vanity-project-for-the-rich-or-the-future-of-humanity.

Wang, Lucy. "Floating Cities: It's an Unsinkable Idea." *Anthropocene*. https://www.anthropocenemagazine.org/2020/07/its-an-unsinkable-idea.

Waste360. "Dr. Luis Aguirre-Torres Resigns as City of Ithaca's Director of Sustainability." November 9, 2022. https://www.waste360.com/sustainability/dr-luis-aguirre-torres-resigns-as-city-of-ithaca-s-director-of-sustainability.

Watts, Jonathan. "From Miami to Shanghai: 3C of Warming Will Leave World Cities below Sea Level." *The Guardian*, November 3, 2017. https://www.theguardian.com/cities/2017/nov/03/miami-shanghai-3c-warming-cities-underwater.

World Ocean Forum. "Itai Madamombe Session Keynote," November 18, 2022. https://www.youtube.com/watch?v=pGRLeh2rvoY.

CHAPTER FOURTEEN: THE ATOLLS' LEGACY

Books

Currie, Ruth Douglas. *Kwajalein Atoll, the Marshall Islands and American Policy in the Pacific.* Jefferson, NC: McFarland, 2016. Kindle edition.

Jetñil-Kijiner, Kathy. *Iep Jāltok: Poems from a Marshallese Daughter.* Tucson: University of Arizona Press, 2017.

Johnson, Giff. *Idyllic No More: Pacific Island Climate, Corruption and Development Dilemmas.* Port Vila, Vanuatu: Pacific Institute for Public Policy, 2015. Kindle edition.

———. *Our Ocean's Promise: From Aspirations to Inspirations; The Marshall Islands Fishing Story.* Majuro: MIMRA Books, 2021. Kindle edition.

Philippo, Jim. *Kwajalein Atoll: The Legacy of Faith and Hope.* Xlibris US, 2015. Kindle edition.

Pincus, Walter. *Blown to Hell: America's Deadly Betrayal of the Marshall Islanders.* New York: Diversion Books, 2021.

Articles

Abella, Maveric K. I. L., Monica Rouco Molina, Ivana Nikolić-Hughes, and Malvin A. Ruderman. "Background Gamma Radiation and Soil Activity Measurements in the Northern Marshall Islands." *Proceedings of the National Academy of Sciences* 116, no. 31 (July 15, 2019): 15425–15434. https://doi.org/10.1073/pnas.1903421116.

Anjain, Jelton. "Spartan Summer Camp Report 2017." Kwajalein Atoll Public School System, Republic of the Marshall Islands.

Babic, Mary. "Big Poultry Finds Workers in an Immigrant Community Known for Its Culture of Forgiving." Oxfam America. November 18, 2015. https://www.oxfamamerica.org/explore/stories/big-poultry-finds-workers-in-an-immigrant-community-known-for-its-culture-of-forgiving.

Bittle, Jake. "Inside the Marshall Islands' Life-or-Death Plan to Survive Climate Change." *Grist,* December 5, 2023. https://grist.org/extreme-weather/marshall-islands-national-adaptation-plan-sea-level-rise-cop28.

Burgos, Annalisa. "'We're Survivors': Marshall Islands Leaders Look to Hawaii for Adapting to Climate Change Crisis." *Hawaii News Now,* March 10, 2023. https://www.hawaiinewsnow.com/2023/03/11/marshallese-leaders-visit-hawaii-raise-awareness-develop-solutions-climate-change-crisis.

Burns, Cameron M. "For the Least Among Us: New Approaches to Refugee Care." *World & I,* January–February 2005, pp. 24–35.

Cappucci, Matthew. "Rogue Waves Slammed U.S. Base in Marshall Islands." *Washington Post,* January 25, 2024. https://www.washingtonpost.com/weather/2024/01/25/rogue-wave-marshall-islands-military-base.

Carr, David F. "Promise Unfilled Case 165 A Dissection." *Baseline,* May 2005. https://www.carrcommunications.com/clips/0505-StrAngel.pdf.

Cleary, Ryder, Will Haga, Max Jenkins, Kelsee Miller, and Kenneth McDonald. *Ebeye 2023: Comprehensive Capacity Development Master Plan.* West Point, NY: Center for Nation Reconstruction and Capacity Development, United States Military Academy, 2012. https://www.westpoint.edu/sites/default/files/inline-images/centers_research/national_reconstruction_capactity_development/pdf%20tech%20reports/Ebeye%2520Report.pdf.

College of the Marshall Islands. https://www.cmi.edu.

Conard, Robert A. *Twenty-Year Review of Medical Findings in a Marshallese Population Accidentally Exposed to Radioactive Fallout.* Brookhaven National Laboratory, United States Energy Research and Development Administration, January 1, 1975. In the collection Office of Scientific and Technical Information Technical Reports, University of North Texas Libraries, UNT Digital Library, Government Documents. https://digital.library.unt.edu/ark:/67531/metadc866112/m1/.

———. *Fallout: The Experiences of a Medical Team in the Care of a Marshallese Population Accidentally Exposed to Fallout Radiation.* US Department of Energy, December 31, 1991. https://doi.org/10.2172/10119323.

Congressional Record. "Urging the Administration to Include the Marshall Islanders in a Government Study of Radiation Experiments." Vol. 140, no. 15 (February 22, 1994). https://www.govinfo.gov/content/pkg/CREC-1994-02-22/html/CREC-1994-02-22-pt1-PgH31.htm.

DOE Openness: Human Radiation Experiments. "Chapter 12: The Marshallese." Capsule descriptions of the *Final Report of the Advisory Committee on Human Radiation Experiments,* 1995. https://ehss.energy.gov/ohre/roadmap/achre/chap12_3.html.

Dutch Water Sector. "Deltares and Marshall Islands Work Together on Battle against Climate Change." July 29, 2021. https://www.dutchwatersector.com/news/deltares-and-marshall-islands-work-together-on-battle-against-climate-change.

EurekaAlert. "Radioactive Contamination in the Marshall Islands." July 15, 2019. https://www.eurekalert.org/news-releases/588860.

Gander, Kashmira. "Marshall Islands, Where U.S. Ran 67 Nuclear Weapon Tests, More Contaminated Than Fukushima and Chernobyl." *Newsweek,* July 16, 2019. https://www.newsweek.com/marshall-islands-u-s-nuclear-weapons-tests-contaminated-fukushima-chernobyl-1449463.

Giardino, Alessio, Leo van Rijn, Ellen Quataert, and Andrew Warren. "SimpleCoast: Simple Assessments of Coastal Problems and Solutions." Conference paper from MEDCOAST 17, Malta, November 2017.

Goldman Environmental Prize. "Jeton Anjain." https://www.goldmanprize.org/recipient/jeton-anjain.

———. "Jeton Anjain: 1992 Goldman Prize Winner, Marshall Islands." YouTube video. https://youtu.be/oMdC_dYwTX4?si=h34pGSZMVvRpg8Co.

Greenpeace. "Murder in the Pacific: The Sinking of the Rainbow Warrior and What Happened Next." March 3, 2023. https://www.greenpeace.org/usa/murder-in-the-pacific-the-sinking-of-the-rainbow-warrior-and-what-happened-next.

———. "Rongelap: The Exodus Project." https://www.greenpeace.org/usa/victories/rongelap-the-exodus-project.

Heine, Hilda, and Patrick Verkoojin. "Don't Let the Rising Seas Drown the Marshall Islands." *Washington Post,* April 10, 2019. https://www.washingtonpost.com/opinions/2019/04/10/dont-let-rising-seas-drown-marshall-islands.

Hofschneider, Anita. "New Study of Marshall Islands Fish Highlights Peril of Using Oceans as Dumping Grounds." *Honolulu Civil Beat,* January 16, 2023. https://www.civilbeat.org/2023/01/new-study-of-marshall-islands-fish-highlights-peril-of-using-oceans-as-dumping-grounds.

Infinitum Humanitarian Systems. https://www.ihs-i.com.

———. "Eric Rasmussen, MD, MDM, FACP." https://www.ihs-i.com/eric-rasmussen.

Innovation City Forum. "Kathy Jetñil-Kijiner: 'Island Cities Under Water: The Case for Marshall Islands.'" Keynote Session 3, October 19, 2018. https://youtu.be/cGsN7tQ_R24?si=eNR_plibXOtgkAC-.

Jetñil-Kijiner. Kathy. "Dome Poem Part I: The Voyage." February 2, 2018. https://www.kathyjetnilkijiner.com/dome-poem-part-i-the-voyage.

———. "Dome Poem Part II: Of Islands and Elders." March 2, 2018. https://www.kathyjetnilkijiner.com/dome-poem-part-ii-of-islands-and-elders.

———. "Dome Poem Part III: 'Anointed' Final Poem and Video." April 16, 2018. https://www.kathyjetnilkijiner.com/dome-poem-iii-anointed-final-poem-and-video.

———. "Why Don't Marshallese People Leave Their Climate-Threatened Islands?" *Climate Home News*, June 7, 2015. https://www.climatechangenews.com/2015/07/06/why-dont-marshallese-people-leave-their-climate-threatened-islands.

Johnson, Giff. "Marshall Islands Celebrate First Coronation of Paramount Chief in 50 Years." RNZ (Radio New Zealand), July 23, 2022. https://www.rnz.co.nz/international/pacific-news/471470/marshall-islands-celebrate-first-coronation-of-paramount-chief-in-50-years.

———. "Micronesia: America's 'Strategic' Trust." *Bulletin of the Atomic Scientists* 35, no. 2 (September 15, 2015): 1–15. https://doi.org/10.1080/00963402.1979.11458582.

Keating, Joshua. "This Program Divided the World into 57 Trillion Squares and Gave Them Names like Usage.Ample.Soup." *Slate*, June 17, 2016. https://slate.com/technology/2016/06/what3words-divided-the-world-into-57-trillion-squares.html#.

Knighton, Hannah. "Navigating the Waters with Micronesian Stick Charts." Smithsonian National Museum of Natural History. https://ocean.si.edu/human-connections/history-cultures/navigating-waters-micronesian-stick-charts.

Kormann, Carolyn. "The Cost of Fleeing Climate Change." *New Yorker*, January 10, 2020. https://www.newyorker.com/news/dispatch/the-cost-of-fleeing-climate-change-marshall-islands-arkansas.

Kwajalein Atoll Sustainability Laboratory. https://www.kasl.earth.

Letman, Jon. "Rising Seas Give Island Nation a Stark Choice: Relocate or Elevate." *National Geographic*, November 19, 2018. https://www.nationalgeographic.com/environment/article/rising-seas-force-marshall-islands-relocate-elevate-artificial-islands.

Marcoux, Shannon. "Trust Issues: Militarization, Destruction, and the Search for a Remedy in the Marshall Islands." *Columbia Human Rights Law Review*, January 2021, pp. 100–145. https://hrlr.law.columbia.edu/files/2021/01/MarcouxJan11.pdf.

Marshall Islands Journal. "Marshall Islands Guide: Enewetak Atoll." https://www.infomarshallislands.com/atolls-a-l/enewetak-atoll.

———. "RMI on 50-Year Clock." November 17, 2022. https://marshallislandsjournal.com/rmi-on-50-year-clock/climate-11-18-22.

Mcdonald, Joshua. "Rising Sea Levels Threaten Marshall Islands' Status as a Nation, World Bank Report Warns." *The Guardian*, October 16, 2021. https://www.theguardian.com/world/2021/oct/17/rising-sea-levels-threaten-marshall-islands-status-as-a-nation-world-bank-report-warns.

McKenzie, Pete. "Marshall Islands Caught in U.S.-China Pacific Race for Influence." *Washington Post*, January 27, 2023. https://www.washingtonpost.com/world/2023/01/27/us-marshall-islands-china-pacific-power.

Miller, Michael E. "Tuvalu's Deal with Australia Addresses Rising Sea Level, but Also China." *Washington Post*, December 26, 2023. https://www.washingtonpost.com/world/2023/12/26/australia-tuvalu-deal-climate-change-pacific.

Moscia, Lorenzo. "Ebadon, Marshall Islands 4K." https://www.youtube.com/watch?v=Rr4ZFGtg4Gg.

Nakano, Gregg. "Failing to Survive: Autoethnography of an Accidental Educator." Doctor of education diss., University of Hawai'i at Mānoa, 2021.

New York Times. "Around the World; 70 Are Moved from Atoll Used in '54 Atomic Test." May 21, 1985. https://www.nytimes.com/1985/05/21/world/around-the-world-70-are-moved-from-atoll-used-in-54-atomic-test.html.

NPR. "We Are on the Front Line of Climate Change, Marshall Islands President Says." September 24, 2019. https://www.npr.org/2019/09/24/763679518/we-are-on-the-front-line-of-climate-change-marshall-islands-president-says.

Nuclear Age Peace Foundation. "Nuclear Zero Profiles: John Anjain." January 15, 2015. https://www.wagingpeace.org/john-anjain.

Pacific and Virgin Islands Training Initiatives. "Jelton Anjain." December 5, 2022. https://pitiviti.org/participant/231-pacific.

PBS NewsHour. "Marshall Islands: A Third of the Nation Has Left for the US." https://youtu.be/ZB8s_Yqp3ko?si=lH8XAYXA5MaqzQsA.

Peak, Robin. "Military and Civilian Student 2019 Cohort Completes Pacific ALLIES Internship." U.S. Indo-Pacific Command. August 5, 2019. https://www.pacom.mil/Media/News/News-Article-View/Article/1926324/military-and-civilian-student-2019-cohort-completes-pacific-allies-internship.

Rasmussen, Eric D. "Oceania Human Security Laboratory Research Expansion." Proposal submitted to the US Office of Naval Research in response to FY22 Long Range Broad Agency Announcement (BAA) for Navy and Marine Corps Science and Technology, 2023.

Rasmussen, Eric, Gregg Nakano, Kitlang Kabua, and Alex Hatoum. "Climate-Focused Field Research within the Kwajalein Atoll Sustainability Laboratory." Submission to Office of Naval Research, November 2022.

Rust, Suzanne. "How the U.S. Betrayed the Marshall Islands, Kindling the Next Nuclear Disaster." *Los Angeles Times,* November 10, 2019. https://www.latimes.com/projects/marshall-islands-nuclear-testing-sea-level-rise.

———. "Radiation in Parts of the Marshall Islands Is Far Higher Than Chernobyl, Study." *Los Angeles Times,* July 15, 2019. https://www.latimes.com/nation/la-na-marshall-islands-radiation-20190715-story.html.

Rust, Suzanne, and Carolyn Cole. "High Radiation Levels Found in Giant Clams near U.S. Nuclear Dump in Marshall Islands." *Los Angeles Times,* November 10, 2019. https://www.latimes.com/science/environment/la-me-marshall-islands-dome-is-leaking-radiation-20190528-story.html.

Spitznagel, Eric. "Biggest US Nuclear Bomb Test Destroyed an Island—and This Man's Life." *New York Post,* November 20, 2021. https://nypost.com/2021/11/20/biggest-us-nuclear-bomb-test-destroyed-an-island-and-lives.

Storlazzi, Curt D., Stephen B. Gingerich, Ap van Dongeren, Olivia M. Cheriton, et al. "Most Atolls Will Be Uninhabitable by the Mid-21st Century Because of Sea-Level Rise Exacerbating Wave-Driven Flooding." *Science Advances* 25 (April 2018): 1–9. https://doi.org/10.1126/sciadv.aap974.

The Trebuchet. "Gregg Nakano." September 25, 2021. https://www.the-trebuchet.org/blog/2021/9/25/gregg-nakano.

UNICEF. "16 Children File Landmark Complaint to the United Nations Committee on the Rights of the Child." October 13, 2019. https://www.unicef.org/pacificislands

/press-releases/16-children-file-landmark-complaint-united-nations-committee-rights-child.

US Citizen and Immigration Services. "Status of Citizens of the Freely Associated States of the Federated States of Micronesia and the Republic of the Marshall Islands." Fact sheet. https://www.uscis.gov/sites/default/files/document/fact-sheets/FactSheetVerifyFASCitizens.pdf.

Wolters Kluwer. "Yacht Registration in the Marshall Islands." https://www.wolterskluwer.com/en/solutions/ct-corporation/yacht-registration-in-the-marshall-islands.

World Bank. "Marshall Islands: New Climate Study Visualizes Confronting Risk of Projected Sea Level Rise." October 29, 2021. https://www.worldbank.org/en/news/press-release/2021/10/29/marshall-islands-new-climate-study-visualizes-confronting-risk-of-projected-sea-level-rise.

Worldwide Wilbur. "Ebeye Island: 'The Slum of the Pacific.'" November 30, 2018. https://worldwidewilbur.com/ebeye-travel-guide.

Zak, Dan. "On the Island of Ebeye, a Nuclear Past and Ballistic Present." Pulitzer Center. December 18, 2015. https://pulitzercenter.org/stories/island-ebeye-nuclear-past-and-ballistic-present.

CHAPTER FIFTEEN: HOPE DIES LAST

Books

Choi, Jae-Eun. *Lucy and Her Time.* Seoul: Kukje Gallery, 2012.

Articles

Aizen, Marcelo A., Lucas A. Garibaldi, Saul A. Cunningham, and Alexandra M. Klein. "How Much Does Agriculture Depend on Pollinators? Lessons from Long-Term Trends in Crop Production." *Annals of Botany* 103, no. 9 (June 2009): 1579–1588. https://doi.org/10.1093/aob/mcp076.

Art It. "The Nature Rules: Dreaming of Earth Project [Hara Museum, Tokyo]." https://www.art-it.asia/en/partners_e/museum_e/haramuseum_e/198093.

Arute, F., K. Arya, R. Babbush, et al. "Quantum Supremacy Using a Programmable Superconducting Processor." *Nature* 574 (2019): 505–510. https://doi.org/10.1038/s41586-019-1666-5.

Bansal, Sarika. "The Plans for the World's Next Largest City Are Incomplete." *New York Times,* January 20, 2022. https://www.nytimes.com/2022/01/20/world/asia/delhi-worlds-largest-city.html.

BBC. "Google Says Its Carbon Footprint Is Now Zero." September 14, 2020. https://www.bbc.com/news/technology-54141899#.

Bearak, Max, Dylan Moriarty, and Júlia Ledur. "How Africa Will Become the Center of the World's Urban Future." *Washington Post,* November 19, 2021. https://www.washingtonpost.com/world/interactive/2021/africa-cities.

Bloomberg Media. "AI Is Helping Mitigate Aviation's Climate Impact." https://sponsored.bloomberg.com/article/google-sustainability/ai-is-helping-mitigate-aviations-climate-impact#.

Climate Defiance. https://www.climatedefiance.org.

———. "Big update . . ." X post, August 21, 2023. https://x.com/ClimateDefiance/status/1693744432393113817.

Correal, Annie, and Genevieve Glatsky. "Tourists to Colombia Warned against Using Dating Apps after Sedatives Fuel Crime." *New York Times*, January 24, 2024. https://www.nytimes.com/2024/01/23/world/americas/colombia-dating-apps-sedatives-deaths.html.

Designboom. "A Two-Part Exhibition at Ginza Maison Hermès Forum." https://www.designboom.com/art/jaeeun-choi-retrospective-ginza-maison-hermes-forum-examines-ecology-art-tokyo-10-25-2023.

Eye on A.I. Episode 20. Featuring Craig Smith and John Platt. Podcast. CCS Media. 2019. https://www.eye-on.ai/podcast-020?rq=craig%20smith%20john%20platt.

Feffer, John. "Colombia Adopts an Unprecedented Energy Policy—but Needs Help to Pull It Off." *The Nation*, April 7, 2023.

Gallucci, Maria. "In a First, a Major Airline Will Cross the Atlantic without Fossil Fuels." *Canary Media*, November 27, 2023. https://www.canarymedia.com/articles/air-travel/in-a-first-a-major-airline-will-cross-the-atlantic-without-fossil-fuels.

———. "The Smelly, Greasy Truth about How Sustainable Aviation Fuel Is Made." *Canary Media*, January 12, 2023. https://www.canarymedia.com/articles/air-travel/the-smelly-greasy-truth-about-how-sustainable-aviation-fuel-is-made.

Gayle, Damien. "More Than 1,000 Climate Scientists Urge Public to Become Activists." *The Guardian*, December 4, 2023. https://www.theguardian.com/environment/2023/dec/04/more-than-1000-climate-scientists-urge-public-to-become-activists.

Gell, Aaron. "Democrats and Climate Activists Are on a Collision Course in 2024." *New Republic*, January 2, 2024. https://newrepublic.com/article/177572/democrats-climate-activists-collision-course-2024.

Google. "White Paper | Google's Carbon Offsets: Collaboration and Due Diligence." https://static.googleusercontent.com/media/www.google.com/en//green/pdfs/google-carbon-offsets.pdf.

Google Deepmind and EMBL-EBI. "AlphaFold Protein Structure Database." https://alphafold.ebi.ac.uk.

Google Research. "Flood Forecasting." https://sites.research.google/floodforecasting.

———. "John C. Platt." https://research.google/people/john-c-platt.

———. "Machine Intelligence." https://research.google/research-areas/machine-intelligence.

———. "Project Contrails." https://sites.research.google/contrails.

Gopal, Keerti. "How a Climate Group That Has Made Chaos Its Brand Got the White House's Ear." *Inside Climate News*, February 11, 2024. https://insideclimatenews.org/news/11022024/how-a-climate-group-that-made-chaos-its-brand-got-white-house-attention.

———. "Like 'Em or Not, Climate Defiance Is Determined to Get in Your Face." *Mother Jones*, February 13, 2024. https://www.motherjones.com/politics/2024/02/climate-defiance-fossil-fuel-protests-radical-extinction-rebellion.

Gotal, H., M. W. Binderbauer, T. Tajima1, S. Putvinski, et al. "Formation of Hot, Stable, Long-Lived Field-Reversed Configuration Plasmas on the C-2W Device." *Nuclear Fusion* 59, no. 11 (June 5, 2019). https://doi.org/10.1088/1741-4326/ab0be9.

Hao, Karen. "AI Is Taking Water from the Desert." *The Atlantic*, March 1, 2024. https://www.theatlantic.com/technology/archive/2024/03/ai-water-climate-microsoft/677602.

———. "Here Are 10 Ways AI Could Help Fight Climate Change." *MIT Technology Review*, June 20, 2019. https://www.technologyreview.com/2019/06/20/134864/ai-climate-change-machine-learning.

Hu, Akielly. "The Overlooked Climate Consequences of AI." *Grist*, July 6, 2023. https://grist.org/technology/the-overlooked-climate-consequences-of-ai.

Jefferson, Nikayla. "Why a New Climate Group Is Heckling Powerful Democrats." *Yale Climate Connections*, December 4, 2023. https://yaleclimateconnections.org/2023/12/why-a-new-climate-group-is-heckling-powerful-democrats.

Kaack, L. H., P. L. Donti, E. Strubell, et al. "Aligning Artificial Intelligence with Climate Change Mitigation." *Nature Climate Change* 12 (June 9, 2022): 518–527. https://doi.org/10.1038/s41558-022-01377-7.

Kristof, Nicholas. "Why North Korea Is Offering New Reasons to Worry." *New York Times*, January 14, 2024. https://www.nytimes.com/2024/01/17/opinion/north-korea-war.html.

Marris, Emma. "Crypto Is Mostly Over. Its Carbon Emissions Are Not." *The Atlantic*, March 22, 2023. https://www.theatlantic.com/science/archive/2023/03/crypto-bitcoin-mining-carbon-emissions-climate-change-impact/673468/#.

McKibben, Bill. "Could Google's Carbon Emissions Have Effectively Doubled Overnight?" *New Yorker*, May 20, 2022. https://www.newyorker.com/news/daily-comment/could-googles-carbon-emissions-have-effectively-doubled-overnight.

Metz, Cade. "Finally, Neural Networks That Actually Work." *Wired*, April 21, 2015. https://www.wired.com/2015/04/jeff-dean.

———. "'The Godfather of AI' Quits Google and Warns of Danger Ahead." *New York Times*, May 4, 2023.

Milman, Oliver. "Alarm as Fastest Growing US Cities Risk Becoming Unlivable from Climate Crisis." *The Guardian*, July 20, 2022. https://www.theguardian.com/us-news/2022/jul/20/us-fastest-growing-cities-risk-becoming-unlivable-climate-crisis.

———. "'Outrageous' Climate Activists Get in the Faces of Politicians and Oil Bosses—Will It Work?" *The Guardian*, April 24, 2024. https://www.theguardian.com/environment/2024/apr/25/climate-crisis-activists.

Mo, L., C. M. Zohner, P. B. Reich, Jingjing Liang, et al. "Integrated Global Assessment of the Natural Forest Carbon Potential." *Nature* 624 (2023): 92–101. https://doi.org/10.1038/s41586-023-06723-z.

Neimann, Emily Saya. "The Nature Rules: Dreaming of Earth Project." *Tokyo Art Beat*, August 2, 2019. https://www.tokyoartbeat.com/en/articles/-/the-nature-rules-dreaming-of-earth-project.

Osaka, Shannon. "Throwing Tomato Soup on Van Gogh: Why Climate Protests Are Getting Weirder." *Washington Post*, October 14, 2022. https://www.washingtonpost.com/climate-environment/2022/10/14/tomato-soup-sunflowers-climate-protest.

Pearce, Fred. "How Airplane Contrails Are Helping Make the Planet Warmer." *Yale Environment 360*, July 18, 2019. https://e360.yale.edu/features/how-airplane-contrails-are-helping-make-the-planet-warmer.

Perez, Andrew, and Tim Dickinson. "The Climate Activists Fighting Off Cane-Wielding Country Club Members." *Rolling Stone*, January 26, 2024. https://www

.rollingstone.com/politics/politics-features/climate-defiance-activists -moynihan-bank-america-biden-1234955026.

Petro, Gustavo. "Intervention within the Framework of COP28—Dubai, United Arab Emirates." https://www.youtube.com/live/3auQstHfHEc?si=AnQLFHJYy6bCL7op.

Petro, Gustavo. "Para atender crisis climática se necesitan 30 veces más los USD 100 mil millones pactados en la COP de París: Presidente Petro." https://petro.presidencia .gov.co/prensa/Paginas/Para-atender-crisis-climatica-se-necesitan-30-veces-mas-los -USD-100-mil-mil-240117.aspx.

Pichai, Sundar. "Our Global Progress Toward a Sustainable Future." *Sustainable with Google*, October 10, 2023. https://blog.google/inside-google/message-ceo/google -sustainable-future-global-progress.

Platt, John. "Keynote Talk: AI for Climate Change; The Context." SlidesLive, June 14, 2019. https://slideslive.com/38917851/keynote-talk-ai-for-climate-change-the-context.

Presidencia de la República—Colombia. "Intervention by President Gustavo Petro before the 77th UN." September 20, 2022. https://youtu.be/F_HJHZd1w2o?si =yqvDcgv6oJLweOUD.

Ransohoff, Nan. "Scaling the Carbon Removal Market." Stripe. https://stripe.com /sessions/2022/scaling-carbon-removal-market.

Riahi, Keywan, Detlef P. van Vuuren, Elmar Kriegler, Jae Edmonds, et al. "The Shared Socioeconomic Pathways and Their Energy, Land Use, and Greenhouse Gas Emissions Implications: An Overview." *Global Environmental Change* 42 (January 2017): 153–168. https://doi.org/10.1016/j.gloenvcha.2016.05.009.

Rolnick, David, Priya L. Donti, Lynn H. Kaack, Kelly Kochanski, John C. Platt, et al. "Tackling Climate Change with Machine Learning," arXiv:1906.05433v2 [cs.CY] (November 5, 2019). http://arxiv.org/pdf/1906.05433.

Rubiano A., María Paula. "How Colombia Plans to Keep Its Oil and Coal in the Ground." BBC, November 16, 2022. https://www.bbc.com/future/article /20221116-how-colombia-plans-to-keep-its-oil-and-gas-in-the-ground.

Sang-Hun, Choe. "70 Years along the Zone Where the Korean War Never Ended." *New York Times*, July 26, 2023. https://www.nytimes.com/2023/07/26/world/asia /korea-dmz-north-south-border.html.

———. "North Korea Says It No Longer Wants to Reunify with South Korea." *New Yorker*, January 16, 2024. https://www.nytimes.com/2024/01/16/world/asia/north -korea-reunification-policy.html.

Shead, Sam. "Google Plans to Stop Making A.I. Tools for Oil and Gas Firms." CNBC, May 20, 2020. https://www.cnbc.com/2020/05/20/google-ai-greenpeace-oil-gas.html.

Solomon, Caren G., and Regina C. LaRocque. "Climate Change—a Health Emergency." *New England Journal of Medicine* 380 (2019): 209–211. https://doi.org/10.1056 /NEJMp1817067.

Stierwalt, Sabrina. "Einstein's Legacy: The Photoelectric Effect." *Scientific American*, August 18, 2015. https://www.scientificamerican.com/article/einstein-s -legacy-the-photoelectric-effect.

Stokel-Walker, Chris. "Data Centers Are Facing a Climate Crisis." *Wired*, August 1, 2022. https://www.wired.com/story/data-centers-climate-change.

Tabuchi, Hiroko. "Bank of America Pledged to Stop Financing Coal. Now It's Backtracking." *New York Times*, February 5, 2024. https://www.nytimes.com/2024/02 /03/climate/bank-of-america-esg.html.

TAE Technologies. https://tae.com.

Temple, James. "Is Carbon Removal Crazy or Critical? Yes." *MIT Technology Review*, March 1, 2019. https://www.technologyreview.com/2019/02/27/136958/one-mans-two-decade-quest-to-suck-greenhouse-gas-out-of-the-sky.

Tigue, Kristoffer, and Keerti Gopal. "Will Biden's Temporary Pause of Gas Export Projects Win Back Young Voters?" *Inside Climate News*, January 26, 2024. https://insideclimatenews.org/news/26012024/todays-climate-biden-trump-pause-lng-gas-export-young-voters.

Torres, Émile P. "Worst-Case AI Scenario? Human Extinction." *Washington Post*, August 31, 2022. https://www.washingtonpost.com/opinions/2022/08/31/artificial-intelligence-worst-case-scenario-extinction.

Vincent, James. "Here's How AI Can Help Fight Climate Change According to the Field's Top Thinkers." *The Verge*, June 25, 2019. https://www.theverge.com/2019/6/25/18744034/ai-artificial-intelligence-ml-climate-change-fight-tackle.

Weisman, Alan. "Turn the Korean DMZ into a Bridge of Peace?" *Los Angeles Times*, December 27, 2017. https://www.latimes.com/entertainment/arts/la-ca-cm-dmz-art-20171231-htmlstory.html.

White House. "Statement from President Joe Biden on Decision to Pause Pending Approvals of Liquefied Natural Gas Exports." January 26, 2024. https://www.whitehouse.gov/briefing-room/statements-releases/2024/01/26/statement-from-president-joe-biden-on-decision-to-pause-pending-approvals-of-liquefied-natural-gas-exports.

Whittaker, Matt. "Google Has One of Big Tech's Most Aggressive Sustainability Plans. Here's Its 3-Step Playbook for Helping the Planet." *Fortune*, September 2022. https://fortune.com/2022/09/12/google-has-one-of-big-techs-most-aggressive-sustainability-plans-heres-its-3-step-playbook-for-helping-the-planet.

Work on Climate. "PAW Climate 2021—Keynote (John Platt, Google): How Can AI Fight Climate Change." https://youtu.be/hRQkUYjPg_c?si=CFumzqucGWDRiqxn.

Woskov, Paul, and Daniel Cohn. *Annual Report 2009: Millimeter Wave Deep Drilling for Geothermal Energy, Natural Gas and Oil.* MIT Report #PSFC/RR-09-11 MIT Plasma Science and Fusion Center, September 2009. https://dspace.mit.edu/handle/1721.1/93312.

Zielinski, Chris. "Time to Treat the Climate and Nature Crisis as One Indivisible Global Health Emergency." *Sexually Transmitted Infections* 100, no. 1 (February 2024): 1–2. https://doi.org/10.1136/sextrans-2023-056026.

Zurich, Eth. "Scientists Begin Building Highly Accurate Digital Twin of Our Planet." *EurekaAlert*, February 24, 2021. https://www.eurekalert.org/news-releases/1035439.

EPILOGUE

Books

Al-Ansārī, Ayatollah Sheikh Mohammad Hussein. *Human Cloning: An Islamic Study on Its Permissibility and Implications.* Translated by Mohammad Basim Al-Ansārī. Najaf: Al-Ansārī Foundation, 2010.

———. *Knowing the Imam of Your Time: Nihayah al-Mayrifah, the End or Recognition (of Imam Mahdi).* Translated by Sheikh Zaid Alsalami. Sydney: Office of Ayatollah Al-Ansārī, 2019.

Ali, Abdullah Yusuf. *An English Interpretation of the Holy Qur'ran*. Lahore: Fine Offset Printing-Urdu Nagar, 1934.

Berlin, Adele. *Lamentations: A Commentary (The Old Testament Library)*. Louisville, KY: Westminster John Knox Press, 2004.

Hattin, Rabbi Michael. *Joshua: The Challenge of the Promised Land (Maggid Studies in Tanakh)*. New Milford, CT: Toby Press, 2015.

Ish-Kishor, Sulamith. *Children's History of Israel*. Vol. 2, *From Joshua to the Second Temple*. New York: Jordan Publishing Company, 1960.

Mervin, Sabrina, Robert Gleave, and Géraldine Chatelard. *Najaf: Portrait of a City*. Paris: UNESCO, 2017.

Naqvi, Alsyyed Abu Mohammad. *The Holy Qur'an Shia Translation*. Amazon, 2016. Kindle edition.

Pruzansky, Rabbi Steven. *Prophet for Today: Contemporary Lessons from the Book of Yehoshua (BIBLE/TANACH 16)*. Jerusalem: Gefen, 2006. Kindle edition.

Tabbaa, Yasser, and Sabrina Mervin. *Najaf: The Gate of Wisdom*. Paris: UNESCO, 2014.

Articles

Adamo, Nasrat, Nadhir Al-Ansari, Varoujan K. Sissakian, Sven Knutsson, and Jan Laue. "Climate Change: Consequences on Iraq's Environment." *Journal of Earth Sciences and Geotechnical Engineering* 8, no. 3 (2018): 43–58, ISSN: 1792-9040 (print version), 1792-9660 (online).

Al-Islam.org. "The Twelfth Imam, Muhammad ibn al-Hasan (Al-Mahdi-Sahibuz Zaman) (The hidden Imam who is expected to return)." https://www.al-islam.org/story-holy-kaaba-and-its-people-smr-shabbar/twelfth-imam-muhammad-ibn-al-hasan-al-mahdi-sahibuz.

Al Jazeera. "Shia Muslim Pilgrims Commemorate Arbaeen in Karbala." September 29, 2021. https://www.aljazeera.com/gallery/2021/9/29/in-pictures-iraq-shia-muslim-pilgrims-commemorate-arbaeen-in-karbala.

Alma'itah, Qais Salem, and Zia ul Haq. "The Concept of Messiah in Abrahamic Religions: A Focused Study of the Eschatology of Sunni Islam." *Heliyon* 8, no. 3, e.09080 (March 9, 2022). https://doi.org/10.1016/j.heliyon.2022.e09080.

Associated Press. "Waves of Shiite Pilgrims Descend on Iraqi City." August 26, 2007. https://www.nbcnews.com/id/wbna20446813.

Aster, Shawn. "Lamentations." *My Jewish Learning*. https://www.myjewishlearning.com/article/lamentations.

"Book of Joshua Commentary." http://www.yeshshem.com/sefer-joshua.htm.

Branigan, William. "The Army's 'Thunder Run' to Baghdad to Oust Saddam Hussein Changed Iraq War." *Washington Post*, March 17, 2023. https://www.washingtonpost.com/history/2023/03/17/iraq-war-thunder-run-baghdad-saddam-hussein.

Britannica. "Book of Joshua." https://www.britannica.com/topic/Book-of-Joshua.

———. "Wheel." https://www.britannica.com/technology/wheel.

Chandrasekaran, Rajiv. "20 Years Later, and Despite U.S. Mistakes, Iraq Is Turning the Corner." *Washington Post*, March 17, 2023. https://www.washingtonpost.com/opinions/2023/03/17/iraq-invasion-anniversary-recovery.

Cohen, Roger. "Between Israelis and Palestinians, a Lethal Psychological Chasm Grows." *New York Times*, November 22, 2023. https://www.nytimes.com/2023/11/20/world/middleeast/israelis-palestinians-conflict.html.

Eden in Iraq Wastewater Garden Project. News release. https://mailchi.mp/393dfdaf68df/meridels-art-on-view-turner-carroll-container-santa-fe-nm-opening-march-10-talk-5pm-13540636?e=1a0aad7fb8.

Frontline Defenders. "Environmental Human Rights Defender Jassim Al-Asadi Released by His Kidnappers after 2 Weeks." https://www.frontlinedefenders.org/en/case/environmental-human-rights-defender-jassim-al-asadi-released-after-2-weeks.

ICRC. "Iraq: Expanding Deserts, Searing Temperatures, and Dying Land: Climate Crises Deepen Struggle of Farmers." November 14, 2022. https://www.icrcnewsroom.org/story/en/2037/iraq-expanding-deserts-searing-temperatures-and-dying-land-climate-crises-deepen-struggle-of-farmers.

Iraqi Civil Society Solidarity Initiative. "Statement from Protect Iraqi HRDs, NOW! Campaign." February 5, 2023. https://www.iraqicivilsociety.org/archives/13872.

Jewish Virtual Library. "The Jewish Temples: The Babylonian Exile." https://www.jewishvirtuallibrary.org/the-babylonian-exile.

Lonergan, Stephen. "Don't Be Afraid for the Marshes." Al Jazeera, May 4, 2024. https://www.aljazeera.com/features/longform/2024/5/4/dont-be-afraid-for-the-marshes-the-battle-to-save-iraqs-waterways.

Ludwig, Mike. "20 Years after US Invasion, Iraq Faces Cascading Climate and Water Crises." *Truthout*, March 20, 2023. https://truthout.org/articles/20-years-after-us-invasion-iraq-faces-cascading-climate-and-water-crises.

Mahmoud, Sinan. "Hope and Tragedy after Unprecedented Heavy Rains Sweep Iraq." *MENA*, March 25, 2024. https://www.thenationalnews.com/mena/iraq/2024/03/25/iraq-rains-floods.

———. "Iraqi Environmental Activist Kidnapped near Baghdad, Family Says." *MENA*, February 5, 2023. https://www.thenationalnews.com/mena/iraq/2023/02/05/iraqi-environmental-activist-kidnapped-near-baghdad-family-says.

Middle East Monitor. "Prominent Iraq Environmental Activist Kidnapped near Baghdad." February 5, 2023. https://www.middleeastmonitor.com/20230205-prominent-iraq-environmental-activist-kidnapped-near-baghdad.

Ministry Voice. "When Was the Book of Joshua Written: A Historical Analysis for Biblical Scholars." https://www.ministryvoice.com/when-was-the-book-of-joshua-written.

My Islam. "Read Surah Ash-Shams Translation and Transliteration." https://myislam.org/surah-ash-shams.

Project on Shi'ism and Global Affairs. "The Hidden Imam and the End of Time." https://shiism.hds.harvard.edu/files/shiism-global-affairs/files/the_hidden_imam_and_the_end_of_time_-_a_primer_on_the_mahdi_islamic_theology_and_global_politics_04.pdf.

Rubin, Alison J. "20 Years after U.S. Invasion, Iraq Is a Freer Place, but Not a Hopeful One." *New York Times*, March 21, 2023. https://www.nytimes.com/2023/03/18/world/middleeast/iraq-war-20th-anniversary.html.

Uruk Project. "History of Uruk." https://staff.cdms.westernsydney.edu.au/~anton/Research/Uruk_Project/History.html.

Warrick, Thomas S. "What Happened to America's Postwar Plans for Iraq. Here's How Israel Should Plan for Gaza." *New York Times*, October 16, 2023.

Weisman, Alan. "The World, with or without Us." *Salon*, May 22, 2022. https://www.salon.com/2022/05/22/the-world-with-or-without-us-increasing-gas-and-oil-production-is-a-disastrously-idea.

WikiShia. "Holy Shrine of Imam al-Husayn." https://en.wikishia.net/view/Holy_Shrine_of_Imam_al-Husayn_(a).

———. "Imam Ali b. Abi Talib." https://en.wikishia.net/view/Imam_Ali_b._Abi_Talib_(a).

World Bank. "Iraq Economy Rebounds but Economic Diversification Remains an Urgent Priority." November 17, 2022. https://www.worldbank.org/en/news/press-release/2022/11/17/iraq-economy-rebounds-but-economic-diversification-remains-an-urgent-priority.

Yee, Vivian, and Alissa J. Rubin. "In U.S.-Led Iraq War, Iran Was the Big Winner." *New York Times*, March 20, 2023. https://www.nytimes.com/2023/03/19/world/middleeast/iraq-war-iran.html.

INDEX

Page numbers in italics refer to photographs or illustrations.

ABOUT THE AUTHOR

Alan Weisman has reported from all seven continents and in more than 60 countries. His books include the *New York Times* bestseller *The World Without Us*, translated into 34 languages and a finalist for the National Book Critics Circle Award and winner of China's Wenjin Book Award; and *Countdown*, winner of the *Los Angeles Times* Book Prize. His work has appeared in *Harper's*, *The New York Times Magazine*, *The Atlantic Monthly*, *The New York Review of Books*, *Vanity Fair*, *Orion*, *Mother Jones*, *Discover*, and *Salon*, among others, and on NPR. A cofounder of the journalism collective Homelands Productions, he also has been a laureate professor of international journalism at the University of Arizona. He and his wife, sculptor Beckie Kravetz, live in western Massachusetts.